THE PRACTICE OF CONSTRUCTION MANAGEMENT

People and Business Performance

THE PRACTICE OF CONSTRUCTION MANAGEMENT

People and Business Performance

Barry Fryer

with contributions from
Marilyn Fryer

FOURTH EDITION
revised by
Charles Egbu, Robert Ellis & Christopher Gorse

Blackwell
Publishing

Editorial offices:
Blackwell Publishing Ltd, 9600 Garsington Road, Oxford OX4 2DQ, UK
   Tel: +44 (0)1865 776868
Blackwell Publishing Inc., 350 Main Street, Malden, MA 02148-5020, USA
   Tel: +1 781 388 8250
Blackwell Publishing Asia Pty Ltd, 550 Swanston Street, Carlton, Victoria 3053, Australia
   Tel: +61 (0)3 8359 1011

First edition published by Collins Professional and Technical Books Ltd
Second edition published by BSP Professional Books 1990
Reprinted 1992, 1994
Third edition published by Blackwell Science 1997
Reprinted 1998
Fourth edition published 2004 by Blackwell Publishing Ltd

Library of Congress Cataloging-in-Publication Data
Fryer, Barry G.
   The practice of construction management/Barry Fryer, with contributions from Marilyn Fryer. – 4th ed.
      p. cm.
   Includes bibliographical references and index.
   ISBN 1-4051-1110-0 (alk. paper)
   1. Construction industry – Management.   I. Fryer, Marilyn.   II. Title.

   HD9715.A2F59 2004
   624′.068–dc22   2004046324

ISBN 1-4051-1110-0

A catalogue record for this title is available from the British Library

Set in 10 on 13pt Times
by DP Photosetting, Aylesbury, Bucks
Printed and bound in Great Britain
by MPG Books Ltd, Bodmin, Cornwall

The publisher's policy is to use permanent paper from mills that operate a sustainable forestry policy, and which has been manufactured from pulp processed using acid-free and elementary chlorine-free practices. Furthermore, the publisher ensures that the text paper and cover board used have met acceptable environmental accreditation standards.

For further information on Blackwell Publishing, visit our website:
www.thatconstructionsite.com

Contents

Author Biographies   x
Preface to the Fourth Edition   xii

1 The Development of Management Thinking   **1**
Early contributions to management thinking   2
Management and the social sciences   4
Systems management   6
Situational or contingency management   9
Dynamic engagement   12
The new Ps of management: post-industrial portfolio, pragmatic,
    post-modern   13
Quality and environmental management   15
Summary   15
Discussion and questions   16

2 Managers and Their Jobs   **17**
The tasks of management   17
How managers spend their time   22
The manager's skills   23
The manager's power   25
Empowerment   27
Professionalism and construction management   29
Summary   30
Debate motions and presentation topics   31

3 Organisation   **32**
Organisational activities   33
Objectives   34
Characteristics of organisations   36
Types of organisation   42
Summary   46
Discussion topics   47

4 Leadership   **48**
The characteristics of the leader   49

473475

vi Contents

624
068
FRY

Leadership style		50
The leader and the situation		53
Leadership, goals and social exchange		57
Task and socio-emotional roles		59
Formal and informal leaders		59
The leader's competence		60
Summary		60
Discussion topics		61
5 Communication		**62**
Communication process		63
Functions of communication		66
Communication structure		68
The direction of communication		68
Why communication fails		70
Communication methods		74
Plain talking and writing		78
Graphic and numerical communication		81
Information management		82
The manager's behaviour: communication and influence		84
Personal skills and interaction		86
Summary		88
Group communication exercise		89
6 Conflict and Conflict Management		**93**
Definition: functional and dysfunctional conflict		93
Conflict emergence and development		95
Managing conflict		99
Conflict and disputes		104
Summary		106
Conflict management exercise		106
7 Individual/Group Behaviour and Teamwork		**111**
Personality and individual behaviour		111
Individuals and groups		111
Developing group performance		112
Observing and analysing group behaviour		117
Summarising individual and group behaviour		122
Teamwork		123
Features of a good team		125
Teamwork roles		127
Team leadership		127
Team leadership and the self-managed team		129
Training in teamwork and team leadership		130

Evaluation of teamwork training   132
Summary   133
Exercise   133

8 Motivation and Human Performance   **134**
People and work   134
Employee performance   135
Motivation   144
Sub-contractors   149
Job design   150
Time management   151
Summary   153
Discussion   153

9 Problem-Solving and Decision-Making   **154**
Types of problem and decision   155
Stages in problem-solving and decision-making   157
Human reasoning and problem-solving   158
Group decision-making   161
Suspending judgement in problem-solving   167
Creative problem-solving   168
Summary   171
Exercise   172

10 Managing Change   **173**
Future studies   174
The process of organisational change   174
Managing change   176
Strategic management   177
Marketing   180
Organisational development   184
Changing people's attitudes   187
Managing creativity   190
Summary   192
Exercise   193

11 Value and Risk Management   **195**
Best value   195
Collaborative agreements   196
Value management   196
Value management and enhanced team-working   198
Value management interventions   198
Risk management   201
Collaborative workshops   204

| | Summary | 204 |
| | Student activity | 205 |

12   Managing Innovation in Construction — **207**

Challenges associated with managing innovations in organisations	208
Knowledge management and innovations: building and maintaining capabilities	209
Organisational innovations and strategies: critical success factors	210
Measuring innovation success	213
Knowledge management and improved innovations: issues of strategy, process and structure	214
Managing knowledge and organisational learning for innovations	216
E-business initiatives and the construction industry	218
Challenges facing organisations in using the Internet for business activities	219
Summary	221
Discussion and questions	221

13   Managing Supply Chains and Construction Networks — **222**

The nature, types and importance of supply chains in construction	222
A model of SCM in construction	223
Mechanisms used to set up successful collaborative relationships – implementing SCM in construction	223
Partnering	224
Prime contracting	227
Public private partnerships (PPP) and Private Finance Initiatives (PFI)	228
Discussion and questions	231

14   Personnel Management and HRM — **232**

Personnel management or human resources management?	232
The personnel function	233
The tasks of personnel management	234
Personnel policy	235
Strategic planning and organisational development	237
Employee remuneration	237
Counselling	238
Administration and records	239
Summary	239
Exercise	240

15   Recruitment and Staff Development — **241**

Employment planning	241
Forecasting and budgeting	242
Producing the plan and an action programme	243
Planning for projects	243

Recruitment 245
Personnel selection 248
Staff development 253
Performance appraisal 253
Education and training 257
Systematic staff development 258
Approaches to staff development 259
Staff development methods 261
Mentoring 263
Management development 264
Continuing professional development (CPD) 265
Construction Industry Training Board (CITB) 267
Summary 267
Mind map tasks 268

16 Health and Safety **269**
Safety 269
Health 275
Effective communication and managing health and safety 279
The common law on health and safety 280
The Health and Safety at Work etc. Act 1974 282
Construction regulations 286
Safety representatives and committees 287
Protective equipment 289
Summary 289
Discussion and questions 290

17 Industrial Relations **291**
Employers' associations 292
Trade unions 293
Collective bargaining 295
Employee participation and industrial democracy 298
Employment and workplace relations 299
Summary 305
Discussion and questions 305

18 Managing Quality and Environmental Impact **306**
Quality management 306
Installing a quality management system 311
Environmental impact 315
Environmental and quality auditing 320
Summary 320
Discussion and questions 321

References 322
Index 341

Author Biographies

Barry Fryer, MA, MSc, MCMI, FRSA was formerly Assistant Dean of Faculty and Professor of Construction Management at Leeds Metropolitan University. He has been an examiner for the Chartered Institute of Building and external examiner for a number of degree and postgraduate courses in construction management. He began his career with Sir Robert McAlpine & Sons and was a Chartered Quantity Surveyor and Chartered Builder.

Charles Egbu, BSc(Hons), PhD, MCIOB, MAPM
Charles Egbu is Professor of Construction and Project Management at Glasgow Caledonian University. He has a background in quantity surveying and worked as a site quantity surveyor with a large UK construction company before obtaining his PhD in construction management from the University of Salford. He was previously a Reader of Construction and Project Management at Leeds Metropolitan University and has worked as a Senior Research Fellow at University College London and as a Senior Research Consultant at the Building Research Establishment.

Robert Ellis, BSc(Hons), MSc, PhD, FRICS, MCIArb, ILTM
Robert Ellis is a National Teaching Fellow and Principal Lecturer at Leeds Metropolitan University. Having worked both in local government and private practice as a Chartered Quantity Surveyor, he moved into higher education where he has been heavily involved in consultancy and research. He has played an active role in the Royal Institution of Chartered Surveyors and is a past Chairman of the Yorkshire and Humber Region.

Christopher Gorse, BSc(Hons), PhD, MCIOB, MAPM, ILTM, CertEd, PGDip (Ed.), Dip (H&S)
Christopher Gorse is a Senior Lecturer in Construction and Project Management at Leeds Metropolitan University. He worked as an engineer and project manager with a large UK contracting organisation before returning to work in higher education. He is Course Leader for a number of undergraduate and postgraduate construction courses and is active in consultancy, research and in-company training programmes.

Marilyn Fryer, BA(Hons), PhD, CPsychol, AFBPsS, GradCertEd, FRSA is Managing Director of The Creativity Centre Ltd. She was formerly Reader in Psychology at Leeds Metropolitan University, before co-founding The Creativity Centre Ltd and The Creativity Centre Educational Trust.

Preface to the Fourth Edition

In the Preface to the Third edition, Barry Fryer acknowledges the 'unprecedented changes' affecting the practice of construction management in the mid-1990s. Ironically, change appears to be the only constant in all our working lives, therefore it comes as no surprise that change has continued unabated in the construction industry during the late 1990s and early 21st century. Witness the growth in partnering, prime contracting and strategic alliances, the emphasis on integrated project teams and supply chain management and the importance of innovation in construction. Moreover the adoption of new techniques, once the preserve of the professional consultant, has now entered the domain of the construction manager. Value and risk management, for example, are *de rigueur* as the contractor becomes central, rather than peripheral, to the project team.

Accordingly, construction managers face new and sometimes far more subtle pressures, as they seek to balance relationships with commercial issues, to develop new skills and to promote a culture of collaborative working. Contractors are no longer 'in the pay' of the professional team, they are the 'paymasters'. So with this new found power, is there the opportunity to shape future development?

Since the last edition of the book in 1997, many new initiatives and government sponsored reports have begun to impact on organisational strategy, organisational culture and the culture of the industry in general.

1998 saw the publication of the Construction Task Force Report – *Rethinking Construction*. The report has in many ways been viewed as an agent for change both in terms of productivity improvements and the meeting of clients' needs. Further initiatives such as the Movement for Innovation (M4I) and 'Respect for People', a commitment to people as our important asset, have also gained widespread recognition. In June 2002, the UK Government launched an initiative entitled 'Revitalising Health and Safety in Construction' (HSE, 2002), a key aspect of 'Respect for People'. This is designed to inject new impetus into the health and safety agenda. The industry continues to work towards a fully qualified workforce and the Construction Skills Certification Scheme (CSCS), and other equivalent training schemes, are already being acknowledged as a minimum industry standard.

In terms of research and how this might contribute to the construction industry, Sir John Fairclough's Report – *Rethinking Construction Innovation and Research* (2002) and Sir Gareth Robert's Report (*Robert's Review*, 2003) on the future of

Research Assessment are likely to shape the future of research, at least in the next five years.

Charles Egbu
Robert Ellis
Christopher Gorse

Chapter 1

The Development of Management Thinking

Management is both a fascinating and frustrating subject. It abounds with exciting and challenging ideas, but even the most promising ideas don't always work. Throughout the twentieth century, managers searched for a set of guidelines for running a business. The result has been a jungle of diverse and often conflicting ideas about what managers are and what they do – or ought to be doing.

People have looked at management in different ways. Some have tried to identify the things managers do, whilst others have looked at how they do them. Some have put forward management principles to apply to all organisations, whilst others are sure there are none.

Despite many attempts to describe management, no widely accepted definition has emerged. Simple definitions include 'running things properly' and 'getting things done through people'. Rosemary Stewart (1997) brings decision-making into her definition of management – 'deciding what should be done and then getting other people to do it'.

To be more precise, we need to say how and why the manager does these things; what tasks or processes are involved. The early management writers, who were mostly practising managers, said that these processes included planning, organising, directing and controlling. This led to definitions like:

> Management is the process of steering an organisation towards the achievement of its objectives, by means of technical skills for planning and controlling operations, and social skills for directing and co-ordinating the efforts of employees.

Although harder to take in, this definition highlights the complexity of management. Yet it still tells us little about *how* managers work. Like the simple definition, it tells us that a manager is someone who plans and gets things done; that the role involves achieving objectives and co-ordinating the work of others. It does not tell a site manager whether to use the same planning techniques as a factory manager, or how to get co-operation from a site team.

Such definitions also give little indication of how management is changing. Management today is harder and less intuitive than in the past. Building and civil engineering firms used to be smaller and simpler. There were fewer specialists and fewer rules. Jobs were more flexible. Managers were closer to the work and communications were better.

1

Today, many construction firms have grown and their activities are more complex. The ratio of managers and specialists to workers has increased. There are more rules and procedures. Roles are more tightly defined and there are many external controls.

Managers need more skills and more information to cope with these changes. In large organisations, the days of the individual manager running things have gone. The efficient organisation of big business now demands *team* management.

In the recommendations of the joint review of the industry, *Constructing the Team*, known as the Latham Report, Sir Michael Latham (1994) drew attention to the wide-ranging scope for improving the construction industry's performance, through improved management practices and procedures, including more carefully thought out project strategies, more systematic quality assurance and productivity measures, and improved teamwork on site between contractors, trade specialists and consultants.

The Egan Report (1998) and the Fairclough Report (DTI 2002a) have called for improved collaboration between industry and academia in research, increased focus on diversity and equality issues, and performance-based approaches to tackle the challenges of modern day management of construction organisations. As such these reports recognise the increasingly competitive environments in which managers in the construction industry have to work due to globalisation, changing construction procurement routes, mergers and acquisitions, the potential arising from the exploitation of information communication technologies, and changing employment practices.

Early contributions to management thinking

The systematic study of management to find out what managers ought to be doing emerged at the end of the nineteenth century. The industrial system was well established. People had migrated to the towns to work in the factories and mills. They worked long hours for low pay. They worked hard – or they lost their jobs. The managers were powerful and this made their jobs easier.

Some of the managers wanted to learn more about their work. They tried to analyse their jobs and the events happening around them. They wondered if there could be principles of management that would work anywhere – a science of management. Their experiences seemed to support this, for managers everywhere appeared to be doing similar things – drawing up programmes, marshalling resources, allocating tasks and controlling costs.

They came to believe that it was possible to devise an ideal organisation using a set of design rules that would apply anywhere. The two main contributors to this line of thinking were Henry Fayol (1841–1925) and Frederick Winslow Taylor (1856–1917). The books they wrote formed the basis of the *classical* or *scientific* management movement. The design rules were later developed and refined by writers like Lyndall Urwick. These rules or principles included:

- *The principle of specialisation.* Every employee should, as far as possible, perform a single function.
- *The principle of definition.* The duties, authority and responsibility of each job, and its relationship to other jobs, should be clearly defined in writing and made known to other employees.
- *The span of control.* No one should supervise more than five, or at most six, direct subordinates whose work interlocks.

How useful are such guidelines to a manager setting up a civil engineering site, or a resourceful joiner wanting to start a small building firm? The answer is that they offer only general guidance rather than a blueprint for designing an organisation.

The principle of specialisation is heavily qualified by the phrase 'as far as possible'. How many people in construction perform only 'a single function'? What is 'a single function' anyway?

The principle of definition is sometimes impractical. How many managers in construction have a clearly defined, set task? Most have to adapt to each new project and cope with constantly changing problems as it moves from start to finish.

The principle of the span of control is very specific and has been widely quoted among managers. Many now believe it is too restrictive. Some writers have modified the principle, saying that a manager's span of control should be limited to 'a reasonable number', but this reduces the principle to a statement of the obvious.

Certain factors clearly affect the size of group a construction manager can handle. They include:

- The manager's character and abilities.
- The attitudes and capabilities of the members of the group.
- The amount of time the manager spends with the group.
- The type of work the group is doing.
- The proximity of the manager and group members.
- The extent to which the manager is supervising direct or sub-contract personnel.

A site manager can co-ordinate a site team fairly easily. Contracts managers controlling projects spread over a sixty mile radius will find it more difficult. They may spend a lot of time travelling!

People have used arguments like these to refute many of the early management ideas, although they probably worked well enough in their day. Applied to modern organisations, the management principles can be justifiably challenged because:

- Conditions have changed radically. Projects are technically and contractually more complicated; legislation affecting businesses is more extensive and demanding; competition is fiercer; people's attitudes towards work and towards their managers have changed. These and many other changes have altered the manager's job significantly compared to that of the tough task-master of the early 1900s.

● Evidence now suggests that there is a divergence between what managers do and what management writers say they ought to do. Henry Mintzberg (1973, 1976) found, in his studies, that managers were not very systematic. He dismissed much of the early management thinking as folklore, saying that managers are not the reflective, analytical planners they are made out to be. Instead they spend their time liaising and negotiating with people and coping with an unrelenting stream of problems and pressures.

Most managers today recognise the importance of people in organisations, but the early management thinkers concentrated mainly on the tasks of the business. They thought the main problem in the factories and mills was to design efficient work-places and control resources tightly. Most of them treated labour as a resource, to be worked as hard as possible.

From the outset of the Industrial Revolution, a few managers showed concern for the well-being of employees, but experience of large-scale industry was limited. No one fully understood the effect the new workplaces would have on people, but some managers quickly sensed that they could not treat people like machines.

Management and the social sciences

During the early decades of the twentieth century, social scientists began to study people in industrial settings. At first, their interest centred mainly on how work practices and working conditions affect people. Later, some of their attention switched to how workers affect organisations. Elton Mayo is regarded as the founder of this *human relations* movement, which brought into prominence the idea that employees must be understood as human beings if organisations are to be run efficiently. Mayo's far-reaching research at the Western Electric Company near Chicago – the Hawthorne studies – generated momentum for other work, including extensive research on group behaviour at the University of Michigan.

In the UK, one of the most determined and practical studies of the relationship between organisational efficiency and employee well-being was initiated at the Glacier Metal Company in London. It involved many years of close collaboration between managers and social scientists. The Glacier team took the view that the manager not only has a technical role, but a social one of creating an organisation with which workers can identify and in which they can participate and exercise discretion (Brown and Jaques, 1965).

Other studies have looked at specific topics, such as:

● Communication
● Worker participation
● Leadership
● Stress
● Labour turnover

- Performance
- Motivation.

Such work is still going on, supported, in the UK, by bodies like the Medical Research Council and the Economic and Social Research Council. The research has yielded many interesting results. For instance, an early discovery was that work groups exercise considerable influence over their members' behaviour and, in particular, over how much work they do. It was found that workers consider pay less important than had been thought. Many of them ranked factors like steady jobs, good working conditions and opportunity for promotion, higher than pay. Other findings suggest, for example, that:

- Satisfaction and dissatisfaction depend not so much on physical conditions, but on how people feel about their standing in the firm and what rewards they believe they deserve.
- Complaints are not necessarily objective statements of fact, but symptoms of a more deep-seated dissatisfaction.
- Giving a person the chance to talk and air grievances often has a beneficial effect on morale and performance.
- Employees' demands are often influenced by experiences outside, as well as in, the workplace.

Whilst these conclusions are fairly simple and clear, many research results are complex, fragmented and difficult to apply. Some construction managers are openly sceptical about the social sciences, arguing that many studies pursue trivial and obvious relationships, whilst findings are often difficult to interpret. Psychology, for instance, is every bit as concerned with the behaviour of building workers on site as it is with the study of mental disorders. Yet the applications of psychology on site have rarely been made clear and busy site managers are left to make their own conceptual leap from theory to application (M. Fryer, 1983).

Nevertheless, psychologists and sociologists have made a substantial impact on management ideas and business practices. There has been a noticeable shift in attitudes over the years (see Fig. 1.1). Managers are more aware of the construction worker's needs and aspirations and take a more humane approach.

Legislation has also compelled managers to give employees a better deal, and collective bargaining between the unions and employers has improved the terms and conditions of employment of most construction workers.

By the 1960s, so much was being written about the relationship between people and organisations that managers came under pressure to modify their leadership styles, get subordinates involved in making decisions and give them more autonomy in their jobs. The work of American writers like Argyris, Herzberg and Likert, and British writers, such as Emery, Trist and Rice at the Tavistock Institute, were brought to the notice of managers through books and business courses. For a time, it seemed that so much attention was being lavished on the worker by management

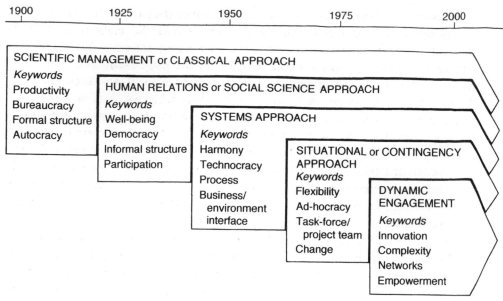

Figure 1.1 The development of management thinking.

writers and educators that managers might forget that their organisations still had work to do and profits to make.

Eventually there was a call for a more balanced approach to management, which would recognise the importance of both people and tasks. Indeed, the Tavistock Institute researchers were among the first to express this view. Two new trends in management thinking started to emerge and gain ground in the 1960s and 1970s, namely that:

- people and tasks must be considered as related parts of an organisational *system*; and
- managers must be more flexible and tailor their approach to the needs of the *situation*.

Systems management

Since the 1960s, people have tried to apply systems thinking to organisations, to see if it could help make them more manageable. The essence of *systems theory* is that the structure and behaviour of all systems, whether living organisms, machines or businesses, have certain characteristics in common. The manager who is aware of these characteristics is better able to predict the behaviour of the system and understand why it sometimes breaks down.

The construction project is a good example of a system that can be studied over its full lifespan. The project can be viewed as a temporary system, set up for a specific purpose, with well-defined tasks and a set timescale (Miller and Rice, 1967).

In systems thinking, the emphasis is not so much on the parts of the organisation – site set-up, head office departments, and so on – but on the relationships between them. There is a *technical* subsystem, the network of activities for erecting the structure or building, and a *social* sub-system, the people who contribute their energy and skills to the project. The human and technical problems cannot be divorced from one another. A change in a site bonus scheme will affect the quantity and quality of work. Changing a work method or introducing new equipment may influence operatives' attitudes and morale. The parts of the system are intertwined.

Moreover, the system is open and is influenced by events outside the organisation. The success of a building project depends not only on the project team, but on the activities of competitors, suppliers, government, clients and local communities. Many of the factors affecting the business are not only external, but are beyond the manager's control.

The project is an input/output system. Inputs of information, materials and mechanical and human energy, are turned into outputs of finished buildings. The inputs are not wholly within the manager's control and depend on the co-operation of many people, including designers, sub-contractors and suppliers. Outputs include profit, wages and job satisfaction. But there are unintended outputs too. They include noise and waste, toxic fumes and other damage to natural systems. People are injured and exposed to health hazards. They may become dissatisfied and alienated. Profits can turn into losses. Taking a systems approach means looking at the bad consequences of the organisation as well as the good!

Systems thinking emphasises the importance of *feedback*. In every organisation, managers and other employees rely on feedback to regulate their performance. For instance, managers have long acknowledged the importance of feedback in the principle of 'management by exception', where the manager puts most effort into tackling problems and breakdowns and keeps a minimal eye on the trouble-free operations. In systems terms, management by exception means that the manager is acting on negative feedback – feedback which shows something is wrong – and devotes his or her energies to bringing the system back on course.

Some of the feedback the manager receives is intermittent, giving an incomplete picture, or delayed (feedback 'lag'), which may mean that by the time the feedback reaches the manager, it is too late to take corrective action.

A contractor made a detailed monthly comparison between unit costs and the unit rates in the bills. One month, the comparison showed that the bulk excavation was making a loss of 98p per cubic metre. By the time the information reached the site agent, some 10 000 cubic metres had been excavated, making an irretrievable loss of nearly £10 000.

Managers need quick and reliable feedback on costs, progress and the quality of materials and workmanship. The time taken to obtain each kind of feedback varies. Feedback on progress can be very fast, providing the manager is keeping a close eye on operations, has a good system for recording work done and finds time to compare this data with a well-formulated programme. Cost feedback is probably the slowest and can also be the most inaccurate, since the information on which costings are based is often distorted. Labour returns are often inaccurate, and managers themselves are not always systematic in their record keeping.

Systems analysis gives a fresh angle on management. Managers have the delicate task of regulating a complex system, maximising the intended goals, whilst keeping unintended effects to a minimum. This requires a high standard of performance. Managers have to strike a balance between the technical and human demands on their time. They must keep the system in tune with the world outside and maintain its internal harmony.

Petit (1967) points out that the job of keeping the firm on course and coping with outside pressures is not the same as running the day-to-day operations of the business. Using a systems approach, he defines three distinct kinds of managerial work:

- *Technical*. At the technical level, managers run the production process. In construction, this takes place mainly on site, although some of the office work is directly concerned with production too. Site managers co-ordinate direct and sub-contract labour, plant and materials in order to achieve short-term project goals. They are protected from some of the outside pressures on the business, because the senior managers cope with these.
- *Institutional*. The senior managers are at the institutional or corporate level and Petit defines their task as relating the firm to the world outside. They cope with the risks and uncertainties caused by events and long-term trends over which they have little or no control. The survival or long-range success of the firm is their prime concern. Technical managers have access to a fair amount of reliable information for solving their problems, but senior managers deal with the unforeseen and rely heavily on intuition and judgement.
- *Organisational*. A third group of managers mediates between the other two groups, co-ordinating and integrating their tasks. These organisational managers often have to search for compromises between the strategic concerns of the top managers and the immediate, operational problems of the technical managers. They have the difficult task of supporting production, making resources available when needed, whilst ensuring that the day-to-day activities contribute to the long-range goals of the enterprise.

Figure 1.2 illustrates these levels of management, although they may merge and overlap. In small firms, the same manager may perform all three roles and will need to understand the demands of each role, know when he or she is performing each and apply the appropriate skills. In larger firms, the three levels are likely to be separate. They will be carried out by different people, often relatively independently of one another.

Viewing the construction *site* as a system in its own right, a rather different picture emerges. The site manager is the top manager of this smaller, 'task-force' system. This job involves welding together an effective team as well as dealing with outside influences, such as the local labour market, competitors, local authorities and suppliers. Site managers may regard the design team and even their own head office as outside forces which make demands on them that are difficult to meet.

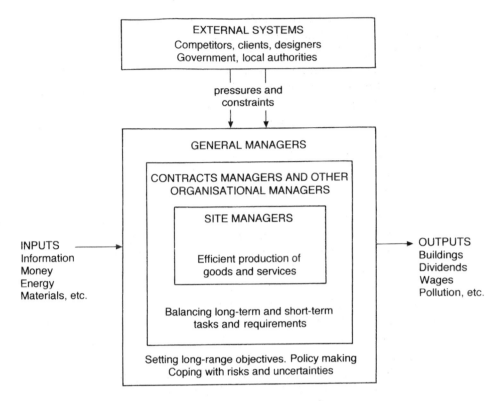

Figure 1.2 Management of a construction firm: a systems view.

They usually lack administrative help and have to perform both technical and institutional roles. To cope with the conflicts between these, they have to be organisational managers as well, using both quantitative methods and judgement to find compromises between the short-range goals of the project and the long-term strategies of the company (see Fig. 1.3).

Some site managers enjoy considerable autonomy in running their sites, but others have a narrower role and are expected to leave some of the tasks to more senior managers – contracts managers or directors – and concentrate their efforts on the day-to-day running of the site.

Clearly, managers with the same job title may not always have the same responsibilities, and managers at different levels in an organisation perform quite different roles. They are responsible for different aspects of the system's performance.

Situational or contingency management

The long search for similarities between managers' jobs, to build up a picture of the ideal manager, has had only qualified success. In the late 1950s, people started to

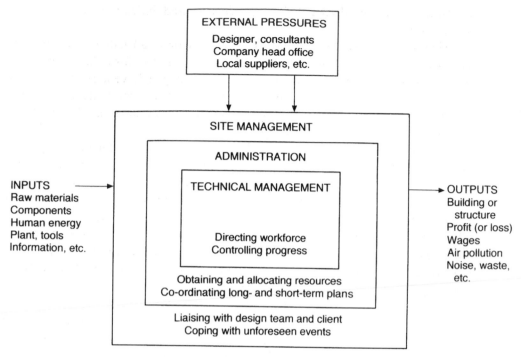

Figure 1.3 Management of a construction site: a systems view.

take a serious look at the differences between managers' jobs and collected evidence that management is really a family of roles in which managers do different things.

A site manager, co-ordinating the contractor's work with that of a dozen or more sub-contractors, may have a very different task from the production manager in a textile mill whose more stable workforce is doing repetitive work. There is increasing evidence that firms vary in their approach to management and that this has a lot to do with their size, the type of work they do, the people they employ and external and market forces.

This *situational* or *contingency* approach to management argues that there is no single best way to run a business and that managers must adapt their style and methods to suit the circumstances. In other words, the way a firm is organised and managed is 'contingent' upon factors like its size, tasks, technology and markets.

Joan Woodward (1958, 1965), in a pioneering study of British firms, showed that there are important variations in the management of different technologies. She identified three technology groupings:

- Unit and small batch production
- Mass production
- Process production.

In processing and mass production industries, Woodward found that there are many managers, many levels of management and more administrative rules. The top managers are rather divorced from production and industrial relations easily become strained.

Businesses like construction, producing one-off or small batch products, have shallower management structures, fewer specialists and less formality in their procedures. With fewer management levels, senior managers have more contact with employees and labour relations are usually better. The informality gives people more opportunity to negotiate their roles and define the boundaries of their jobs. In other words, they have more freedom.

Woodward's fresh approach showed that many of the so-called *principles* of management were derived from experience of mass production or process operations and may not apply to industries like construction. She showed that it is possible to compare large numbers of organisations and draw conclusions about management that are firmly based on and specific to the situation.

Burns and Stalker (1966) looked at the management of 20 British firms and found two kinds of management structure which they called *mechanistic* and *organic*. The more rigid, mechanistic approach to management seems to work best in firms operating in relatively stable conditions, where the technology and markets are changing slowly. The more flexible, organic approach to management works well for firms operating in unstable conditions, where markets are unpredictable or technology is changing fast. Both kinds of management structure work well, providing they 'fit' the markets and technology concerned.

Studies in the United States produced similar results. For instance, Morse and Lorsch (1970) found that the manufacture of standardised containers in a stable technology and market was most efficient when organised along fairly rigid lines, with managers exercising tight control. But companies carrying out research and development work in the fast-changing, unpredictable field of communications technology were most successful when organised flexibly, with employees having considerable freedom and with few rules and managerial controls.

The construction industry does not have to cope with rapid technical change but the market for buildings is changeable and unpredictable. Designers, builders and civil engineering contractors are likely to find a flexible style of management more effective.

A similar conclusion was reached by Lansley *et al.* (1975), who measured the commercial and human performance of 25 building companies. They pointed out that for general contractors, every new project poses fresh problems, making programming difficult. Flexibility is essential and co-ordination and teamwork are vital. However, for specialist contractors, projects tend to follow a pattern, making programming easier. Managers can exercise tighter control. The situation differs again for small works contractors (where little co-ordination is needed) and for firms that sub-let most of their work. Each type of work involves its own problems and constraints.

These and other results add weight to the argument that there is no single ideal way of organising construction and no best style of management. It seems that managers must look critically at what they are trying to achieve and the prevailing conditions, and adapt their organisations accordingly. In a period of rapid change, the need for flexibility becomes vital. Peter Lansley has contrasted the relatively stable construction environment of the 1950s and 1960s, where the passwords were productivity and efficiency, with the unstable conditions of the 1970s and 1980s, where the password became *flexibility* (Lansley, 1981).

Dynamic engagement

The ideas summarised above are still influencing managers and will continue to do so, but the backdrop is changing faster than ever and managers are having to ask themselves what will happen to their organisations in the next century. Stoner *et al.* (1995) use the term *dynamic engagement* to describe this rethinking process and capture the mood of current debate about management and organisation. They argue that managers are having to rethink their activities in the face of unprecedented external changes – changes which are causing the boundaries between cultures and nations to blur; changes in which the world is becoming a global village, as international and intercultural relations expand rapidly. The term dynamic engagement stresses the vigorous and intense involvement managers are having, in order to deal with new and changing human relationships and constant adjustment to change over time.

Dynamic engagement builds on the underpinnings of contingency management but recognises more fully the implications of the extent, type and rate of change affecting business and society. Stoner and his colleagues identify six management themes emerging from this approach:

- *New organisational environments.* These consist of complex, dynamic networks of people interacting with one another, competitors, customers, suppliers, subcontractors, specialists and so on.
- *Ethics and social responsibility.* Shaping new corporate cultures which reflect the needs and aspirations of individual employees, groups, clients and others outside the organisation in the community and wider society.
- *Globalisation and management.* The expansion of business opportunities into markets and production activities which transcend national boundaries in a 'borderless' world, where global competition also features.
- *Inventing and reinventing organisations.* To find ways of unlocking the creative skills of managers and their teams, so that they can discover innovative organisation structures and ways of operating as conditions change.
- *Cultures and multiculturalism.* In a global economy and multicultural society, cultural differences create fresh challenges for managers, who must capitalise on the values and strengths of various cultural traditions, synthesising the benefits of each of them.
- *Quality.* Organising every aspect of organisational activity to meet the higher standards demanded by clients and to maintain competitiveness in an ever tougher business environment.

Peters (1992) identifies a matching concept, *liberation management*, which focuses on escaping from existing approaches to problems of organisation and striking out to find creative solutions. Hammer and Champy (1994) also argue for a dynamic approach, saying that managers need to make a fundamental reassessment of their organisations, questioning and 're-engineering' the very processes through which their firms function. Hannagan (1995) recommends an integrated approach to management, where emphasis is on the synthesis of best practice found in organisations and cultural traditions around the world.

The new Ps of managment: post-industrial, portfolio, pragmatic, post-modern

Management in a post-industrial society

Consistent with the ideas of dynamic engagement is the work of such analysts as Warren Bennis, Tom Peters and Charles Handy who are noted for their original and perceptive insights into management and their analyses of the future of business and work. Many of these contemporary analysts recognise that the *post-industrial* era has arrived and that business practices will never be the same again. Because they are having a marked effect on managers' attitudes and behaviour, these future-oriented thinkers are often labelled *management gurus* (Kennedy, 1993). They put a lot of emphasis on change, especially with respect to organisational cultures, strategies, structures and processes and the effects these have on people's work and lives.

As organisations realise that the post-industrial society has finally arrived, a lot of attention is being directed at the role of strategic managers and the impact of delayering the middle management levels, as firms try to become leaner and more efficient to combat growing global competition. But, importantly, there has been a sharp divide between those analysts who describe future organisational success in terms of improved strategies, structures and processes, and those who believe that success is ultimately rooted in more effective *people management.*

Portfolio management

Handy (1994) successfully bridges this divide, stressing that both organisation structures and employment patterns are changing fundamentally and permanently. We will have to get used to the fact that many younger people will not achieve the 'permanent' jobs their parents mostly had and that future careers will often consist of a series of mini-careers, successive short-term contracts and/or periods of self-employment, often involving different sets of skills – and interspersed with periods of unemployment.

Handy refers to these as *portfolio* careers. They require a different kind of management and self-management and necessitate new attitudes within organisations and society. In the portfolio world, employment has to be redefined. Voluntary work, self-employment and unemployment all have new meanings. Handy even suggests that 'agents' may become more commonplace. As well as finding work for their clients, these agents will also act as mentors, helping their clients to organise their lives, develop their skills and build their portfolios. So, a new breed of independent construction managers may emerge, people who turn to their agents to help them find 'contracts', just as many writers, actors and sports people already do.

Pragmatic management

Since the mid-1980s, there have been new fashions in presenting management ideas and approaches to managers and aspiring managers. One of these has been a flux of books by well-known and successful managers, people like Sir John Harvey-Jones (see, for instance, Harvey-Jones, 1993). Other influential books include *The Change Masters* and *When Giants Learn to Dance* by Rosabeth Moss Kanter (1983 and 1990, respectively). They are refreshingly practical books that express the seasoned management experience of their authors and (in the main) reflect well the innovations, culture changes and problems of the times. Collectively, these works cannot be ignored; they have been so successful. They could be said to represent a new *pragmatic* approach to management thinking. One drawback is that each work reflects the analysis and conclusions of only one manager, something which makes the seasoned management researcher a little uneasy.

Another development has been the explosion of 'How to' management books dealing with almost every conceivable aspect of the subject from performance appraisals to presentation skills, from time management to teamwork. These are also pragmatic in that they provide accessible and easily assimilated information for managers and some of them are very good. But the advice provided can be simplistic, often summarising major areas of the manager's work in bullet-point checklists. Management is a difficult business and is becoming more so. Managers must guard against the temptation to adopt simple 'off-the-peg' solutions to complex problems.

Post-modern management

If all these changes are taken together, management needs to take a leap into a post-industrial business environment which is bound up with what Donald Horne has called 'the public culture'. Describing this new culture as part of the *post-modern* condition, Hewison (1990) aptly refers to it as a managed, official culture, supported by both public and private corporations. This culture has become enmeshed with commerce to the extent that culture itself has largely been turned into a commodity, often mediated most effectively by television – but perhaps, in future, by computer-based information networks.

Quality and environmental management

Quality management and environmental management became major issues for managers in the 1990s. Organisations in almost every sector, public as well as private, came under pressure to deliver their goods and services not only more efficiently, but to higher quality and environmental standards. To improve one of these isn't too difficult, but to produce more added value (better quality using less resources) in environmentally acceptable ways became a central challenge for managers in almost every field of activity.

These dual issues of quality and environment have become linked. There are significant parallels between the British and European standards covering these topics and the remit of a CIRIA project, started in 1996, was to examine the possibilities for integrating the management of quality and environmental impact (as well as health and safety).

The quality movement took off first, but the explosion of information about environmental damage in the late 1980s and early 1990s has forced many managers to think about building the concept of sustainable development into their business objectives and then manage responsibly. Although environmental problems are global and need to be tackled on a broad front, much of the detailed work of overcoming them needs to be done locally, often by individual companies (Roberts, 1994). Construction managers must share in this process and urgently review processes and practices. Even small construction firms must take the environment seriously, because there are so many of them and their impact is therefore substantial (Fryer, 1994a).

Summary

This chapter has looked at the main strands of management thinking, how they have developed and how they apply to construction. Some of the early guidelines were based on experience of mass production and processing industries. They were useful in their time and laid the foundation for a more systematic approach to management. However, they have not offered the best guidance for managing construction. As technology and markets have changed, the relevance to all industries of the early management principles has been seriously questioned.

Managers have also recognised that they must give more attention to the social aspects of work. In the long run, an organisation cannot be successful in economic terms unless it is also a success in human terms, providing meaningful work and proper rewards.

There are similarities in managers' jobs because some of the tasks they perform are the same. But the differences between managers' jobs tell us more about the process of management. Managers' roles vary with their level in the organisation. Some managers deal mainly with technical matters whilst others spend their time on strategic issues.

Construction must be managed flexibly because some of the problems facing construction managers differ from those found in factory-based industries. The separation of sites from head office and the temporary nature of project teams, taking their members from many different organisations, demand a special 'task-force' manager, who is self-reliant, adaptable and capable of inspiring teamwork under difficult conditions.

More importantly, the need for flexible management arises from far-reaching changes in society and its economic and value systems. The globalisation of business means that the construction industry will increasingly work in an environment of tough, international competition, adopting innovative organisational structures and work practices in order to achieve new standards of quality, safety and environmental protection set by knowledgeable, demanding clients from many parts of the world.

Discussion and questions

A construction director of a large building company, who is also a part-time lecturer in the School of Built and Natural Environment of a UK university told his students that 'the purpose of a management course is to teach students about management, not to teach them to be managers'. Do you agree or disagree with this statement? Discuss.

Based on your experience at college or from working in the industry, describe some ways in which the principles of Scientific Management and the Contingency Theory approach to management are still used in organisations. Do you believe these principles and theories will ever cease to be a part of organisational life? Discuss.

Chapter 2
Managers and Their Jobs

The tasks of management

Asked what they do, most managers will answer with words like *planning, organising, directing* and *controlling*. A handful of words like these have dominated management thinking since they were introduced in the early 1900s by Henri Fayol, the French industrialist. Yet these words are too vague to tell us much about what managers actually do and they take little account of the differences between managers' jobs. Everyone agrees that managers make plans, but so do other people. Moreover, a site manager's plans are quite different from those of a board of directors.

Many management writers add *communicating* to the list of tasks, but it seems more useful to regard this as a management skill. After all, managers are communicating when they give orders (directing) or arrange for materials to be delivered (organising). Moreover, communication is a two-way process and the manager should often be on the receiving end.

Some managers also rank *co-ordinating* as a management task. This is a useful term but Stewart (1997) argues that it is too general to be called a separate management function. It is hard to distinguish co-ordinating from organising and it seems to overlap with directing and motivating too. Co-ordination involves planning, as in deciding who should do what and when. Swedish researcher Sune Carlson was perhaps the first to seriously question the validity of co-ordination as a separate management task, arguing that it does not describe a particular set of actions, but rather all operations which lead to unity of action. In other words, co-ordination is another word for management itself.

Peter Drucker has added a further task which he feels is a vital part of management – *developing people*. The effectiveness of an organisation will depend on how well managers counsel and support their teams. Managers can bring out the best in people or they can frustrate and stifle them.

Planning

All managers plan. They set objectives, try to anticipate what will happen and devise ways of achieving their targets. However, some planning is long-term, extending over a period of years, whilst other plans cover immediate targets, achievable in a week or less. Long-range planning involves more risk and uncertainty, for it is difficult to know what will happen; short-term plans are usually based on more reliable information.

Planning has grown exponentially both in construction and in other industries for the following reasons.

- There are more large, complex projects, with a lot of people and resources to co-ordinate.
- The increasing reliance on sub-contracting means that the work of many organisations has to be co-ordinated.
- There are greater external controls over business activity, additional constraints have to be met and approvals obtained.
- Markets are more turbulent and economic and social change has accelerated, making the future less predictable.

The main features of a good plan are that it is realistic, flexible, based on accurate information and readily understood. The stages in planning are:

- Set clear performance objectives, usually in terms of time, quality, safety, cost and environmental impact.
- Identify accurately the resources needed and action to be taken to achieve objectives.
- Decide on the best action to take and the most effective use of resources.
- Set up procedures for monitoring the implementation of the plan.

Operational research has developed with the growth of large, technically complicated projects. Computer-aided project planning has come into increasing use. Despite such tools, planning remains difficult. The information needed for such thorough planning is seldom available and the manager rarely has enough time to plan properly.

Some construction firms have set up planning departments in which specialists prepare plans on behalf of, or in collaboration with, management. In other companies, plans are drawn up by the managers themselves. The latter may be less skilful in planning techniques, but they are likely to be more committed to a plan they have produced themselves. Centralised planning used to have the advantage that it made possible the use of a computer as a planning tool. Companies are taking up computer-aided project management (CAPM) software quite widely. They are tending to use it initially for planning and scheduling and, as they become more familiar with the software, are gradually exploiting the package's full potential (Sturges et al., 1997).

Plans are usually converted into bar charts – visual statements backed by descriptive notations. They show what the targets are, how they are to be achieved and, ideally, what activities are critical to completion. Plans should be expressed as simply as possible. They must be flexible to allow for unforeseen events. In construction, every new contract is a period of uncertainty for the construction manager.

Organising

Managers are organising when they put plans into action – allocating tasks to people, setting deadlines, requesting resources and co-ordinating all the tasks into a working system. Questions arise about how far to go in splitting the total operation into individual tasks. An easy task will not provide a challenge and the operative may become bored. A task which is too demanding may cause frustration. In either case, the task will fail to motivate and will not be performed efficiently.

Another problem is to decide how fluid to make the boundary of each job. In the building trades, jobs have been quite tightly defined by custom and job boundaries may be vigorously defended. Technical and supervisory jobs may be more flexible, with scope for the manager to vary the individual's job to create interest, improve motivation and skills and meet changing demands.

However successfully managers divide up the total operation into individual jobs and match them to people, they still have to co-ordinate them, so that one work group is not held up by another and materials are there when needed. Activities like these take up a lot of the manager's time and it may be difficult to analyse clearly what the manager is doing. The process of organising becomes inseparable from planning, directing and controlling.

People are the manager's major asset, but some costly resources must be managed too. Plant has to earn its keep; materials and components must be stored, handled and used efficiently to avoid waste. This is all part of organising and a good plan will indicate when plant and materials are needed, what stocks to hold and when to call up deliveries.

The task of organising is very specific to the manager's role. For the personnel manager, whose concern is people, organising will not be the same as it is for the site manager, who has to co-ordinate a diverse range of material, plant and labour inputs and integrate the work of sub-contractors with that of the company's own labour.

Directing

This task involves leading, communicating and motivating, as well as co-operating with people and, sometimes, disciplining them. So central are these to the manager's work that some definitions of management put directing people at the focus. The most carefully prepared plans are useless unless people are effectively directed in implementing them. At the same time, if plans have not been made and resources not organised, work will be misdirected. People will pursue the wrong goals or their efforts will not be properly co-ordinated. Clearly, the manager's tasks are inseparable.

To direct people effectively, the manager must:

- have some influence or authority over them;
- develop a style of management acceptable to them;
- earn their respect and co-operation;
- empower them.

Delegation is an aspect of directing people which has been widely discussed. It involves passing authority down the management hierarchy. Techniques like Management by Objectives involve directing people by setting them targets rather than tasks. The manager tells subordinates what must be achieved but gives them some freedom to choose how to go about it. Self-managed teams are a development of this approach.

Many managers find it hard to delegate, with the result that they are overworked and their subordinates become frustrated. Delegation means giving people more control over their work. There may be limits to how far this can be taken, but there is little doubt that managers could do a lot more to involve employees in deciding work methods and allocating tasks within their groups. The building trades have always provided more scope for giving operatives discretion over their work than is possible in machine-paced factory work.

Delegation is vital to staff development for it provides subordinates with new experiences at a measured pace, suited to their abilities and ambitions. Carefully monitored delegation of tasks and responsibilities – *coaching*, in other words – has been recognised by construction managers as one of the most potent methods for developing managers (Fryer, 1994b).

Controlling

This task involves comparing performance with plan. The plan is the yardstick, without which the manager cannot control anything. If the manager does not control performance, the plan is of no value. So, planning and controlling are dependent on each other and the manager must appreciate this. Apart from environmental impact, the main factors to be controlled, whether on site or in a contractor's or designer's office, are time, safety, cost and quality of work. Time is monitored by assessing progress against programmes, whilst quality and safety yardsticks are provided by specifications and regulations. Priced bills of quantities, subcontractors' quotations and the estimator's unit rate analyses contain the information for controlling project costs.

Because the term controlling can have connotations of punishment and censure, some managers and writers prefer to talk of *reviewing*, *monitoring* or *measuring*. After all, much of the manager's controlling work consists of obtaining information (feedback) and comparing it with various documents. A variance between performance and plan may lead to censure, but more often will simply result in the manager taking corrective action. One could say that controlling includes both, (a) reviewing and monitoring operations, and (b) taking decisions to correct variances.

The site manager is controlling when he/she decides to bring more personnel on site after bad weather has delayed progress.

The difficulty over the definition of management words is endless. For instance, when a site manager decides to fence a storage area on site, or use the services of a security patrol at night, is the manager controlling or organising?

Developing staff

Many writers have argued that people are an organisation's most important asset, particularly in a labour-intensive industry like construction. The effective use of a company's resources, whether on a building site or in a designer's office, depends on the motives, abilities and attitudes of people. Good managers have recognised implicitly the importance of staff development for a long time. But it required legislation to make many firms take a serious look at the problems of training and development. This was spearheaded by the Industrial Training Act 1964, which led to the setting up of the Industrial Training Boards, although many have since been disbanded.   *C. I. T. B*

There has been a spectacular growth in formal training in construction and other industries, but many managers are sceptical of much of its value – and rightly so. Formal staff development programmes which take people out of the organisation to acquire new knowledge and skills have certain advantages over learning 'on the job', but the methods used have inherent problems. Many training activities are costly and do not produce the results expected.

Managers and training tutors have put a lot of effort into finding more realistic ways of developing staff. The emphasis has gradually shifted from teaching people facts (which they can usually find out for themselves) to helping them learn skills. Some of the more exciting training activities now focus on practical problems rather than subjects, and managers are realising that people learn best when they work at their own pace, using the study methods they prefer. Managers can play an important part in developing their subordinates by giving them increasingly difficult tasks, mentoring and counselling them, and arranging periods of job rotation in which they experience other parts of the company.

Many managers and management tutors are most interested in helping people learn how to learn. This emphasis on the process of learning rather than its content is most appropriate in a time of rapid change, when knowledge quickly becomes obsolete but what the individual discovers about how to acquire fresh information and skills equips him or her to be an adaptable, life-long learner.

Indeed the concept of life-long learning has been embraced by government through the University for Industry and Learndirect, by professional institutions such as the Chartered Institute of Building and the Royal Institution of Chartered Surveyors and by the vast majority of higher education institutions.

How managers spend their time

The early management writers tried to build a model of the ideal manager. Their focus was on what managers ought to do. A few studies, mainly since 1960, have looked at what managers actually do. Henry Mintzberg (1976, 1980) concluded that the manager's work is characterised by brevity, variety and discontinuity and that managers prefer action to reflection. Several researchers, including Mintzberg, found that managers spend a lot of time in informal, face-to-face communication with people. This can often account for 80 per cent of their working day.

Much of the information collected in the 1970s supported these ideas. Managers were not very systematic and preferred informal, 'unscientific' methods. They would rather talk than write and kept many of their plans locked up in their heads. Their decisions were often intuitive and political, their motives private and hard to define. Mintzberg challenged the traditional, formal management ideas and argued that managers perform an intricate set of overlapping roles, which he called:

- *Interpersonal*. The manager is the group's figurehead, leader and liaison officer, performing rituals and ceremonies, motivating and directing people, and developing a network of contacts and relationships with people outside the manager's group.
- *Informational*. The manager, as monitor, disseminator and spokesperson, is the nerve centre of the unit, reviewing data and events, giving and receiving information, and passing information from the group to others outside. He or she may have to deal with the public and people in influential positions.
- *Decisional*. The manager is a resource allocator, entrepreneur, negotiator and trouble-shooter, searching for opportunities, initiating change and coping with crises. The manager makes decisions and argues about priorities, allocates materials to people and people to tasks.

To perform these roles effectively, managers find ways of gaining control over their time; they use some of this saved time to decide priorities and the rest to discuss information and courses of action with their subordinates.

The roles identified by Mintzberg have largely remained valid, but his comments on managers' use of informal, unsystematic and intuitive methods have become less applicable.

Whilst many managers would still prefer these informal methods, this is not possible in a climate characterised by disputes and litigation. As managers come under pressure to meet tougher performance targets, they have resorted to using computer software to assist in the analysis and synthesis of information, formalising their tasks and committing plans and decisions to paper.

The manager's skills

During the 1960s and 1970s, some of the interest focused on the manager's skills. Robert Katz (1971) identified three broad classes of management skills:

- *Human skill*. This is the manager's ability to work as a group member and build co-operative effort in the team, to communicate and persuade. Managers with good human skills are aware of their own attitudes and assumptions about people and are skilled in understanding and influencing people's behaviour.
- *Technical skill*. Most managers have previously occupied a craft or technical role and are proficient in some aspect of the organisation's work. They have acquired certain analytical abilities, specialised knowledge and techniques, much of their training having centred on developing such skills and knowledge.
- *Conceptual skill*. This is the ability to see the organisation as a whole, how the parts affect one another and how the firm relates to the outside world. The manager with conceptual skill appreciates that a marketing decision must take account of local conditions, the state of the industry, competition and other political, social and economic forces. Such a manager recognises that the decision will affect production, people, finance and other aspects of the business.

Managers use different combinations of skills for different kinds of management work. Katz argues that human skill is important at all levels of management, but especially for junior managers, who have wide-ranging and frequent contacts with people. Junior managers also rely heavily on technical skill, but this becomes less important for senior managers, who depend more on conceptual skill. Indeed, they may get by with little technical or human skill if their subordinates are competent in these.

Each of Katz's skills is really a family of skills, which can be further analysed if managers want a closer insight into their jobs. For example, human skill encompasses the ability to deal with peers and colleagues, bosses and subordinates. It includes skills for negotiating, persuading, empowering, gaining support, encouraging and counselling – and, sometimes, disapproving and giving a reprimand. Many management tasks demand the use of several skills. Resolving a technical problem may require more than technical skill, if it affects people.

These three groups of skills can be matched against Mintzberg's analysis of management roles.

Interpersonal skills

A number of researchers have sought to analyse, and in some cases classify, the skills of the construction manager. Love and Haynes (2001), for example, suggest that construction management graduates should possess problem-solving skills, specialist knowledge, an ability to communicate and an understanding of 'how' information and communication technologies can be used to improve business practice – skills that reflect Katz's classification.

Discussions with 22 practising construction managers and the responses of 142 senior, mid-level and junior managers in a postal questionnaire, revealed that human skills are highly valued (Egbu, 1999). Asked to identify those skills and areas of knowledge which were significant in refurbishment work, construction managers prioritised:

- Leadership
- Communication
- Motivation of others.

Such skills help to create good relationships with colleagues, subordinates and sub-contractors' personnel. They help the manager to develop a network of contacts for securing action and the prompt exchange of information and instructions.

On site, the manager has to deal with sub-contract personnel. They have a contractual duty to co-operate with the main contractor, but their main loyalty is to their own employers. Their priorities, attitudes and values may differ from those of the main contractor's team.

Managers must think carefully about how to influence people and must adapt their social skills to meet the situation. Technical and conceptual skills are essential, but their potential cannot be realised if the manager fails, through lack of human skills, to bring together a cohesive team.

Decision-making skills

Construction managers, like managers in other industries, attach a lot of importance to skills used in decision-making and problem solving. They have been taught, or have found that others expect them, to make quick decisions, thereby showing their competence and resolution. Failure to make a 'snap' decision is often thought to show weakness or lack of self-confidence.

Clearly, this can be dangerous. It is true that quick decisions are often called for and delay can cost money, but a bad decision is sometimes more costly than no decision at all. The manager has to recognise that solving a problem and reaching a good decision sometimes takes time. It relies on more than intuition and judgement. A mix of technical, human and conceptual skills is often needed to achieve a satisfactory outcome. A poor decision rarely brings credit to the manager or the firm in the long run.

Another common problem is that managers spend far too much time making short-term decisions and neglect the long-term issues. This is hardly surprising, since they are judged on the success of current operations. However, many commentators have called for a better balance between immediate and long-term decisions. The attention paid to long-range issues will ensure that the firm keeps pace with developments and survives in difficult times.

Information handling skills

Handling information has become more central to the manager's work as projects and organisational procedures have become more complicated. Managers need a combination of human, technical and conceptual skills for locating and interpreting information, judging its importance and accuracy, sorting facts from opinions and displaying data in various ways. The ability to pass on information clearly, concisely and in an acceptable form is vital nowadays. One problem is that a vast amount of information is available to managers – more than they can possibly absorb. Much of it is not presented concisely or in a suitable form. Managers waste a lot of time sifting through information, picking out important points. IT should have reduced this problem, but it often makes matters worse instead.

The manager's skills are so important that they are discussed in detail in later chapters. But managers' tasks and skills are not the distinguishing feature of their work. Most people make decisions, handle information, draw up plans and organise resources. What distinguishes managers from others is their organisational setting and authority for getting things done. To do their jobs properly and meet objectives, they need *power* over others.

The manager's power

To perform their work of getting things done through people, managers need to exert influence or power over them. This presents special difficulties for contractors and project managers because many of the people working on a project are employed by other organisations, such as professional practices and specialist contractors. These employees owe their allegiance mainly to their companies and not to the project. There are, however, many reasons why people will co-operate with a manager. They may do so because it leads to some reward or removes the threat of unemployment. They may seek the manager's respect or simply like the manager as a person. Managers must know why people co-operate with them, because this is the basis of their power. Four main power bases recognised in management are discussed below.

Resource or reward power

The manager controls some of the resources and rewards that others want and can influence the salary increases, bonus earnings and promotion prospects of his/her team. A site manager may sanction payments to sub-contractors, exerting some indirect power over their site personnel. The manager may have some influence over whether a sub-contractor is used again.

   Resource power is seldom popular. People dislike the idea that their co-operation can be bought and dislike it even more when they have to co-operate with an unpopular manager. But there is no doubt that managers can secure a partnership of effort by rewarding good performance (and perhaps by penalising bad). Reward power will only work, however, if the employee wants the kind of reward the manager is offering and believes that it is conditional on meeting the manager's performance targets. There is a subtle relationship between this and other sources of influence. Sometimes rewards work only in conjunction with other forms of power.

Position power

Managers have some power because of their positions in the organisation. Sometimes this is called *legitimate* power or *role* power. Other people recognise that the manager has the right to give orders, control progress, inspect work and sometimes reject it. Position power is strongest when the manager has explicit backing from senior management. Even when this support is weak, other employees often reinforce the manager's power by expressing group norms about behaviour and attitudes to work, which put pressure on individuals to conform to standards which have become accepted in the organisation. On the other hand, group norms may work the other way, undermining the manager's power.

Position power gives the manager access to, and control over, certain information. Managers are a focal point in the communications network and control the dispersal of information within their teams and to other departments and organisations. The pieces of the information jigsaw often have little value until they are put together. Information displays synergy – the whole is greater than the sum of the parts. Sometimes, people withhold information from managers and deliberately or unwittingly reduce their power.

Managers also have access to people outside their work groups and to other organisations, and can tap their expertise and resources. Above all, they have the acknowledged right to decide how work should be organised and what should be done when things go wrong. These are powerful influences over people's behaviour. However, sub-contract employees may be more impressed by the role power of their own managers. The main contractor cannot rely solely on position power to gain their co-operation.

Personal power

Some managers have the personality, presence or charisma to influence others without recourse to other methods. Such influence may stem from the manager's appearance, manner, poise, confidence or warmth, dominance or decisiveness. More often, it depends on a combination of such factors.

Some managers rely heavily on personal power to get co-operation, but such power can be elusive and temporary. It works sometimes and with some people. It can disappear in a crisis and can seldom be relied on to consistently replace position power. Nevertheless, it is important and managers use it wherever possible to supplement their authority.

Expertise or expert power

Special knowledge and skills give the manager power over those lacking them. Most managers have some expertise which their subordinates lack and this reinforces their position. A project manager, responsible for co-ordinating the design and construction of a complex project, will depend on such expertise.

However, many architects and construction managers lack expertise in the other disciplines of the project team and can be at a disadvantage, particularly when dealing with specialist designers and contractors. For instance, an electrical sub-contractor's site supervisor will be able to exercise expert power over the main contractor's manager by virtue of his or her specialist knowledge and skills. To retain control, the contractor's manager minimises this counteractive power by strengthening other power bases and by becoming more knowledgeable about the sub-contractor's specialism.

There is a further power base – coercive power – based on threat or the fear it induces. Few managers rely on such power, although in some day-to-day situations they may use it very temporarily to achieve a quick result.

Managers should review the kinds of power they use and watch how others react to them. People respond in different ways. They may accept power, ignore it or rebel against it. Some comply with the manager because they think it is worthwhile to do so. Rewards and company rules often lead to such compliance. Others adopt or accept the manager's suggestions because they admire or identify with him or her. Some subordinates may model their behaviour on a manager they admire. Charismatic managers can get co-operation in this way, but people may become too dependent on them. Some people develop such commitment to the task that they carry out their duties with little supervision. Managers have only to keep a watching brief. These subordinates have adopted the goals and have internalised the values and attitudes of the manager as their own.

Most managers achieve their objectives using a combination of rewards, contractual procedures, rules, sanctions, expertise and personal qualities. The methods chosen will depend on the task, the people and the support the manager gets from the organisation.

Empowerment

The early 1990s saw a switch of emphasis from the importance of the manager's power to the need for employees to exercise power. Established notions of employee participation and involvement gave way to the concept of empowerment. Empowerment of employees is based on the premise that the people who actually do a job are in the best position to learn how to do it better. Empowerment aims to eliminate close management control and unnecessary rules, procedures and other restrictions. It gives employees more control over their work (individually and as groups) and the authority to make many of the decisions without asking the manager's approval.

It also means that managers must, to some degree, give up being *in* authority and spend more time being *an* authority – giving employees support and guidance so that they can exercise their empowered status effectively (Stewart, 1994). Seen in this way, empowerment does not lessen the manager's power, but changes the way it is applied. Moreover, empowerment is not just about giving authority to employees, but about providing them with the knowledge and resources to achieve work objectives (Stoner *et al.*, 1995).

Total quality management systems have embraced the concept of empowerment because it offers some significant benefits. Used effectively, it makes far better use of employees' skills, experience and commitment, leading to higher productivity and job satisfaction.

Dainty *et al.* (2002) suggest that empowerment strategies, used selectively, could play an important part in helping construction organisations to address increasing performance demands. However, they also identify many barriers to individual and team-based empowerment. They recommend:

- flatter organisational structures that embrace employee ownership;
- formal support networks to provide empowered individuals and teams with assistance, guidance and leadership;
- the apportionment of appropriate responsibilities for aspects of project delivery and overall strategic performance throughout the supply chain;
- placing employee participation within the context of the strategic need to secure employee commitment to organisational goals;
- a genuine and not merely cosmetic commitment to the use of empowerment.

Self-managed teams

As traditional hierarchies are broken down and organisations adopt flatter, leaner structures, the need for empowered teams increases. Sometimes called self-managed teams or autonomous work groups, the empowered team has great potential for releasing creative action, improving performance and building employee commitment.

But these teams need to be skilfully developed – they don't just happen. Training is needed so that the rationale and benefits of empowerment and team working are understood. Individual employees need to learn how to exercise power within their team: how to do this by communicating effectively and influencing others using appropriate behaviour; how to improve their skills for analysing situations and solving problems creatively; how to be assertive but not aggressive or domineering; how to negotiate compromise and reach consensus. Managers also need training. They must learn how to adapt their behaviour to interact with their new, dynamic teams: how to relinquish power; how to exercise authority when it is needed; how to act as a mentor to the group and its members; how to ensure in a non-controlling way that the team's achievements remain in harmony with the organisation's wider goals.

Empowering individuals and setting up self-managed teams is not always easy. For example, there can be problems associated with staff calibre – a manager who is unable to adapt to the new role, an employee who can't cope with problem-solving – and attitude problems – an employee who feels it is the manager's job to make decisions. Such problems can often, but not always, be overcome by training. Clearly, there are big advantages in training a team together. Attitude changes which are difficult to bring about in an individual are often easier to achieve within an established group. In the main, case studies of organisations which have tried empowerment of individuals and teams have shown impressive results.

Professionalism and construction management

Until perhaps the 1980s, few people would have described construction management as a profession. But the discipline has steadily gained in status and recognition in the eyes of clients and other built environment professions. Of course, the definition of profession has also broadened and includes many more occupations than it originally did. Murdoch and Hughes (2000) refer to the four defining characteristics of a profession as:

- A *distinct body of knowledge*. A special competence or identifiable corpus of expertise.
- *Barriers to entry*. Professional bodies which regulate entry through qualifying mechanisms.
- The goal of *service to the public*. The true professional places the public good before other objectives. This concern is exemplified by the profession's code of conduct.
- *Mutual recognition*. The profession is recognised by other professions and it recognises them.

There is no doubt that construction managers have behaved in an increasingly professional way. They have had to, in order to keep on top of the growing complexity of projects, increasing sophistication of clients and major changes in contract procurement. But, if one applies the above criteria strictly, construction management falls some way short of being a profession in the traditional sense.

There is no shortage of a corpus of knowledge, but the barriers to entry are somewhat ill-defined. This stems from the fact that there isn't a single professional body regulating entry or a single qualifying route. Engineers, architects or quantity surveyors could (and indeed do) perform the role of the construction manager if they have the necessary experience and skills. Engineers are frequently promoted to construction management. QSs who have worked extensively for contractors can progress to project or contract management roles. Organisations like the Chartered

Institute of Building and the Association for Project Management are about the closest thing to a professional institute for construction managers, but they have wider categories of membership. This also creates some difficulty in meeting the criterion of mutual recognition among professions.

A further difficulty is that most construction managers could not say, in all honesty, that public service is uppermost in their minds. Most of them would probably rate the interests of their employer or the client as their first priority. Of course, one can speculate that the established profession's members do not always put the public interest first. But at least they have codes of conduct and the threat of censure.

There are, of course, problems inherent in professionalism itself. The institutions it creates can easily become bureaucratic and resistant to change. Professions try to guarantee reliable service but cannot always succeed – all organisations contain a mix of people of varying competence. Professionalism can create rigid role demarcations which are not always in the best interests of specific projects, where flexibility is essential and interdisciplinary teamwork is needed (Murdoch and Hughes, 2000).

Summary

Managers' jobs are demanding, complex and varied. There are certain common features in the role of manager, but individual jobs differ markedly. Most managers, regardless of their field of operation, have to manage people, information and decision-making processes. They perform these roles using varying combinations of human, technical and conceptual skills to plan, direct, organise and control people and resources. The amount of time they spend on each role and its associated skills depends on their function and level, and on the abilities and motivation of team members. Staff development and mentoring is becoming an important part of most managers' jobs.

Research in the 1960s and 1970s showed that managers were not always the systematic, analytical thinkers that early management theories thought them to be. However, the development of information technologies has meant that managers now have better information with which to manage and powerful techniques for analysing information, planning and decision-making. Even though many managers prefer informal, intuitive methods, they have been forced to adopt more structured techniques in their work because commercial pressures are demanding ever greater efficiency and this has necessitated a more rational and systematic approach to management, and greater professionalism.

Most managers in the construction industry rely heavily on social skills. Good leadership and effective communication are needed in a wide range of situations. These skills are vital in site management, where the work of many organisations has to be co-ordinated.

Some of the power exercised by managers is based on their personal characteristics and behaviour. However, personal power cannot always be relied on to get results and the manager must utilise several power bases to maintain effective control, especially where sub-contractors are involved.

There is continuing interest in empowerment, a process which shifts some of the power from managers to employees, individually and as self-managed teams. Employees, being closer to the workface and having superior knowledge of the work and its environment, are often in a better position to make decisions; empowerment gives them the opportunity to use and develop their talents more fully.

Debate motions and presentation topics

Debates provide a valuable learning opportunity – allowing participants to develop and express their own ideas and to listen to the opinions of the opposition. As Northledge (1993) observes, in an effort to explain a point, you may frequently hear yourself expressing ideas in a form that you had not been quite aware of before – in short, discussion helps you think.

Motion 1

Expertise in the use of computer-aided planning techniques is more important to the construction manager than are people management skills.

Motion 2

The construction manager is unable to influence the motivation of individuals within the project team.

Motion 3

All construction managers should have a professional qualification in construction management.

Chapter 3
Organisation

Many small businesses work well without formal structure or rigid rules. The enthusiasm of the owners or managers keeps these firms on course. But as organisations grow, the work of more and more people has to be co-ordinated. Special attention has to be given to how tasks and relationships are organised and communications maintained.

Organisations as we know them today have only emerged in the last century or so, with the growth of industry and commerce. Many of them have become *bureaucratic* – that is to say, hierarchical, impersonal and controlled by a system of rules.

An organisation can be seen as a set of roles or positions rather than a collection of people. Employees can be replaced by others with similar knowledge, skills and attitudes. The posts or roles are arranged in a hierarchy and those higher up have authority over those lower down. Difficult problems are referred up the hierarchy to a level at which they can be solved, whilst decisions are passed down to the level at which they are implemented. Activities are broken down into manageable, specialised tasks. The number of subordinates each manager has is limited, so that effective control is maintained.

This approach gives rise to the classic 'family tree' organisation structure. Specialist advisers are needed and this leads to the *line* and *staff* distinction present in many companies. Line people are the generalists, responsible for production. Staff are specialists, who give them technical and administrative support.

There has been much criticism of bureaucratic organisations and managers have tried to improve them. A fundamental objection is that they become rigid and inefficient and do not take enough account of human behaviour. The result is breakdown and failure – people do not always comply with orders or accept company goals. They may challenge the power bases in the firm.

The idea that an organisation should be continually restructured and its rules altered as circumstances change was not widely accepted until after World War II. Change and uncertainty are now forcing organisations to be more flexible. Managers in many industries have experimented with temporary task forces or 'project teams', which can adapt quickly to new challenges and conditions.

Cicmil (1997) believes that the emerging paradigm of project management represents a renaissance of the discipline in a contemporary business context. Project management is perceived to have something new to offer organisations that are keen to improve their efficiency and effectiveness (Thoms and Pinto, 1999). Maylor (2002) concurs, stating that modern business is characterised by change and that a project-oriented approach to management could be used to transform client need into reality. Moreover, he recognises that the discipline is no longer dominated by the construction industry, believing that project management is applicable to all organisations.

Managers in the construction industry have wide experience of setting up temporary project organisations of considerable size and complexity, lasting months or even years. Project organisations are created for a specific purpose, have clearly defined goals, change their composition over their life span and are disbanded when the work is done. However, the need to improve performance within the construction industry is an inescapable conclusion of many government and professionally sponsored reports (Latham, 1994; Egan, 1998). Egan, some authors argue, is not asking the construction industry to reflect upon what it is doing already and do it better. Seemingly he wants the industry to operate in an entirely different way.

Organisational activities

One way of analysing an organisation is to consider it as a system and identify its sub-systems. In broad terms, the organisation can be split into a decision sub-system and an action sub-system, but a more detailed analysis suggests four major activities:

- *Deciding on objectives and policies.* An organisation must have a sense of direction and purpose. A high level sub-system works out priorities, sets standards, lays down codes of ethics and gives overall guidance.
- *Keeping operations going.* There is a sub-system for the routine productive tasks of the business necessary to achieve its purpose. This includes most of the production function, office administration and accounting system. Selling comes under this heading, but not the whole of marketing.
- *Coping with crises and breakdowns.* Things will go wrong. A 'trouble-shooting' sub-system deals with problems. Failures can occur anywhere in the organisation. A routine production task may break down because materials are delivered late. A marketing decision may fail because trading conditions change unexpectedly.
- *Developing the organisation.* Some activities are aimed at changing the organisation or its methods. Research and development and parts of production, personnel and marketing contribute to the organisation's development. For instance, the personnel function of staff development is a key aspect of organisational change.

In construction, deciding policy and developing the organisation are mainly the province of the parent companies. The project task-force will be largely concerned with keeping things going – getting the job built on time and within budget – and coping with operational problems.

Objectives

Whatever form an organisation takes, the ultimate measure of its success is whether it meets the needs of the people who have an interest or stake in it. Yet most industrial organisations have fairly limited objectives and have not always catered well for the needs of employees or local communities. Only exceptionally have such organisations attempted to take over some of the functions usually performed by society. In Japan, this approach is more common.

The objectives or goals of those contributing to a construction firm or project are not always clear. Yet managers need to know these goals to measure how well the organisation is doing.

Traditionally, managers have stressed *economic* goals like profitability, high productivity and expansion. Typical economic objectives are:

- To provide a fair return to shareholders.
- To satisfy clients' requirements.
- To utilise resources efficiently.
- To improve the company's position in its markets.
- To develop products which can be sold profitably.

Profit has been the main measure of business success, although it has come under attack from time to time. Changing attitudes have forced profit into a less central role in management thinking, where it is viewed in the context of other objectives.

The modern view is that an organisation is a coalition of people. The organisation, being mindless, cannot have goals – only the people in it can. Therefore, all objectives are really *social*. The so-called 'organisational objectives' are the goals laid down by the more powerful or influential people in the business. Increasingly, these goals have been challenged by groups within and outside the organisation. Unions, governments and other bodies have scrutinised organisations and put pressure on senior managers to modify their actions and expectations. Managers have had to make changes to meet statutory demands and to ensure the continued co-operation of the workforce.

The Business Impact Task Force (2000) argues that organisations should operate their business in a socially responsible way and uphold the following principles:

- To treat employees fairly and equitably.
- To operate ethically and with integrity.
- To respect basic human rights.
- To sustain the environment for future generations.
- To be a caring employer in the community.

Not only will adherence to these principles be of benefit to society, but it will also build business sales, build the workforce and build trust in the company as a whole (BITF, 2000). Corporate social responsibility (CSR) is here to stay as debate within the EU focuses on the introduction of mandatory regulations.

Environmental objectives

Since the 1950s there has been growing concern about the impact of human societies on the global environment. The energy crisis of the early 1970s highlighted the potential vulnerability of developed economies to changes in energy supply (mainly oil) and triggered the development of efficiency improvements. During the 1980s much of the impetus for energy efficiency, at least in the UK, was dissipated as energy supply increased and prices fell in real terms. However during this period much more fundamental concerns emerged about the impact of the burning of fossil fuels on the global climate and ecosystems. At the same time, it became increasingly clear that tackling the environmental problems would require action that reached every part of the global community and dealing with economic and social issues as well as technology. The need to understand and tackle the problems involved prompted the United Nations to set up the Brundtland Commission in 1983, giving rise to the notion of *Sustainable Development*, which was defined as 'development which meets the needs of the present without compromising the needs of the future' (World Commission on Environment and Development, 1987). This idea is fundamental to the environmental challenge that faces all societies.

The construction industry has a crucial role to play in ensuring that the goals of sustainable development are met. The buildings and infrastructure it constructs consume resources, impact on ecosystems through activities such as material extraction and increase the use of energy (and consequent greenhouse gas emissions) during construction and in occupation. Increasingly, the industry as a whole, designers, constructors and upstream suppliers, will be required to respond to changing demands, designed to minimise environmental impact. The need to mitigate climate change through increased energy efficiency of buildings is already a central plank of UK and European policy (DTI, 2003; European Commission, 2003) and is framing the development of building regulations (Bell and Lowe, 2000, 2001; Lowe and Bell, 2000) and impacting on technology (Lowe *et al.*, 2003a and Lowe *et al.*, 2003b). In addition, it is recognised that some climate change is inevitable and that the industry will need to adapt to the changing climate in the way developments are designed and constructed.

The changes required in response to environmental objectives will have far reaching impacts. Not only will technological change be necessary but construction industry organisations will have to adapt to an environmentally aware market place, assess the financial risks of development in a climate prone to increased flooding and extreme weather events, and develop commercial strategies that recognise the impacts of environmental requirements (maintenance of habitat, increased flood protection and the like) on such things as land values (Hertin *et al.*, 2003). Perhaps the most fundamental change that organisations will be required to make in the first decade and a half of the 21st century will be the need to marry the long time horizons involved in achieving the goals of sustainable development with short term commercial imperatives. This is likely to lead to fundamental shifts in the way organisations see themselves and in the way that they operate.

Underlying objectives

However, many so-called objectives are not objectives at all. They are the means by which underlying goals are achieved. For example, profitability can be viewed not as a goal, but as a way of ensuring that organisations survive, wages are paid, shareholders are rewarded and, perhaps, managers' self-images are satisfied! Similarly, the social objective of secure employment is not a goal in itself, but a means of giving employees satisfaction and self-respect from having a place in society and the ability to supply their needs.

If any goal can truly be said to be organisational, *survival* is perhaps the only one. The survival of an organisation affects owners, employees, their families, shareholders and the community. In many organisations, profit is a prerequisite for survival and, for this reason, is important.

The purpose of setting up project organisations is to build buildings and structures. Construction can be thought of as a strategy for achieving a variety of goals for the people involved. Ideally, these goals will be achieved by completing projects on time, at the right cost and quality, but in practice some of the objectives conflict.

Managers use time, quality and cost to measure project performance. These criteria are more quantifiable than social objectives and therefore easier to use. They include cost targets, dates for starting and finishing each operation and specifications of materials and work.

However, it does seem important that the economic goals of the companies contributing to a project should help achieve social and environmental objectives. Organisations should ultimately serve people, both the stakeholders in the business and the members of society at large. People should not be the slaves of organisations, nor should their environment be seriously degraded.

Characteristics of organisations

Organisation structure

Most organisations are not designed, they grow. They eventually reach a size where it becomes necessary to write down who does what, otherwise the managers lose sight of the whole picture and jobs are forgotten, or done twice. The purpose of organisation structure is to ensure that work is allocated rationally, that there are effective links between roles, that employees are properly managed and that activities are monitored.

Structure is the skeleton of the business: it creates enough standardisation of roles and procedures to allow work to be performed economically and to keep the organisation in tune with the procedures of the firms with which it does business. It facilitates control by creating a communications network of instructions and feed-back.

When designing or improving an organisation, senior managers must ensure that:

- tasks and responsibilities are allocated to groups and individuals, including discretion over work methods and resources;
- individuals are grouped into sections or larger units and the units integrated into the total organisation;
- formal relationships are set up, spans of control considered and the number of managerial levels decided;
- jobs are clearly defined, but are not too rigid or specialised;
- authority is delegated and procedures are set up for monitoring its use;
- communication systems are created, improving information flow and co-ordination;
- procedures are developed for performance appraisal and reward.

Structural weaknesses in organisations lead to many business problems, including too much paperwork, people overloaded with work, poor or late decisions, inability to cope with change, low morale, industrial conflict, increased costs and lack of competitiveness.

Specialisation

Most organisations have introduced specialisation in the belief that it leads to better use of people and resources, but it has drawbacks too. It leads to fragmentation and the need to control and integrate tasks more tightly.

In construction, the fragmentation is very marked. Parent firms contribute only a specialised input to projects, and jobs within that limited input are themselves specialised. Specialisation leads to isolation and can cause co-ordination problems. For instance, the R&D laboratory of a heating and ventilation contractor may be annexed in a country house, whilst top managers occupy a high-rise office in the capital. Production takes place anywhere the firm is willing to work, perhaps over an area of hundreds of square miles.

In professional and technical jobs, specialisation can create challenge; in clerical and manual jobs it can lead to boredom. Writers like Friedmann and Argyris have argued that highly routine jobs, requiring little learning, are not a humane use of people because their full potential cannot be tapped. Some managers have recognised the need to adapt work to meet employees' needs and their companies have successfully introduced schemes to enlarge and enrich jobs, making them more satisfying. Many firms, however, have not come to grips with the problem.

Drucker (1968) offered three simple guidelines for improving routine jobs:

- A job should be a distinct step in the work flow, so that the worker can see the result.
- The design of a job should allow workers to vary their pace.
- A job should provide an element of challenge, skill or judgement.

It may be impossible to achieve this in every job, but it can often be done for small work groups, where roles can be swapped, provided that rigid job demarcations are dropped.

There are arguments for and against closely-defined jobs. Drawing up a precise job description forces management to think clearly about the purpose and content of the job and both management and employee know where they stand. On the other hand, job descriptions can be inflexible and unrealistic when conditions are changing fast.

Indeed, the future success of organisations will depend less on traditional jobs and more on the creative use of information, ideas and intelligence – things that don't fit neatly into old specialisations. Work will *have* to be more flexibly defined and organisations will have to be even more adaptable. There are many reasons for this and Handy (1995) explains them well. New roles that we never heard of before will (and already have) come into existence and many of them will need to be organised and managed in new ways.

Information technology is just one of the factors affecting organisation structures. Computer-based decision-support and information systems can lead to different choices of structure and influence the extent of de-centralisation of decision-taking and control (Mullins, 2002). Regrouping of tasks may result from developments in information management, creating new specialisations.

Hierarchy

Most organisations are hierarchical. They are made up of a series of tiers, each having authority over the levels beneath them. The number of levels in the hierarchy may vary from two in a small building firm to a dozen or more in some large organisations. The size of the firm largely dictates the number of tiers, although management may decide to widen spans of control to limit the number of levels.

Where spans of control can be widened successfully, there is a strong case against tall organisation structures, which increase overheads, create communication problems and weaken senior management control. The more levels in the hierarchy, the harder it is to distinguish between the duties and responsibilities of people at different levels. This can restrict the scope for subordinates to show initiative, thereby reducing their motivation and job satisfaction. A small organisation can opt for a flatter structure with few levels of management or it can keep spans of control small, making the structure taller. It will usually choose the former.

A large organisation has a more difficult choice. It is necessary to maximise the span of control to prevent the structure becoming too tall, but clearly there are limits beyond which effective supervision becomes very difficult.

Large organisations employ more specialists who relieve the line managers of some of their tasks. This makes it possible to increase spans of control to levels which would otherwise be impractical. The manager's span of control will also depend on the work and the people involved. Routine, repetitive jobs may need less supervision than complex, non-routine tasks, but this also depends on the capabilities of employees. The span of control can be widened if the manager is very able, if subordinates are competent and willing, and if they share the same workplace. An area manager may have subordinates spread over a wide radius.

Downsizing

A phrase which became popular in the 1990s, downsizing refers to the trend among many organisations to reduce their overall size, often by decreasing the number of levels in the hierarchy, producing a flatter structure. Businesses have done this partly to create flexibility, so that they can respond more quickly to change, and also to achieve improved efficiency to satisfy new quality management systems.

Downsizing has become a competitive imperative for many organisations, but it creates an ethical challenge for managers, who have to cope not only with redundancies but with problems of retaining the loyalty, motivation and sense of security of the employees who stay (Stoner *et al.* 1995). Moreover, downsizing has in many cases resulted in a loss of key skills, knowledge and organisational learning (Littler and Innes, 2003).

Centralisation v. decentralisation

An important structural feature affecting an organisation's efficiency is the degree to which it is centralised or decentralised. This can be measured by:

- The extent to which managers delegate authority and decisions from the top to the lower levels in the business.
- The extent to which the administrative functions of the firm are carried out at head office, rather than being spread through the organisation. For instance, some contractors have a central buying department for all material purchases. Others allow managers in different areas or divisions to organise their own purchasing.

Decentralisation can be based on area or product. If a contractor is working over a wide area, regional decentralisation may be vital to cope with local conditions. If a company builds hospitals and factories and also undertakes speculative housing work, product decentralisation may improve organisational efficiency. In speculative housing, policies and procedures for accounting, estimating, buying and so on, will differ from those suited to contract work.

However, no organisation is likely to be totally centralised or decentralised. Most firms strike a balance between the two. What this balance should be depends on several factors:

- *The size of the organisation.* This is important because the larger it gets, the harder it becomes to control everything from the top, without depriving junior managers of authority and autonomy. Since the 1950s, when the problems of large-scale organisation were becoming clearer, managers have become keen on decentralisation because it permits more realistic control and greater flexibility.
- *The type of work the firm undertakes.* This is important for two reasons – diversity and pace of change. If its operations are diverse, it is difficult for the top managers to keep track of everything. If conditions are changing fast, it is better to leave more of the judgements and decisions to people on the spot. Indeed, some technical decisions have to be delegated because the junior staff are more technically up-to-date.
- *Staff capabilities and motivation.* A decentralised organisation is often more satisfying for people to work in, but staff must be competent and willing to make the necessary decisions. This means that the organisation must have good calibre employees in key positions in its decentralised units.

Centralisation and decentralisation each have their strengths and weaknesses, so a compromise between them is usually best. The advantages of decentralisation are the drawbacks of centralisation, and vice versa, so it is only necessary to consider one of them. Table 3.1 summarises the points for and against decentralisation.

Table 3.1 Advantages and disadvantages of decentralisation.

Advantages	Disadvantages
Makes junior posts more challenging	Makes overall control more difficult
Decisions are taken by those who have to live with the results	Difficult to keep track of decisions taken
Encourages people to show initiative and creates greater commitment among employees	Difficult to keep an overall perspective and safeguard the interests of the whole organisation
Easier to judge the performance of a manager who is responsible for a decentralised unit	Creates higher administrative costs owing to duplication of specialists

Rigidity v. flexibility

Some of the differences between firms were highlighted in Chapter 1. For instance, Burns and Stalker (1966) contrasted the rigid, mechanistic organisation with the more flexible, organic one. As with centralisation, it is unlikely that any firm will adopt an extreme policy. Most will opt for a structure somewhere between the extremes. Size is again important. The larger the firm, the more formal and inflexible it is likely to be, although the degree of rigidity can vary a lot from department to department. For instance, the production part of a firm is often more formal than the sections dealing with marketing or research. In a construction firm, the buying department is likely to be more rigid than the estimating department, whose workload is usually varied and unpredictable. There are several indicators of rigidity in a company, as described below.

Rules and procedures

Rigidity often shows up in the number of rules and procedures used and the extent of written, rather than spoken, communication. All firms have rules governing who is allowed to authorise cheques, sign contracts, buy materials, and so on. The rules are not always written down and this can give a false impression of informality in a formal set-up. Up to a point, procedures and rules are necessary to ensure that tasks are allocated and performed systematically. They underpin the authority of managers and help reduce the number of decisions to be taken. But they can become an end in themselves instead of a way of improving efficiency. Rules and procedures should be kept under review to ensure that they still apply.

Some formality is imposed on the organisation from outside. For example, a contractor's disciplinary procedures are partly dictated by legislation and codes of practice. Similarly, the statutes impose many site safety rules on the contractor.

Organisation charts

Many firms draw up some form of organisation chart, a kind of map of the firm. The chart gives an overall picture of how roles are allocated and helps senior managers to identify organisational problems and develop procedures and succession plans. It gives new employees a better idea of the 'shape' of the organisation.

But organisation charts have their limitations. They give only a crude picture, unless there are detailed explanatory notes to accompany them. Even then, they tend to oversimplify relationships because there is a limit to the amount of information they can show. They tend to emphasise vertical relationships in the organisation, rather than horizontal. They stress the formal links, rather than the informal. They give little indication of status differences between managers on the same tier in the hierarchy. Most important of all, they are static and can quickly become out of date. When this happens, organisation charts are not simply useless but misleading.

Job descriptions and organisation manuals

These documents set out the functions or duties of individuals and departments and the relationships between them. They can be quite detailed. They are intended to make the organisation more efficient, but they can create rigidity, making it hard for people to respond sensitively to unexpected changes.

Paperwork and committees

Paperwork and meetings are a feature of most organisations. The extent to which firms use forms, reports, memoranda and committees, and the diligence with which files and minutes are kept, give a measure of the firm's formality. Many committees meet regularly, even when there is little to discuss. Forms are often filled in, even though the information is little used. Reports are written and considered at length but, all too often, no action is taken. Such waste of time and resources must be eliminated.

Some records, such as accident report forms and records of disciplinary meetings, are kept to comply with legal requirements and codes of practice.

Types of organisation

Line and staff organisations

Most construction firms have an organisation structure of the line and staff type which has dominated management thinking for many decades. The 'line' managers are responsible for production. They pass instructions and information down the hierarchy and monitor what happens. 'Staff' are the functional specialists – engineers, accountants, estimators and so on – who provide a back-up service to the line managers. Some of the specialists run departments and therefore have both line and staff responsibilities. Their authority is, however, limited to their own specialism. Senior planners, for instance, have line relationships with their bosses and subordinates, and staff relationships with the operations managers for whom they provide planning services.

In its basic form, the line and staff structure is split into *functions* as shown in Fig. 3.1, but there are many variations. When a firm widens its scope, it may split into product divisions, each specialising in a type of work or market, such as housing, refurbishment or road construction. A company which expands geographically is more likely to become area-based. Here it makes sense to decentralise some of the administrative functions and perform them locally.

In both cases, divisions are usually fairly autonomous and are responsible for their own profitability. The parent company retains a headquarters, mainly for strategic planning, policy-making and overall financial control. The divisions have their own estimators, project planners, buyers, etc.

In both area- and product-based organisations, the problem of how best to group activities remains. Each division may be split into functional specialisms, so that it appears to be a microcosm of its parent firm. However, the division can respond more quickly and flexibly to the demands of its product or area, than can its parent. Complications arise when a company both expands and diversifies. It may need some of the features of product and area organisation and must operate a blend of functional, area and product organisation.

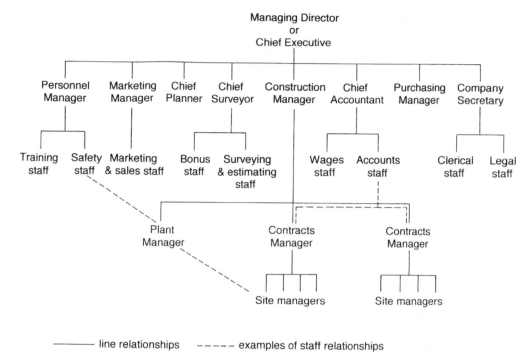

——— line relationships   – – – – examples of staff relationships

Figure 3.1 Line and staff organisation structure: construction firm.

Matrix organisations

Unlike the parent firms, project organisations do not evolve over a period of years, but have to become operational in weeks. Special attention must be given to the design of large or complex project organisations for power stations and other heavy engineering works. These temporary, task-force organisations may be better served by the matrix organisation structure, which first attracted widespread attention in the 1970s. The traditional management hierarchy – the chain of command – is partially replaced in the matrix structure by a network of lateral and vertical role relationships better suited to the need for teamwork and integration (Fig. 3.2).

In the matrix organisation, managers and supervisors responsible for the various trades and specialisms report vertically to their 'line' bosses in the parent firms and laterally to the project manager. This separates the roles of managing people and managing tasks. Project staff have both a functional boss, who runs their careers and tries to balance the demands of the project and the parent organisation, and a project boss, who 'bids' for their services. Clearly, this can create problems of loyalty and commitment. Ideally, individuals remain loyal to their company, but are committed to the project. A number of construction firms and professional practices have tried the matrix approach because they were dissatisfied with traditional methods. There are still problems. The success of this form of organisation depends on people's willingness to break away from established methods and attitudes, shifting their allegiance from specialist group to task group. A buyer has to see him/

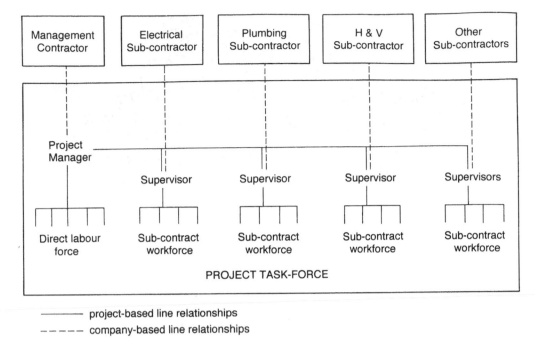

Figure 3.2 Matrix organisation structure: construction project.

herself not primarily as a member of the buying department, but as part of a project team.

Wood (2001) recognises the organisational complexity of projects, the inherent conflict between function and projects and the propensity to fracture along professional lines. He cites an alternative interpretation of these structures (Figure 3.3).

A project organisation is further complicated because its structure changes over its lifespan. This is a major difference between projects and factory-based manufacturing. The skills and resources needed within the project team alter sharply over a period of weeks or months. The team members have to collaborate closely, but their backgrounds and skills are quite different.

A project organisation should be flexible. It should respond to the type and complexity of the job. It will vary, for instance, with the ratio of specialist engineering and services work to main contractor's work. The traditional line and staff organisation may not encourage the close co-operation and good communication that are essential to the success of projects. Rigid roles, captured in job descriptions, can create problems. Loosely defined, overlapping roles can encourage the kind of teamwork needed in construction. The organisation structure must provide for ideas and information to flow in all directions, so that people are better informed and become more supportive of one another. Informal, lateral communications are legitimised in the matrix structure because they are essential for bringing the task-force members together and focusing their attention on mutual problems.

Conventional management

Matrix management

Figure 3.3 Matrix humour (Wood, 2001).

The characteristics of a matrix organisation for a construction project are summarised in Table 3.2. The main variables are its goals, timescale, tasks, people and environment. These alter considerably from project to project, so it is important to adopt a contingency or 'best-fit' approach. The resulting project organisation may not always be tidy, but what matters is whether it works. Adherence to time-honoured principles of organisation is pointless unless, in the end, the project goals are achieved.

Writers like Harrison (1992) have emphasised the special features and problems of project organisations. The characteristics of complex, one-off projects are summarised below.

- Decisions are not repetitive and a bad decision early on can affect the rest of the project. It may be impossible to recover from an early mistake.
- The learning time for those involved is limited. A manager may only experience each stage of a complex project once every few years.
- It is difficult to define suitable work patterns, planning and co-ordination methods, and control systems.
- Project personnel are drawn from many organisations and some contribute to the project on a 'part-time' basis. Their work must be thoroughly integrated.
- The companies and departments involved usually work simultaneously on other projects, each of which is probably at a different stage in its life-cycle.
- As work progresses, the emphasis shifts from design to procurement, then to site organisation and construction, and finally to commissioning and operation. No single firm or department is the most important over the whole project lifespan. No single manager (except a project manager) can assume the leading management role for the entire project period.

Table 3.2 Characteristics of a project organisation.

Goals	Clearly defined and short-term, in comparison with those of the parent firms. Stated as cost targets, time deadlines, quantities and standards of performance, quality and materials. Most project goals are quantifiable and progress towards them can be measured.
Timescale	Relatively short-term. The project lifespan is finite, with specific dates for commencement, completion and key stages of the project.
Tasks	Variable in scope and technical complexity. Less repetitive than most manufacturing tasks. Assembly of a wide range of raw and partly processed materials and components. High level of task specialisation, reinforced by trade practice and custom.
People	Wide range of backgrounds, knowledge and technical skills. Mixture of specialists, craft workers, semi-skilled and unskilled. Many involved for only part of the project duration. Willing to tolerate job mobility, low job security and poor working conditions.
Environment	Comparatively stable for the duration of the project, except for the weather, which is highly variable, and the labour market, which fluctuates in response to local competition and changes in workload.

Summary

We live in an organised society, depending on organisations to satisfy most of our needs. Yet the activities of organisations do not always contribute to people's well-being and there is a need to balance economic, environmental and social objectives of business. Moreover, attitudes to work and to organisations are changing and employees expect a fairer deal from their employers.

There are many types of organisation, but no single ideal one. A well-designed organisation enables tasks and resources to be allocated efficiently and provides a system for co-ordinating and controlling them. Rules and procedures ensure that tasks get done and are carried out efficiently. Good organisation ensures that information flows and decisions are taken.

The size and complexity of organisations has encouraged a shift towards decentralising some organisational functions. At the same time, there is growing support for reducing the level of specialisation in some jobs, where it has been taken too far. Jobs are being re-examined with a view to making them more varied and interesting.

When a client decides to build, the construction industry has to create a temporary, project organisation and make it operational in a very short time. Construction projects have special characteristics and the kind of structure which suits them may not suit their parent firms. The success of a project relies a lot on effective co-ordination of design and production and of main contractors and specialist sub-contractors. The task-force or matrix structure offers some advantages for organising construction projects.

Some organisations need to be more flexible than others, but flexibility is a vital dimension in project work. The ability to adapt to change may be the most important factor affecting the success and survival of many organisations. Downsizing is purported to be one of the ways used to improve organisational efficiency and competitiveness.

Discussion topics

The Association for Project Management's Body of Knowledge (BoK) details the knowledge and experience that people involved in the formal management of projects should possess (www.apm.org.uk/pub/bok.htm).

Revision 4 of the BoK contains the following sections:

- General
- Strategic
- Control
- Technical
- Commercial
- Organisational
- People

Rank the importance of these skills to the construction manager and highlight the principal differences between the roles of the project manager and construction manager.

The complete document can be downloaded from the APM web-site, subject to the stated copyright agreement conditions.

Chapter 4
Leadership

Management and leadership are not the same thing. Management evolved with the growth of formal organisations, but leadership is one of the oldest and most natural relationships in society. Managers have to be appointed, but leaders emerge naturally, whenever people get together to do things. But if the manager is not the person the group would choose as its leader, there could be problems. One of the fascinating debates in management is whether managers can learn to be better leaders and if so, how.

Leadership has been a popular subject for over half a century. It was eclipsed for a while by new ideas about worker participation and group decision-making, but has re-emerged with a new focus which recognises that the leader's role varies with the circumstances.

Leadership is hard to define for it is a complex process. There have been countless leadership studies, but almost all have looked at only a small part of the picture. Few studies have pulled together all the features of leadership in a comprehensive way. Moreover, much of the research overlaps with other areas like power, motivation and group processes. The piecemeal approach to leadership has meant that much of the work is inconclusive and some of the most exciting ideas, put across with conviction and enthusiastically received by many managers, have no sound empirical basis.

One of the many attempts to distinguish management from leadership defines management as 'ensuring effective and efficient operations' and the core of leadership as 'direction setting' (Novelli and Taylor, 1993).

The idea of direction setting is underscored by Schmidt and Finnigan (1992) who cite research that stresses the importance of the leader's ability to create and communicate a vision that inspires the team. These authors also remind us of Warren Bennis' witty remark that while managers give their attention to doing things right, leaders focus on doing the right things!

Without doubt, the concept of management has become debased in the 1990s 'now that everyone claims to manage something' and future-oriented leadership may supersede management as we understand it (Thomason, 1994).

Measuring the leader's behaviour and performance is difficult. One can measure the group's output, but this will depend on many factors, of which the leader's behaviour may be one of the least important. One can ask subordinates, peers or superiors to rate a leader's effectiveness, but they will have the same problem. Moreover, they will find it hard to be objective, because of their personal feelings about the individual. Many biases creep in.

Taking a broad view, the ideas about leadership fall into three categories, focusing on:

- The leader's personal characteristics or traits.
- The leader's behaviour or leadership style.
- The setting or situation.

The characteristics of the leader

For a long time, the popular view was that certain people make good leaders because of their personality traits. Indeed, there have been hundreds of studies of leaders' traits. As personality was thought to be inherited, it was believed that leaders were born not made.

The evidence from psychology now strongly indicates that personality is only partially decided by hereditary factors. Good leaders are not just born, their personalities develop through experience.

Researchers have looked for links between personality and effective leadership. Knowing the ideal personality, firms could then select good leaders, even if they couldn't train them. Long lists have been produced of desirable leadership qualities, like intelligence, good judgement, fairness, insight, self-confidence and imagination. Others include honesty, courage, perseverence, imagination, reliability and industriousness. Yet some of the most successful leaders in history have not had certain of these qualities. Indeed, some have been unjust, neurotic, narrow-minded and even insane!

Certainly, good leaders can be above average in intelligence and may have been chosen for this reason. But many intelligent people never become leaders and research has shown that the correlation between intelligence and effective leadership is low. At best there is merely a *tendency* for leadership and intelligence to go together. Similarly, personal characteristics like dominance and extroversion only correlate weakly with the leader's effectiveness.

The personal traits and qualities of leaders influence their success – but only partially. The writer has found that employees in the construction industry look for qualities like fairness, competence and decisiveness in their leaders. Research findings do not deny the value of such qualities, but suggest that they cannot wholly explain the leader's success or failure.

A common objection to the trait approach is that it labels people as good or poor leaders on the basis of rather subjective measures of leadership performance and fails to take account of other factors which affect the leader's behaviour. To demonstrate whether or not personal traits affect leadership ability, one would need valid and reliable measures of:

- The traits themselves.
- The criteria on which a leader can be considered successful.

So far, this has proved difficult, because both personality and leadership behaviour are dynamic. People often exhibit different characteristics in different situations. A manager who is a good leader when things are going well may not be successful in a crisis.

The overall picture does not suggest an ideal leadership personality. Rather, good leaders come from many backgrounds and the personal qualities they need depend on the circumstances.

Leadership style

The search for an ideal style of leadership was spurred on by the belief that people work harder under the right style of leadership. Styles are commonly classed as *authoritarian* and *democratic*. The difference reflects the personality and attitudes of the leader and the power structure of the firm. Handy (1985; 1993) uses the less emotive titles of *structuring* and *supportive*.

The structuring leader retains most of the power for controlling rewards, settling disputes and making decisions in the group. The supportive leader shares power with the group, so that they have control over what happens.

The extreme authoritarian leader decides objectives and gives orders without consulting the group. The democratic leader seeks the group's views and keeps members informed. The authoritarian tends to be aloof and concentrates on the task. The democratic leader participates as a team member and shows an interest in the group's well-being.

The style a manager adopts reflects his or her attitude to people and assumptions about authority. Negative attitudes lead to a more autocratic style. The author-itarian manager believes that people are basically lazy and need firm control. The democratic manager has a positive attitude to the team, seeing them as responsible, keen and capable of exercising initiative and self-control. The democratic manager listens to their ideas and gives them encouragement.

In the 1950s and 1960s, democratic leadership became very popular and was thought to produce better results. Many people prefer a democratic leader and such a style can improve morale and reduce labour turnover and disputes. But there is little evidence that people will work harder for a democratic leader. Cause and effect are difficult to separate. An efficient, happy group may permit a democratic style rather than result from it. Moreover, some people prefer an autocratic leader and will work harder for one. Some managers believe that in the world of business, democratic leadership is unworkable.

Another way to describe the leader's style is as 'task-centred' or 'employee-centred'. The two terms need not be mutually exclusive. Indeed, construction managers cannot afford to neglect either task or people. This opens up the possibility that managers need to combine the best features of task-centred and employee-centred leadership.

After all, the leader is responsible for certain activities and will have to schedule the group's work, instruct and train subordinates, check finished work and give subordinates feedback on their performance. The manager must decide how closely to get involved in tasks and how much to delegate. Close supervision can cause output to drop and adversely affect job satisfaction and labour turnover. People don't like too much interference!

At the same time, the manager must look after employees' needs. This includes helping them achieve personal goals, dealing with their problems and establishing warm, friendly relationships.

There is some evidence that considerate leaders get better results from their groups, and lower labour turnover and absenteeism. However, the relationship is complex. Peter Smith (1984) cites studies of Japanese firms operating in the West, which show that task-centred managers, who stress efficiency, quality control and good time-keeping, have hard-working, willing subordinates who accept exacting standards. White and Trevor (1983) suggest that such employees co-operate because managers are not aloof, work the same hours and wear the same uniforms. These managers convey a sense of unity of purpose.

Some studies of leadership style

Likert: employee-centred leadership

In a series of studies of morale and productivity, Likert (1961; 1987) concluded that the best supervisors are employee-centred. They concentrate on building cohesive work groups and focus on the human aspects of their groups. They exercise general rather than detailed supervision and are more concerned with targets than methods. They allow maximum participation in decision-making.

There have been some powerful criticisms of the way Likert's data were collected and interpreted. Most of the data was based on surveys. The research did not attempt to change the leader's behaviour experimentally, but merely recorded the relationship between supervisor behaviour and worker performance. It is very difficult to establish cause and effect from such studies.

Indeed, attempts to replicate Likert's findings have produced inconsistent results. Employee-centred leaders sometimes get poorer results than task-centred ones. Likert's work resembles the trait approach, looking for an ideal leader for all occasions. Nevertheless, his work has stimulated managers' interest in leadership style.

Tannenbaum and Schmidt: leader style continuum

In one of the best-known discussions of leadership, these writers identified a spectrum of leader styles ranging from totally autocratic or task-centred to fully democratic or employee-centred. Between these extremes are a number of style variations, one of which may be the most suitable in a given setting. A simplified spectrum of leadership styles is shown in Fig. 4.1.

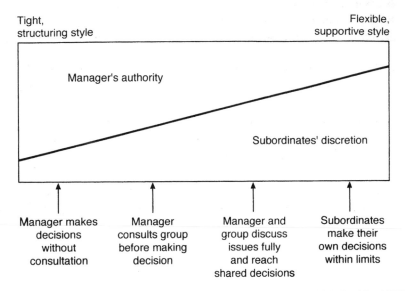

Figure 4.1 Some styles of leadership (adapted from Tannenbaum and Schmidt, 1973).

Tannenbaum and Schmidt (1973) maintain that choosing the right style depends on a careful assessment of the leader, the followers and the situation. Leaders must be sensitive to the needs of the situation and flexible enough to adjust their styles to suit.

Tannenbaum and Schmidt identify some of the factors influencing the leader's style but, like many other leadership studies, do not suggest how managers might assess and improve their own styles.

There has been a renewal of interest in employee-centred leadership, with its flexible, supportive style. This is a result of interest in empowerment and self-managed teams which managers hope will lead to efficiency gains.

Vroom and Yetton: leadership and decision-making

Drawing on previous research on group decision-making, Vroom and Yetton (1973) have developed a prescriptive model of leadership which would provide managers with definite guidelines on leader style. Their focus of attention was on the problems leaders face.

The leader must analyse the problem situation before choosing the right approach for dealing with it. The factors include:

- The qualitative importance of the decision.
- The amount of information the leader and group have about it.
- How structured the problem is.
- Whether subordinates need to be committed to the decision.
- Whether an autocratic decision is acceptable.
- How much subordinates want to solve the problem.
- How much subordinates might disagree about the decision taken.

Rules are given for relating these factors to leadership style. For instance, if the problem is serious and the manager lacks the information or expertise to solve it, participative leadership should be chosen.

Vroom and Yetton's model identifies three main styles for arriving at a solution to a group problem – autocratic, consultative and group – and identifies within this span seven leadership styles ranging from highly authoritarian to totally participative.

The authors recognise that in some settings more than one style might work equally well. In such cases, time constraints or the leader's preference should dictate the style. The effectiveness of a style is judged by:

- The quality of solution reached.
- The time taken to reach it.
- Its acceptance by subordinates.

This approach tries to offer managers a practical framework for leading. Some commentators claim that there is little empirical evidence to support the validity of the model, although Handy (1985) claims there is a lot of pragmatic evidence to support it. The approach uses a decision tree and appears rather mechanistic.

Most of the research evidence suggests that the amount of attention the manager should give to task and people depends on many factors. All in all, it seems that there is no single ideal style of leadership. For example, a style that will work on site for direct labour may not be effective in controlling sub-contractors.

The leader and the situation

Most of the evidence suggests that leadership is specific to the situation. Faced with a difficult work problem, a group may turn to someone tough, clever or experienced. In a routine or social setting, they may follow the lead of someone friendly. After a serious accident, a first-aider may temporarily become leader.

Managers need to know what kind of leadership will work in a specific situation. The main variables are:

- The leader
- The subordinates

- The task
- The setting.

The leader

The success of leaders depends on many factors, including their personalities, values and preferred styles of management, their level of competence and self-confidence. It also depends on how much they trust their teams and their ability to cope with stress.

Whether leaders choose structuring or supportive roles depends on such factors. Some leaders will trust their teams more than others do, others will feel it is their job to make the decisions. Leaders who give their teams more of a free rein must be able to live with uncertainty – and not all leaders can.

The subordinates

The success of a group partly depends on how competent its members are, how interested they are in their work, their attitude towards their leader, how much freedom they want in their jobs, their goals and how long they have worked together. On construction sites, the composition of groups can change frequently. Groups will try to balance task demands with their own needs.

The more competent they feel, the more they will want control over their work, especially if it is important or challenging. Past experience will affect the kind of leadership they find acceptable. Younger people, reared in a more permissive and democratic society, expect more involvement than many of their elders.

The task

The kind of operation is important – whether it is well defined, long term or short term, important or trivial. Mass production often needs tight supervision and control because the job has to be done in a certain way. On the other hand, research work cannot be strictly controlled. Much has to be left to the researcher's discretion, because the manager may not know what the end result will be.

Construction falls between these extremes. Some work is repetitive and inter-linked and has to be tightly controlled, but other tasks are one-off and must be loosely programmed and left to the initiative of those involved.

Key issues are whether the task requires obedience or initiative, whether it is routine, problematic or pioneering and whether it is urgent. If a task has to be performed in a hurry, this may push the leader towards tighter control. Participation takes time.

The complexity of the task will affect the leader's style. Technical complexity may necessitate a supportive style if the leader lacks expertise, or may demand a tight rein because operations are closely interrelated. Organisational complexity can have similar effects. In the construction of a power station, novel technical problems may force managers to be flexible, relying on their teams to come up with fresh answers. Conversely, a large contractor, experienced in commercial contracts, will have evolved many set procedures which staff are expected to follow.

An added difficulty is that work groups often have a variety of jobs to undertake, ranging from well-defined routines to ill-defined and long-term tasks. The leadership demands may be different for each task and this calls for a good relationship between leader and group. It is understandable that many managers give up trying to cope with such complexity and simply fall back on their habitual style.

The setting

The leader's behaviour is affected by his or her position in the firm, the extent to which the work is important and closely related to other activities in the business, and the organisation's norms and values. No manager or worker is entirely free from organisational pressures or from systems and procedures. The power the manager wields is not static; it changes from one setting to another and this affects leadership behaviour too.

Some studies of situational leadership

Fiedler: situational leadership

Fiedler's leadership studies in the late 1960s provided a much needed new focus. Arguing that leadership varies with the situation, he identified three factors which seem especially important (Fiedler, 1967):

- Whether the leader is liked and trusted by the group.
- How clearly the group's task is laid down and defined.
- The amount of power and organisational backing the leader has.

In his view, the relationship between leader and group is the most important of these factors. A leader who is liked and accepted by the team, and has their confidence and loyalty, needs little else to influence their behaviour. If the leader is unpopular or rejected, the group will be difficult to lead.

Fiedler found that the style which worked best depended on how 'favourable' all three factors were to the leader. The most favourable situation is where the task is well-defined and the manager is liked and respected and has good position power. The situation is most unfavourable when these conditions are absent. Fiedler concludes that in very favourable or unfavourable conditions, a structuring approach is better. When conditions are only moderately favourable, the leader will find a supportive approach more effective.

If the task is confused or the construction manager is unpopular or lacks power, a firm stand is needed to keep control. If the leader does not take charge, the group may fall apart. If the task is well-defined, or the manager is popular or powerful, he/

she is expected to take a firm lead, giving clear information and instructions. Under these conditions the passive construction manager may lose the group's respect.

A supportive style of leadership seems to work best in two situations. One is where the task is unstructured but the leader is popular. Here, a people-centred approach is needed to elicit the team's help in finding the answer to the problem. The stricter, directive style will not elicit the group's co-operation, for they will be afraid that their ideas will be judged unfavourably. The second is where the task is structured, but the manager lacks popularity or power. Here the leader must tread softly and be diplomatic to avoid being rejected by the group. Here the democratic leader is likely to get better performance than the tougher, controlling leader.

Fiedler's model suggests that we may have paid too much attention to selecting and training leaders, whilst neglecting the needs of the situation.

Hersey and Blanchard

Hersey and Blanchard (1982) and Hersey *et al.* (1996) put forward a variation to the situational leadership approach in which the leader's style changes over time, as the employee develops. In their model, the task-centredness of the leader starts high and diminishes as the employee becomes more experienced, skilful and willing to take responsibility. The leader's relationship behaviour (such as giving support and encouragement) starts low but increases in the early stages, eventually diminishing again as the employee achieves high levels of skill, motivation and autonomy. This approach uses a four-sector grid, reminiscent of Blake and Mouton's managerial grid. But Hersey and Blanchard's theory differs in proposing shifts in the leader's style, whereas the Blake and Mouton model argues for a single best style.

Charles Handy: 'best fit' approach

Handy (1985; 1993) puts forward a contingency approach to leadership in what he calls the 'best fit' approach. This puts the style preferences of the leader and subordinates and the demands of the task along a continuum, ranging from tight (structured) to flexible (supportive). There is no fixed measuring device for this scale – it is rather subjective.

Handy suggests that effective performance depends on some changes being made so that the three factors 'fit' together on the scale. How the leader or organisation achieves this depends on the group's *setting* – such things as the leader's power or position, organisational norms and relationships, and the kind of technology the business uses. Unless the match between the factors is improved, the group will cease to be effective. Leaders who have strong organisational back-up may pull the group and task towards their preferred ways of working. Leaders who lack this may alter their own behaviour.

Handy's approach recognises that the leader has two main roles vital to the performance of the group – ambassador and model. As ambassador, the leader represents the team in dealings with others at the same and higher organisational levels. As a model, the leader must recognise that some subordinates will copy his or her successful behaviour.

Leadership, goals and social exchange

Tolman (1932) showed that most human behaviour is goal-directed. To achieve their objectives, people often have to co-operate with others and have to choose between different courses of action. Their choices depend on many factors. Belonging to a work group is a way of achieving some of their goals and they see the leader as someone who can help or hinder them in this process.

Evans (1970) and House (1971) laid the foundations for *path–goal* leadership theory, which argues that the leader's job is to define a path along which subordinates expend effort to achieve a group goal. The approach assumes that:

- subordinates will accept the leader's behaviour if they believe it is helping them achieve immediate or future goals; and
- rewards are made conditional upon subordinates achieving the work targets set.

The effectiveness of leaders depends on their ability to help subordinates clarify their goals and see ways of achieving them. If employees feel that the leader is giving this help, their motivation will increase.

Like Fiedler's approach, the theory stresses that the leader's behaviour is influenced by subordinates, as well as task demands and environmental factors. For example, subordinates who need to work independently or feel competent at their jobs may show their dislike of having a structuring leader.

Workers carrying out routine tasks, for which the rewards are clearly identified and related to performance, would not require an authoritarian leader, because behaviour is goal-directed and the path to it clear. This conclusion differs from that of some other researchers.

Path–goal theory makes some plausible statements and House found it held up in studies in seven organisations. Research is needed to look more closely at how subordinates' expectations affect, and are affected by, the leader's behaviour.

Hollander (1978) argued that there has been a tendency to view leadership as something static, with leader and group in fixed positions. Realistically, leadership is a process in which leader and followers influence one another and their situation.

Viewing leadership as a social exchange puts the emphasis on the impact of all group members, not just the leader. Initiatives and benefits are seen to come not just from the leader but from the other team members too. Being a leader and a follower are not mutually exclusive roles.

An effective leader does things that benefit group members, but makes demands on them too. The team provides the leader with status and other privileges of position, but influences and makes demands on the leader as well. Both leader and group must give and take for the relationship to work. They are parts of a system that takes time to develop.

The leader often defines standards, sets objectives, maintains the group and acts as its spokesperson. But many situations are ambiguous, with the goals, tasks and procedures not clearly defined. Here, the leader's help is especially sought because the group wants guidance on what to do, how to do it, or why.

Trust and fairness are important. If leader and group trust each other, they are more willing to take risks. Without trust, the leader may have to resort to position power or authority. Fairness is essential in the social exchange. Even a friendly and unthreatening manager may not help the group achieve job satisfaction or meet its demands for fair play, if he or she is not fair. Members may feel they are being exploited.

Leadership: Goal setting

Before setting goals, which serve to guide, motivate and encourage people, there are a number of factors that should be considered:

- *Set task that is appropriate for the employee*. Identify the task, then consider the necessary skills, competencies, work ethics and qualities required to deliver the task. Select an employee with suitable attributes.
- *Difficulty*. Most people are motivated by tasks that are slightly more difficult, complex and interesting than their usual mundane tasks. Difficult tasks often present a challenge, encouraging people to work harder to complete the task. A heightened sense of achievement and satisfaction is often experienced when a difficult task has been completed. However, if a person considers that a task is too difficult they may avoid the task, or the thought of not being able to do the task may increase stress and anxiety, resulting in mental health problems.
- *Workload*. High levels of work can help performance. People who are busy may be less distracted than people who only have a few things to do. However, if people are given so much work that it is impossible for them to achieve all that is required, personal levels of stress and anxiety may be heightened.
- *Context*. Some people may be more willing and capable of working with different tasks, people and problems in different fields, others may not be capable or willing to work on tasks that are removed from their normal context.
- *Monitor and record*. Where people are aware that their performance is monitored and checked they are likely to increase their effort to ensure that the correct standard is maintained. The act of recording behaviour can encourage people to adopt appropriate behaviour. Quality, time, safety, absenteeism, cleanliness, co-operation or obstructive behaviour can be recorded.
- *Feedback*. Ensure that parties are aware of their performance. If improvement is required the parties need to know. Where work is performed to a satisfactory standard appropriate feedback should be given. Identifying aspects that could be improved and praising good practice are both important.
- *Reward and transparency*. All payment schemes should be clear, easy to understand and honoured. Payment should be related to effort and skill required for

the task. Bonus for work delivered on time, work that is free from defects and delivered safely can serve a useful purpose. Absenteeism has become a problem for many organisations. Offering bonuses for employees who have little unauthorised absence has been shown to reduce absenteeism.

● *Trust.* Where employers or managers do not honour promises to pay bonus it is difficult to make bonus schemes work in the future.

Task and socio-emotional roles

In most groups, two leadership roles are present – a *task* role for co-ordinating the work, and a *socio-emotional* role, for looking after the well-being of the group.

Two people may even share these roles where, for example, one person is seen as competent in the task, whilst the other is more popular and recognised as having skill in holding the group together.

The leader cannot do everything. There are many roles to be performed in a group such as trouble-shooter, negotiator, advocate and counsellor. Some of these may be delegated to group members, or they may take them on uninvited. Some individuals have more status than others and will be closer to the leader. They exert more influence over the leader and the others.

Formal and informal leaders

Every work group has an appointed leader – its supervisor or manager. If there isn't one, a leader will almost always emerge, because groups need task and social leadership.

Most organisations expect one person – the *formal* leader or manager – to perform both roles. The manager has to allocate work, show people what to do and make sure they do it properly, in addition to dealing with human problems and ensuring that group members work together as a team. The successful formal leader does both things well, achieving high productivity and group satisfaction. But this is a tall order.

Blake and Mouton (1964, 1978) recognised this in their managerial grid which measures, on separate scales scoring from one to nine, the leader's concern for production and for people. The 9.1 manager concentrates on the task and shows little concern for the group; the 1.9 manager does the opposite; 5.5 is a compromise – the middle-of-the-road manager – whilst 9.9 is often regarded as the ideal, to be aimed for, but rarely achieved. Perhaps more important is that the leader knows when to concentrate on the task and when to focus on the group.

If the formal leader tends to be task-centred and fails to meet the social needs of the team, an *informal* leader may emerge within the group – someone the members turn to with their work problems and personal worries. Similarly, a group whose formal leader concentrates on the social aspects of the group may accept an informal, task leader, especially if success at the task is vital to the achievement of their own goals. A group can operate successfully with two leaders, an official and an unofficial one, but this can lead to conflicting goals and loyalties. The group member accepted as informal leader may vary with the situation.

A further complication arises if the group's manager is ineffective in both leadership roles. In this case, informal task and social leader roles may be adopted by one or more members of the group.

The leader's competence

Many kinds of leadership study have taken account of the leader's competence or ability, either in the limited sense of technical ability or in the wider sense of competence to lead. One factor which has sometimes been underestimated is the group's view of its leader's competence. The group's perceptions of the manager's ability can account for much of his or her success as a leader. Although subjective, judgements about the leader's ability to get results carry a lot of weight with the group (Hollander and Julian, 1970).

Of course, competence may be attributed to someone who has been lucky or who has been helped by others. It may rely on a reputation built on earlier success. A successful site manager will not necessarily be an effective contracts manager.

Clearly, numerous factors affect leadership performance, yet the leader may be unaware of many of them. Researchers have often focused on only some of these variables when studying the leader's behaviour. They should not do so if they wish to discover what leaders could do to improve their performance.

Summary

The search for the ideal leader has led to the conclusion that one does not exist. There are no specific traits which can be relied on to make a manager an effective leader. Personality is not a fixed commodity. People change. A manager's confidence, decisiveness, judgement and so on, will vary over time and with circumstances. A leader may display good judgement on one occasion and poor judgement on another; be confident about some matters, unsure about others. The leader will handle some people skilfully and make enemies of others. At best, there may be certain combinations of personality factors which give the manager an advantage in some situations.

There is no mode of behaviour or ideal style which can be relied on to be effective. Leaders must learn to be flexible and alter their behaviour to suit the circumstances. Some managers are better at this than others. The evidence shows that both autocratic and supportive leaders can get good results, but attitudes to authority are changing. Many people today have been brought up to expect a better deal in their jobs and want more involvement and autonomy. Employees today will not always accept without question what managers tell them to do. Many of them want and expect to be involved in the management process.

Leadership depends on a dynamic relationship between the leader, the group members, the task and the setting in which they operate. Good leaders know the right behaviour to match the circumstances. They know when to be tough and when to be friendly. They understand that when the task is non-routine and ill-defined, as it often is in construction, they must be flexible and encourage group participation. They also know that there are times when the group may need them to take a firm lead. There are always many factors to consider, not least the abilities and preferences of the people who work under the leader's direction and how willing they are to take responsibility.

How ideas about leadership will change in the future is difficult to predict, but shifting attitudes to authority could have a big impact on the kind of leadership that will be acceptable. Nevertheless, every unique situation will still produce leaders for the job in hand. How effective they are will depend on their skills for dealing with the variables they can control and on being lucky with those they cannot.

Discussion topics

Leaders are born, not made. Discuss.

Task leaders are ineffective if they are not supported by socio-emotional leaders. Discuss.

True leaders survive whatever the situation. Discuss.

Chapter 5
Communication

Poor communication has long been a problem in the construction industry. Part of the trouble is the way the industry is organised. The project team is made up of people from many different firms. Their contributions vary and a lot of information has to pass among them. This requires a well-organised network of communication using the latest technology. Even when this network exists, communication still breaks down at a personal level, because people fail to keep their messages simple; they pass on too much information or too little; the information they give is inaccurate or misleading.

On the receiving end, people are flooded with paperwork they haven't time to read, yet often they cannot get the information they want. Estimates may be wrong, drawings out-of-date, descriptions ambiguous. Meetings go on for too long and people stop listening.

The size of the firm matters. In small organisations, communication is often good. There is more face-to-face contact, so if people don't understand what is being said, they are more likely to say so and the problem is cleared up straight away. Communication is more direct. Those making the decisions are closer to those who have to implement them.

Larger firms rely more on the written word. This puts the message on record, but misunderstandings cannot easily be cleared up. Information can be delayed and distorted as it goes up and down the hierarchy. People are separated by divisions and departments, sometimes by shifts.

Formal communication channels can be slow and impersonal. The faster 'grapevine' takes time to develop and is often discouraged anyway. The larger the firm, the more acute the communication difficulties tend to be.

Poor communication skills make matters worse. Most people, including managers, are poor communicators and don't even realise it. Yet, improvements can easily be achieved through training or simply by making people aware of the main pitfalls and giving them feedback on how well they are communicating.

Communication process

Communication is one of those skills that we all use, yet few of us give any real consideration to how complex the process is. It is an aspect of the human that sets us aside from almost any other species on earth.

Face-to-face communication is not one way, it is transactional (Figure 5.1).

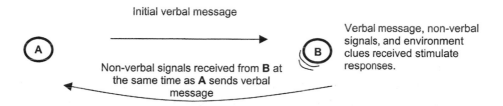

Initial verbal message

A

B

Verbal message, non-verbal signals, and environment clues received stimulate responses.

Non-verbal signals received from **B** at the same time as **A** sends verbal message

- A sends a message to B
- While A is sending the message, B is sending back communication signals, e.g. facial expression, eye movement, body language, etc.
- Both parties communicate at conscious and subconscious levels.
- Although B's body language and facial expressions suggest that B understands, other signals, such as speed of reaction and actions, suggest that B's interpretation of the message is incorrect.
- A recognises that B does not understand even though B thinks that s/he understands.
- The interesting phenomenon here is all of this can be done without B speaking and can occur before A has finished sending his/her initial message.

Figure 5.1 Non-verbal responses during interaction.

During interaction it is possible to recognise whether the people we are talking to are following and understanding what we are saying, even before we have finished a sentence. During conversation we may recognise that a person thinks that they understand what we are saying, yet their body language, facial reactions and other signals inform us that they have not actually properly understood the message. This helps us to change our sentence and add further information to help the person understand.

Because communication skills are both hereditary and developed from a very early age many of the interactions sent and received are processed at a subconscious level. We give little thought to the information being received and sent, although we do react and process the information subconsciously.

Although people are often told that they must consider what they say before they say it, during face-to-face interaction people do not process speech in their conscious mind. Conscious processing is too slow. During interaction, speech and grammatical structure, sentences and words are processed in the subconscious mind (LeDoux, 1998). While it is possible to prepare for meetings and rehearse speeches, once people react to others in a natural communication environment they will respond through their subconscious processing. If people had to think through exactly what was to be said before verbalising each sentence the natural flow of speech would be broken up and slow. This does not mean that people cannot alter the way they communicate.

Using training, education and experience people develop a repertoire of skills that the subconscious draws upon when initiating communication and constructing responses. What this means is that the subconscious may have a library of responses and actions that it has previously used successfully in certain situations. If the situation encountered is similar to that previously encountered the subconscious quickly processes a reaction before the conscious mind has a chance to consider it. The conscious mind may be aware of the interaction as it occurs, but may take little part in the processing of the communication.

If, however, the situation is not similar to any previous encounters the reaction may be a result of the subconscious and conscious mind. The subconscious may prevent us reacting straight away with an incorrect response while the conscious mind thinks over the matter, attempting to understand and contextualise the situation.

An example of this would be when people meet and greet each other. If a person meets a close friend they do not normally take time to think about how they will greet their friend, they will normally rely on one of the ways that they have used so many times in the past. However, if a person is introduced to someone they have never met before, as soon as they encounter the person they may start to consciously and sub-consciously assess the person, considering how the person is acting, the type of person they are and what would be an appropriate response. The greeting may still be one that has been used before, or one that has previously been used in similar situations. However, greater effort will be used to think through and consciously process information, trying to determine what to say or do next.

Why experience, education and training are important for effective communication

To ensure that speech is successful, professionals should rehearse and train for different events that they are likely to encounter so that when the situation arises the speech and reaction come naturally. A natural reaction is one that is processed in the subconscious, one that is given very little thought. Although many people do not overtly train their communication skills, they may often engage others in low risk situations. During such encounters they may subconsciously practise different interaction approaches. Some approaches will not work and others will be more successful in gaining the required outcome. Once successful interaction techniques are developed in low risk situations a person's confidence may increase and they may safely enter more risky situations, drawing on the skills learnt in the low risk environment.

- *Conscious processing* is slow – when a person processes something in their conscious they are having to think things over, or rehearse, what to do or say before doing it.

● *Subconscious processing* is quick – it relies on a person's inbuilt survival skills and patterns of behaviour, which are hereditary and have evolved. Subconscious processing draws on experiences and skills learned from an early age. Experiences gained through training and education may become so well rehearsed that they no longer need any real thought and are quickly processed in the subconscious mind.

Group communication – task and relational interaction

Within this diverse and complex industry it can be difficult to form and maintain inter-organisational relationships. However, without an organised social system, individuals are limited to their own efforts; the accomplishment of major projects is achieved through interlocked co-ordinated activities. It has long been recognised that communication can be divided into two distinct categories: communication aimed at achieving the group goal (task based) and interaction that is used to maintain relationships (socio-emotional) (Bales 1950, 1970; Frey, 1999; Keyton 1999, 2000).

Successful work groups balance task and relational communication (often called positive and negative socio-emotional interaction). As team members work through tasks, differences of opinion emerge and a level of conflict develops. Thus, task-based discussions result in tension that is expressed as negative socio-emotional interaction. Negative socio-emotional interaction threatens relationships. When dealing with a problematic task, group members diffuse negative emotion with positive emotional discourse (e.g. by showing support, joking, etc.), returning to the task issues once the tension has been dissipated. If negative socio-emotional talk occurs, tension is released in stages, first through task related discussion (e.g. finding common ground, explaining and reasoning, giving logical or rational explanation, etc.), then by positive socio-emotional acts (e.g. by agreeing on issues, showing support and making friendly gestures or comments).

The total communication of healthy groups is said to contain several times as much positive socio-emotional as negative socio-emotional acts (Shepherd, 1964). Group members prefer positive feedback (Jacobs *et al.* 1974); interaction that suggests the group is effective increases morale (Frye 1966). Although groups may prefer positive emotional feedback, Cline's (1994) research found that too much emphasis on agreement resulted in unsuccessful outcomes. High levels of agreement (positive socio-emotional interaction) and a very low level of disagreement and conflict are characteristics of groups that are subject to 'groupthink'. Groupthink occurs where individual members of the group feel unable to show their concern with suggestions or disagree with others, thus the group seems to be in unanimous agreement, yet, for a number of reasons, individuals may suppress their dissent. While it is clear that positive relational (socio-emotional) communication should be greater than negative interaction, the amount that it should be greater is disputed. Gorse's (2002) research into construction meetings between the management and design team found that the difference between positive and negative interaction in successful teams was only 1–2% greater than those teams which did not achieve their objectives. The important theme that runs through all of the research is that task based discussions will result in tension (negative socio-emotional interaction), but that the tension is dispersed through positive socio-emotional interaction. Successful working relationships are maintained in a climate that has more positive than negative socio-emotional interaction. However, the occurrence of negative emotional interaction is important to encourage diversity and remove groupthink.

Figure 5.2 provides an indication of how task based and socio-emotional interaction are used.

TASK BASED INTERACTION
Ensure different specialist
perspectives are exposed on the tasks

GIVES TASK BASED	**ASKS FOR TASK BASED**

Task based interaction

Provides information on group goals. Offers ideas, opinions, perspectives, beliefs, suggestions, directions	*When groups first meet members develop relationships and tentatively discuss task issues. As members become more confident they exchange opinions and beliefs on the tasks.*	Requests information on goals, activities and tasks. Ask questions, requests ideas, opinions, suggestions

Socio-emotional interaction
Disagreements then emerge resulting in

SHOWS POSITIVE SUPPORT   *tension. Successful groups discuss the*   **SHOWS NEGATIVE**
differences and rebuild the relationship with   **EXPRESSION**
positive emotional expression

Agrees, shows support, encourages, praises, give assistance, makes friendly gesture, attempts to relieve tension, shows solidarity	Disagrees, show tension, disputes others proposals to expose weakness and risks associated with argument

SOCIO-EMOTIONAL INTERACTION
Negative exchanges ensure
different specialists challenge
others and then use positive
interaction to rebuild and maintain
relationships

Note: While more positive than negative socio-emotional interaction is required to help groups perform, too much emphasis on positive socio-emotional interaction can be detrimental. Failure to engage in critical discussion reduces potential to identify risks.

Figure 5.2 Use of task based and socio-emotional interaction (Emmitt and Gorse, 2003, p. 177).

Functions of communication

Communication serves many functions, all of which are important in construction management. The list below is not exhaustive and most of the manager's tasks involve several of these functions.

Information function

Information is being exchanged all the time. A manager explains a company policy to an engineer; a joiner tells an apprentice how to prepare a joint; a senior estimator tells a junior how to build up a unit rate.

But information passes both ways. The engineer will tell the manager about a problem with a sub-contractor. The joiner's apprentice will talk about a grievance over bonus.

Instrumental function

Communication is used to get things done. Good communication is vital in organisations, where groups undertake discrete tasks and depend on one another to achieve mutual goals. People need to know what they are expected to do, how quickly and how well. In construction, most of the targets are available in drawings, programmes and specifications, but the manager needs skill to communicate them clearly and make sure that they have been understood.

Social relationships function

Much of the communication which circulates round an organisation is aimed at maintaining relationships between individuals and groups, so that they continue to work as a team. The larger the organisation, the more important this social contact becomes. The contact itself is not directly productive, but it facilitates the kind of communication that is the life-blood of the business. On site, where communication channels have to be created from scratch, social contact helps create co-operation between members of the team.

Expression function

Communication enables people to express their feelings. This may happen spontaneously, as in an argument during a site meeting. But it may be carefully planned, for instance, to create a favourable impression at an interview. A grievance procedure is an example of this function operating at a formal level.

Attitude change function

Simply giving orders is not always enough. Managers may need to change employees' attitudes to get the best work from them. This would apply if, for instance, employees felt that the firm was treating them unfairly.

But this can be difficult. Some kinds of attitude are resistant to change. Others are easier to influence and personal discussion is often the best way. The manager may use group discussion to achieve certain kinds of attitude change, especially where several people are affected.

Role-related or ritual function

Sometimes people communicate because they are expected to. An operative who talks little may be labelled unsociable. The manager is often expected to give a speech or have a few words with a retiring employee.

Communication structure

An effective system for passing on information and instructions, and for receiving feedback, is essential for management control. In construction, this system must work both within and among the many firms – consultants, contractors, sub-contractors, suppliers, and client – who contribute to the design and production of the finished structure.

In large organisations, it becomes necessary to use recognised channels of communication to ensure that people get the information they need. Even in small groups, studies have shown that a communications 'free-for-all', in which anyone talks to anyone, can be less effective than a network which directs information through specific channels. In a business, these channels are:

- A leadership or line hierarchy, linking people who decide policy with those who implement it.
- Functional and lateral relationships, linking people in different sections, some of whom contribute specialist knowledge and skills.
- Procedures through which managers and workers can consult and negotiate with one another to resolve conflicts and increase commitment and co-operation.

Yet the existence of these information channels is not enough. Communications must not only reach the right people, they must be accurate, timely and clear. This demands reliable sources of data, prompt action and skilful communication.

To produce reliable information, firms need procedures for recording and storing data systematically and retrieving it in various forms to suit different needs. For instance, some of the data needed by contracts managers, estimators and planners are similar, but they want the information for different reasons and in a different form.

Information and telecommunications technologies have made the information generated during design and construction more reliable. Cheap, portable PCs have made it more accessible. But technology alone will neither make people understand a communication nor make them willing to act on it.

The direction of communication

Communication within companies and project organisations can be classed as upward, lateral or downward, although the distinction is not always helpful. Some lateral communication is between people of roughly equal status (e.g. consultant to contracts manager), whilst some is between people with functional relationships (e.g. plant manager and site supervisor).

Within a work group, a lot of lateral communication takes place and is expected to take place, as people swap information and advice about the job. Much of the information which passes informally along the grapevine is lateral and travels fast. It can be vital for getting work done quickly and efficiently.

Upward communication provides essential feedback to management. It is used for reporting progress, making suggestions and seeking clarification or help, although people often seek help from their peers before going to their bosses.

Managers may have difficulty in getting feedback on progress and costs when things are not going well. Bad news often reflects on someone's ability, possibly the manager's, so no one is in a hurry to break the news. Upward communication for control purposes is often delayed and distorted. Supervisors and managers are told what they want to hear, or what subordinates want them to hear – and only when they are in the mood to take it! Upward communication can become distorted when the sender wants promotion. People are reluctant to take suggestions or complaints to their bosses if it means admitting to failure.

Traditionally, management discouraged upward communication, but modern organisations encourage it. This is achieved through participative management, joint consultation, disputes procedures and empowerment. The employment legislation has put pressure on firms to make sure that employees can express their grievances and get a sympathetic hearing.

Downward communication is used not only to give instructions and explain strategies and objectives, but to give people information about their progress, as in appraisal interviews, and to give advice, as in contacts between head office specialists and site personnel.

More firms are recognising the importance of keeping the workforce informed about policies and activities, although some companies don't even tell their managers what is happening! However, it is widely accepted that employees ought to know about the firm's background, objectives and plans, and should be kept up to date on their prospects. Most people want to know how their work fits in with the organisation's overall goals, otherwise a sense of isolation and alienation from the task can set in.

Communication with sub-contractors demands special attention. Sub-contract site personnel have responsibilities both to their own company and to the main contractor, so that lateral and downward communications 'compete' for priority. This is a problem in any task-force or matrix organisation and there is heavy reliance on contract documents to define the duties and obligations of the contractor and sub-contractor.

It is vital that good communications are established at the outset and that contractor and sub-contractor have continual, direct contact throughout the sub-contract period. Special problems arise with engineering services on complex projects and main contractors sometimes have to appoint services co-ordinators to liaise with services sub-contractors and consultants.

Why communication fails

Many organisational problems are caused by communication failure. Breakdowns occur because of faulty transmission and reception of messages and because people put their own interpretation on what they see and hear. And, of course, the computer is often blamed! Common causes of communication failure are given below.

Poor expression

The communicator does not encode the message clearly because of difficulty in self-expression, poor vocabulary, lack of sensitivity to the receiver or, perhaps, nervousness.

People often fail to speak and write directly and simply. Obscure and redundant words clutter messages and hide their meanings. This problem shows up clearly in many formal communications such as reports and standard letters.

Reluctant communicators

People who avoid communication are often reluctant communicators. An individual's willingness to speak may lie outside the direct influence of the group (Wallace 1987); however, an individual's reluctance to communicate may affect the ability of the group to make a fully informed decision. McCroskey (1977; 1997) found that, under virtually identical situations, some people will initiate communication and others will not. Shyness may occur due to communication discomfort, fear, inhibition or awkwardness.

In groups, apprehensive individuals talk less, avoid conflict, are perceived more negatively and are less liked by other members who are not apprehensive about communicating (McCroskey and Richmond, 1990; Haslett and Ruebush, 1999). Highly apprehensive people also have a tendency to attend fewer meetings (Anderson et al., 1999). Anderson et al., reporting on the findings of group research, found that the degree of communication apprehension diminishes with group experience.

Communication dominance

As well as reluctant communicators, in most groups there are individuals who interact more frequently than other group members do. In decision-making groups, those who talk the most 'win' the most decisions and become leaders (Bales, 1953), unless their participation is excessive and antagonises the other members (Hare, 1976). The more proactive interactors have a greater influence on socialisation and the development of group norms. Those who dominate communication can use their influence on the group to direct questions to members; however, they may also suppress members. When less knowledgeable members dominate interaction they may suppress specialist contributions from the expert members of the group. If dominant communicators are not aware of the specialisms within the group or want to ignore other members, they may be successful in preventing experts who are less active from participating and giving their valuable knowledge.

Failing to ask questions

Asking questions is the single most effective way to extract ideas and information, yet most people are not very good at asking questions (Ellis and Fisher, 1994). The level of question asking within groups is often low compared with other types of communication activity. People have a tendency to give information, opinions and suggestions rather than ask questions (Hawkins and Power, 1999). Also, some questioning approaches can be perceived as accusation (for example 'who did that?') and can result in defensive arousal that can reduce the effectiveness of group discussions, although questions without emotional overtones do not ordinarily result in defensive behaviour (Gibb, 1961). While it would be expected that the most inexperienced or least skilled people would tend to ask the most questions, research has shown that it is often those who are more experienced or capable who ask the most questions (Gameson 1996; Gorse, 2002). Asking questions is different from asking for help. Studies suggest that where professionals do not understand a situation they are often reluctant to ask for help. This phenomenon seems to increase as the status of the professionals increase (Lee, 1997).

Asking closed and open questions

Sometimes closed questions, i.e. questions that have a limited or specific answer (e.g. 'yes' or 'no') can be useful to get a quick answer without irrelevant or misleading information. Other times, open questions, which allow people to explain their answers, are more useful.

Based on information collected during focus groups, many project managers claimed that their subcontractors exaggerate the truth or lie about performance. One problem that seemed to occur was that people would offer 'yes' or 'no' answers to questions rather than supplying detail. Thus, when subcontractors are asked if they will complete their work by the weekend, they often respond with a simple 'yes', when in fact only part of the work will be completed. One way of avoiding such answers is to vary the method of questioning. With trustworthy people we can ask simple questions, which may only require a 'yes' or 'no' answer. Where we have learnt not to trust people we can ask for more detail, or rather than asking a question, we can give an instruction e.g. 'Tell me what needs to be done to finish the work', or ask an open question e.g. 'Can you explain what work you have left to do?', then follow up with a more delving question e.g. 'What plant, equipment and workforce have you ordered to complete the works?'. Always take a strong interest in what is being said. Ask them to clarify dates and numbers, let them know you are recording the detail and will follow up with a visit to site 'just to check everything is going OK'. Such attention may galvanise the person into action, if it does not, at least as a manager you are aware of the problem and need to seek a remedy.

Failing to seek help

Professionals may not seek help, even when help is required, as help-seeking behaviour implies incompetence and dependence. Research on help-seeking behaviour suggests that as the status of the professional increases they become more reluctant to seek help from others (Lee, 1997).

Research has also shown that costly errors made in multi-disciplinary projects could have been prevented by seeking expert help that was available at the time, for example, Capers and Lipton's (1993) research into the behaviours of engineers involved in the development of the Hubble Space Telescope. During the development of the telescope the engineers were monitored using surveillance equipment. The engineers were found to avoid interaction with the specialist employed to provide expert optical advice. The engineers' behaviour showed that they did not want faults to be seen by others and wanted to resolve problems on their own, even though they did not have the knowledge to resolve the problems. The result was that the telescope was launched into space with faults, and the expert who was employed to provide help and advice, but was blocked during development, was used to help correct the faults when the telescope was in space. It would have been much more effective to correct the problems before the launch by seeking help and consulting with the experts.

Help-seeking behaviours are fundamentally interpersonal; one person seeks assistance from another (Lee, 1997). Seeking help from others often occurs simultaneously with information and feedback-seeking (Morrison, 1993). Individuals are more likely to seek help from equal status peers (Morrison, 1993; Lee, 1997) and others who have helped them earlier; co-operative patterns are reciprocal (Patchen, 1993).

Failing to disagree

Disagreement is often seen as a negative term, yet it is found in most observations of group interaction. Moreover, Cline (1994) found that when groups avoid disagreement the vulnerability of a proposal may be overlooked. Conflict during discussions can have positive effects on decision-making, challenging and evaluating proposals and exposing risks of decision; however, if conflict results in a dispute (allocating blame and fault), outcomes of a satisfactory nature are substantially reduced.

A certain amount of challenge, evaluation and disagreement is necessary to appraise alternatives and reduce the risks. Furthermore, Averill's (1993) review of anger based research found that a typical angry episode would often result in change which had positive benefits, and typically the relationship within which the anger was expressed was strengthened more often than it was weakened.

Overloading

Managers often give and receive too much information at once. This causes confusion and misunderstanding. Research has shown that the amount of information a person can cope with at one time is quite limited, especially when the subject matter is unfamiliar and several communication channels (spoken, written, graphical) are being used.

Poor choice of method

People don't always stop to think how to get their message across. Sometimes the spoken word is best, but what is said is usually quickly forgotten. The written word is often preferred and it leaves a semi-permanent record. A simple sketch may be clearer than a lot of words. The method must suit the communication.

Disjunction and distortion

Sender and receiver may not share the same language, dialect, concepts, experiences, attitudes and non-verbal behaviour. Non-verbal cues can have different meanings in different cultures. A message can be misinterpreted because receivers see it in terms of their own experiences, expectations and attitudes. Their outlook and what they think is important will influence how they interpret the message.

Communicators may also 'shape' the message, sometimes unconsciously, to protect their own position or through lack of trust. People often edit information when they feel their credibility is threatened.

Distance

Designers are separated from contractors, sites from parent companies. This limits face-to-face communication and non-verbal signals, like facial expression, which help the communicator and receiver to judge each other's responses.

Status differences

People in relatively junior positions may find it difficult to communicate with those in more senior positions. The opposite can happen too. People may be reluctant to report difficulties or lack of progress to their managers, yet they often like to be consulted and given the chance to air their grievances.

Feelings

How a person feels about a message or about the sender can distort or overshadow its content. In face-to-face communication, the sender may be able to detect this problem, often through the body language of the other person. If a message is received unfavourably, a negative attitude may be provoked in the sender and this in turn affects the receiver. Positive feedback has the opposite effect. If people are aware of this problem, they can avoid setting up a chain of negative reactions. People sometimes totally ignore negative or critical communication to protect their self-esteem.

Skilful managers recognise that each communication is more or less unique. They judge the situation and use all their skills to ensure that people understand what they are trying to convey, accept it and are willing to act on it.

Communication methods

People communicate through language and pictures. Language is conveyed through speech, writing and symbols; pictures are communicated by graphical means, such as drawings and photographs. Managers seldom give enough thought to choosing the best means for conveying an instruction, idea or piece of information. Each method offers a range of options, but has drawbacks as well as strengths. One, or a combination of, methods will usually provide the manager with the right vehicle for conveying a message.

Spoken communication

This can be direct, face-to-face conversation or an indirect telephone call or recorded message. Face-to-face communication is a powerful method, although many people do not use it skilfully. It takes several forms:

- Individual directives, such as a work instruction.
- One-to-one discussions, as in staff appraisal.
- Manager to group, as in a briefing.
- Group discussions, as in site meetings.

Spoken communication needs careful planning, clear expression and the ability to arouse the listener's interest and support.

With indirect conversation via telephone or two-way radio, lack of non-verbal feedback can cause problems. With recorded messages, the sender gets no immediate feedback at all.

If the manager wants to give the same information orally to many people, it usually pays to call them together. But if the manger wants to gauge individual reactions or understanding, the group should be small.

Spoken communication leaves no permanent record. This encourages people to speak more freely, but they soon forget most of what they hear.

Meetings

Organisations use meetings to exchange information, generate ideas, discuss problems and make decisions. Some meetings, like company annual general meetings,
are required by law.

Site meetings are used to inform, co-ordinate, allocate tasks, update plans and
check progress. They create commitment and enable people to get to know and trust
one another. They help people to understand one another's viewpoints and problems. Problem-solving meetings have become more common because the manager
seldom has all the information and skills needed to find a solution single-handed.

However, meetings can fail. They can be so formal that time is wasted on rituals.
They can be so casual that they lack direction and purpose. In meetings, people
seldom build on one another's ideas. Instead, they wait for the chance to make their
point, ignoring what was said earlier. They often criticise and antagonise one
another before ideas have been properly debated.

A good chairperson avoids competing with the others, encourages everyone to
contribute, listens to what they say, keeps the group on course and makes sure all
ideas are considered.

However, chairpersons can unwittingly stifle creative suggestions and discourage
the positive thinking that is needed to throw up new ideas. Also, they are usually
senior employees and have influence outside the meeting, so people are careful what
they say.

Some people believe that unchaired meetings are more productive, but others
claim that even a reasonably competent chairperson can increase the value of a
meeting. He or she acts as a conciliator, controlling aggressive and defensive
behaviour; and sums up, stating clearly the agreements and decisions reached.

Before calling a meeting, a manager should ask:

- Is the meeting necessary?
- What will it achieve?
- How can it be effectively managed?

Meetings and action points: make people act on action

It is common for meetings to identify issues and then decide for action to be taken;
however, in many meetings the specific action agreed fails to be delivered on time.

To help instil responsibility for action:

- Discuss the issue thoroughly.
- Identify action and responsibility.
- Confirm that the action point is agreed with the person responsible.
- Ask the person responsible to specify a time when the action is to be undertaken
  and completed.

- Record the action, person responsible and the date that the action will be undertaken and completed in the minutes. Such information can also be recorded by the chair or project manager in their diaries to help them remind parties and check progress.
- Between meetings, remind people of action points using written communication, save evidence of reminder for the next meeting.
- Nominate someone to check on the progress of the action and report.
- If at a subsequent meeting the action agreed is not delivered by the specified date, find out why and ask for a new date of delivery. Both the previous and new date should be recorded in the minutes.
- If parties continually fail to deliver, the meeting minutes and reminders become embarrassing for the individuals and provide strong supporting evidence for employment, contractual or legal disputes. However, the act of recording dates agreed and any slippage often results in action and prevents issues developing into disputes.

Project meetings

These meetings, attended by members of the project team, are used to:

- ensure that the contractor and other team members understand the project requirements and have an opportunity to check contractual, design and production details and ask for clarification or information;
- ensure that proper records are kept and contractual obligations met;
- compare progress with targets and agree on any corrective action;
- discuss problems like delays or sub-standard work which may affect the quality, safety, cost or timing of the project;
- ensure that contractors and sub-contractors agree on action necessary to meet their obligations;
- check that changes are confirmed in writing and that work is recorded and agreed.

The designer, quantity surveyor and main contractor normally attend project meetings, together with those consultants and sub-contractors involved at each stage of the project. Normally, meetings are held at regular intervals.

Site meetings

The main contractor will hold regular site meetings, some of which will be attended by sub-contractors and key suppliers. The designer may be invited. A meeting will often be used for several purposes. These may include:

- *Internal control*, to review progress, cost, safety and quality against targets and contractual commitments; to update plans.

- *Co-ordination*, to ensure that the work of the main contractor and sub-contractors is properly co-ordinated.
- *Problem-solving*, to identify and discuss problems such as delays, materials shortages and labour difficulties, and to take action to remedy them.
- *Contract administration*, to identify any information needed; to check that proper records are being kept; to monitor the documentation and agreement of variation orders.
- *Labour relations*, to discuss problems relating to work methods, working conditions, safety, incentives, etc.

Written communication

Written communications range from a hand-written note on a scrap of paper to a formal, word processed report. They can be transmitted manually or, as is increasingly the case, by electronic means using systems like fax, e-mail or the Internet. Technology has made it possible to transfer a written communication, in hard copy, to someone's desk the other side of the world, in seconds.

Written communications can be carefully planned and leave a permanent (or at least, semi-permanent) record. On the other hand, an effective written message demands considerable skill and can take time to produce. Once published, it is difficult to retract. People are therefore careful what they write. Their readers can quickly see any contradictions when the message is on paper!

Reports

There are many kinds of report. On site, they give feedback on costs, progress and other aspects of performance. At head office, they may precede a policy decision or change of procedure, or simply give an account of something happening in the organisation. Reports don't necessarily result in decisions or action, but frequently do because they show a deviance from intended standards or targets.

Business reports can be oral, but are usually written because they deal with matters needing careful consideration. They are often supported by figures and diagrams.

A good report is clear, accurate, concise and timely. It should:

- contain everything the reader needs to know and nothing more;
- present the subject matter accurately and logically, giving sources of data, where appropriate;
- make sense to anyone intended to read it;
- clearly summarise the key points, conclusions and any recommendations.

Most reports are structured to help the reader obtain information easily. The exact arrangement depends on the purpose and subject of the report, but typically includes an introduction, the body of the report and a terminal part.

Introduction

This states the aims and terms of reference. It may explain the format of the document and give an outline of the findings. A good introduction focuses the reader's attention on the theme and purpose of the report. There may be a title page and contents page, depending on the length and formality of the report.

Body of report

This contains the subject matter and discusses the data and findings. It need not necessarily be lengthy. Some of the best reports set out the main points in short, crisp paragraphs. Sub-headings make the arrangement clearer, but should be short and self-explanatory.

If the data are bulky, they should be put into appendices at the end of the report. This keeps the body of the report short and clear and readers need only refer to the appendices if details are needed.

Terminal part

This ranges from a *Summary*, if the report has simply gathered data, to a lengthy *Conclusions* section, if advice has been sought. Some reports contain *Recommendations*, where stipulated in the terms of reference.

Busy managers welcome brevity and often rely on reading the summary or conclusions of a report. The terminal part of the report should contain nothing new, apart from any appendices and, if necessary, references and an index.

Plain talking and writing

Business communication is about getting information and ideas across to people. So much information flows through the organisation nowadays that neither manager nor team has time to waste on elaborate communications. Messages must be put over as clearly and succinctly as possible.

Engineers may wish to know that 'transmissions containing formal gearing require detergent lubricants of high viscosity range', but the fitter wants to know whether to use green label oil in the lower gear box (Maude, 1977).

Writing and speaking skills have been neglected. Few managers are trained in the use of language beyond their school-days. The following extracts from construction publications show how much improvement is possible:

> Drawings are all too rarely fully available at this stage of the proceedings, but now is a good opportunity to initiate a comprehensive drawing register and index. [27 words]

The author was trying to say: 'Start a drawing register and index now, even though some drawings are missing.' The main point comes across here in half the words.

A building magazine reported:

> It is difficult to approach the subject of the possible takeover and rehabilitation of failed housing from the public sector by entrepreneurs from the private sector with any confidence, simply because there is not a single case where this has actually happened. [42 words]

In other words: 'As no private developer has ever taken over failed council housing, it is difficult to comment'. (16 words)

Vague, general words should be driven out in favour of 'concrete' words. Key words should be near the beginning, so that the receiver knows what the message is about.

Another building publication had this to say:

> The more optimistic among us might have expected that post-war housing, taking advantage of new building techniques, would be less troubled by condensation and damp than pre-war housing. Unfortunately the reverse is the case. [34 words]

What the author meant was: 'Post-war houses have more condensation and damp troubles than pre-war housing, despite new techniques'. (14 words)

Some might argue that the original versions had more style. Harold Evans (1972) cites Matthew Arnold's advice: 'Have something to say and say it as clearly as you can. That is the only secret of style.' One of the beauties of the English language is that clarity, vigour and economy of words can go hand in hand.

Evans says that people should write positively, prune ruthlessly, and care about the meanings of words. His advice is given below.

Limit the ideas in sentences

Sentences should communicate one idea. Short sentences make for clarity. Too many compound sentences make the message heavy-going. The following sentence contains too much information:

> Three bricklayers who between them had more than twenty years' continuous service with the company and who, until now, had given no cause for complaint, were ordered off the site today by the angry supervisor, after two verbal warnings and a written warning about their bad behaviour and poor workmanship.

Be more direct

Use the active voice. 'The manager called a meeting' is more vigorous and economical than the passive version: 'A meeting was called by the manager'. A succession of passive sentences can ruin a communication.

Be positive. Make sentences assertive. 'The manager has abandoned the new bonus scheme' is more effective than the negative statement: 'The manager is not now going ahead with the new bonus scheme'.

Evans argues that government officials, reports and ministers are the worst per-petuators of the passive: 'It was felt necessary in the circumstances; it should perhaps be pointed out; it cannot be denied', and so on.

Communicators should avoid double negatives. 'It is unlikely that annual bonuses will not be paid to site staff' means that they probably will! Look at the improvement that is possible:

> At its meeting last month, the Board of Directors decided that it was highly unlikely that there would be no deterioration of the housing market and that the company could not be expected to maintain its present market share unless a drastic change of policy was agreed by all concerned. [50 words]

> The Board of Directors warned at last month's meeting that a drastic policy change is needed to maintain the company's workload in a declining housing market. [26 words]

Avoid monotony

Messages can become monotonous if the suggestions above are too rigidly followed, but there is plenty of scope for variety. The structure and length of sentences can be varied without losing vigour and directness. The function of a sentence can be changed between statements, questions, exclamations and commands.

Avoid unnecessary words

Every word should earn its keep. If a word doesn't add something to a message, it should be left out. Redundant words waste the reader's time and obscure meaning. Driving out abstract words often saves on length and aids clarity. Abstract nouns like issue, nature, circumstances and eventuality are often mere padding:

> In the circumstances, the plasterers should be paid last week's overtime, even though the issue cannot be resolved to the entire satisfaction of the manager because of the faulty nature of their work. [33 words]

> The plasterers should be paid last week's overtime, even though the manager is still dissatisfied with their work. [18 words]

Economy has to be used intelligently, but writing with concrete words is usually shorter and more interesting. As Harold Evans points out, words stand for objects, ideas and feelings. Failure to match words with objects leads to vagueness.

Car parking facilities	Car park
Adverse climatic conditions	Bad weather
The canteen has seating accommodation for 80 people	The canteen seats 80

Like words, signs and symbols also stand for objects and information. They have become popular and important in communication. When they make use of icons, as they often do, they become graphic communication.

Graphic and numerical communication

Written communication can be unsuitable when information is extensive or complex. Text ceases to be effective when:

- whole paragraphs have to be read before meaning can be understood;
- individual facts or numbers are difficult to single out from the mass of data; or
- trends are hard to identify and comparisons difficult to make.

In construction, there is heavy reliance on graphic and numerical communication, mostly as drawings, diagrams, schedules and charts. A single drawing often conveys a great deal of information in a much clearer way than would be possible using words alone. Drawings are very useful as long as they are accurate, easy to understand and supplied at the right time. Bills of quantities use numerical data linked with tightly structured text to give condensed information. They are expected to fully and accurately describe a project. Bar-charts and network diagrams are good ways of presenting information which is partly numerical and partly written. They are a valuable tool for management control.

These communication methods are not always satisfactory. A designer's drawing may be supplied late or may be unclear. Bills of quantities don't always describe the work as fully as they should. Programme charts are based on approximate information and may not be kept up-to-date.

However, charts, tables and graphs are powerful methods of communicating certain kinds of information. They are often regarded as an aid to text communication, but can in fact do the main work of communicating (see Fig. 5.3). Tabulated information:

- makes the information clearer by presenting it in a logical way;
- communicates more concisely than would be possible using words alone;
- makes comparisons much easier, by arranging data in columns and rows.

Graphic presentation is especially useful for:

- highlighting key trends or facts in complex information;
- showing relationships and differences;
- displaying information that can best be understood against some visual scale.

On the other hand, graphic information takes time to produce and can only effectively show a limited amount of information at one time, without causing confusion.

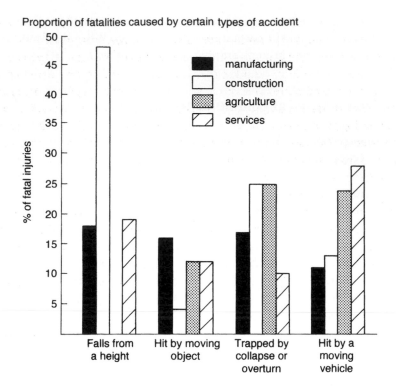

Figure 5.3 Example of graphic communication.

Information management

Communication is about moving information around and processing it in various ways. Some of the information may be in the form of ideas or expressions of feelings, but it still affects organisational performance. Because communication is the life-blood of an organisation, managers now recognise that creating effective systems for managing information is crucial to their success. Information technology (IT), with its associated fields of microelectronics and telecommunications, has revolutionised information management (and therefore communication) in several ways, in particular by:

- speeding up enormously the processing of information (collection, collation, analysis, synthesis, presentation and transfer);
- making available to organisations much more information about their own performance, knowledge of their competitors, and data about other external bodies, events and trends;
- improving management information systems through computer-based systems, giving managers faster access to better information, leading to more effective planning, decision-making and control.

It is easy to decry IT as merely a tool of management, but it is much more than that. IT does not simply improve communication, it performs work for the organisation. For example, computer programs can simulate dozens of project management decisions the manager might make, and present and compare the outcomes. Such a task, performed in minutes or even seconds using a PC, can produce results which might have taken the manager weeks or months to achieve (if at all). Database technology can be used to manage the large quantities of data generated on projects. Flowers (1996) describes the operation of relational databases and shows how data can be structured to give maximum benefit to the manager on a construction project.

Expert systems already exist, programs which perform tasks using artificial intelligence to simulate human expertise. These systems can make diagnoses and judgements, cope with unreliable and unclear information and handle probabilities ('it seems as if...') and possibilities. Developments in IT are so far-reaching, they will require managers to rethink words like information and system, and perhaps even the concept of management itself (Harry, 1995).

The implementation of computer-based management information systems can create problems, typically:

- *Negative attitudes to change.* Many employees resent having their tried and tested routines overturned.
- *Lack of employee commitment.* Employees have not been consulted or involved in the design of the new system.
- *Disruption of organisation structure.* The system disrupts established departmental boundaries.
- *Disruption of informal communications.* New systems alter communication patterns and destroy the informal networks which existed.

In addition, some communications, at a more personal level, depend for their effectiveness on face-to-face contact and body language, vital to the richness and success of interaction. Here, electronic communication remains inadequate.

Management information systems create new posts, including the chief information officer (or MIS manager), whose roles include change agent, overseeing the design, introduction and monitoring of MIS and its surrounding technologies; and 'human link' with senior management. Unlike conventional data processing managers, who concentrate on the day-to-day tasks of their departments, MIS managers focus on planning and developing creative solutions to the organisation's changing information needs.

In the future, it seems that self-managing computers and robots will learn about the organisation and its activities, teach themselves to perform tasks, repair and update themselves as situations change and, of course, communicate with and learn from one another.

The manager's behaviour: communication and influence

One of the most influential works on this subject is Dale Carnegie's highly readable book, the title of which has become a catch phrase: *How to Win Friends and Influence People*. The book has sold more than ten million copies in dozens of languages.

Carnegie's underlying message is simple enough – how you behave towards others must be based on what you hope to achieve and how people will react to you. If your behaviour makes other people feel upset, this will more often than not limit your chances of achieving what you want to achieve in working with them. On the other hand, if your behaviour makes other people feel good, they tend to feel positive towards you and are more likely to be co-operative – and this improves your chances of achieving your goals. In the context of managing yourself, the important point here is that managers who learn to control their own behaviour understand the impact they are having on others and adapt their behaviour to get the results they want.

Such managers learn how to make other people feel good, so that they are more likely to be motivated. Achieving this is not simple, but some of the following suggestions can help to get the best from people:

● Make people feel important.
● Show that you value them and recognise their abilities.
● Be a good listener and show an interest in them.
● Show that you can see people's points of view.
● Be sympathetic to their ideas and needs.
● Give plenty of praise and encouragement.
● Be sincere and fair with everyone.

Recognition

A special reason for wanting to make people feel important and for recognising their capabilities and achievements is that it often helps in getting the most out of them – spurring them on to greater success. To achieve this, the manager must behave in such a way that the individual's confidence is built up and this means seeking opportunities for giving the person praise and recognition. Many managers are quick to criticise, but slow to congratulate people on a job well done. Yet the praise – or positive reinforcement – can produce an improvement in the individual's performance.

People will, of course, see through false praise – or flattery – but the manager should be able to find some basis, however small, for complimenting people on their work. Many employees want to be seen to be competent and want to maintain their self-esteem, so even a word of praise for a minor job can have a beneficial effect on their future performance and motivation.

Of course, there are times when subordinates have been careless or lazy – or for some other reason have done a bad job. How does the manager criticise such people? This depends on the individual – but what the manager must guard against is the negative effect that direct criticism can have on many employees. If criticism damages their self-esteem or creates bad feeling between them and the manager, the net effect of criticism is negative – and some long-term harm can be done to the relationship between manager and subordinate. In extreme cases, the manager may cause bitter resentment or even become hated for handing out criticism.

Empathy

Carnegie emphasises the importance of trying honestly to see things from other people's viewpoints. But most managers are somewhat self-centred. They are mainly interested in their own problems and achievements. The trouble is, everyone else is the same. So the manager who can break out of this mould and show a real interest in others will make a big impact. Such a manager will really try to understand people's aspirations, feelings, ideas and worries – and show that these are as important as his or her own. To show empathy with another person, the manager should pause before starting a conversation and think 'if I were the other person, what would I want to hear now?'. This requires considerable sensitivity on the part of the manager, a quality well worth developing.

Empathy involves not only trying hard to understand what another person is saying or thinking, but responding in a way which *shows* that you understand or are trying to understand. So, the many signals the manager gives to the other person – verbal and non-verbal – can be very important.

Listening

Being a good listener is an important skill, often lacking in managers and non-managers alike. In fact, most people much prefer talking to listening. The manager who can listen not only conveys a message to others that they are *worth* listening to – but also learns a lot from what they have to say. Of course, there are exceptions and managers generally haven't got time to waste on irrelevancies. But there is scope for a lot of useful listening, if the manager has the skill to do it properly and be selective about it.

Among the skills of listening are:

- Interpreting what is being said to understand its meaning (this involves 'decoding' non-verbal as well as verbal signals from the person talking).
- Giving feedback which shows you really understand what the person is saying, but without interrupting.

When the other person is being long-winded and taking up too much of the manager's time, then action is needed to curtail the listening. Here the manager must signal to the other person that the exchange must be brief and to the point.

Encouraging a long encounter	*Signalling a time limit*
'Come in, Joan. How are you? Has Henry recovered from his operation yet? Did you enjoy your trip to France...'	'Joan, I have an appointment at ten but I'm happy to spend ten minutes with you now, if we can solve the problem in that time.'

Assertiveness

Assertiveness has not been given much attention in management, probably because it has rarely been thought of as a problem. But in the last few years, the value of assertive behaviour has been recognised and taken more seriously. Training in assertiveness has become quite common and there are even self-help guides for those who want to assess or improve their assertiveness (see, for example, Lloyd, 1988). An insight into assertiveness shows that many managers are *aggressive* rather than *assertive* – and the two are not the same.

Aggressive managers convey an impression of superiority and often disrespect, their wants and rights being placed above those of others and therefore tending to infringe the freedom and rights of others. Aggressive people tend to stand their ground, are often inflexible and obstinate, belittling others and making them angry or humiliated. They can be sarcastic, accusatory and rude.

Compare this with assertive behaviour. Assertive managers encourage honesty and directness – and do so *by example*; they communicate a feeling of self-respect and respect for others. They try to help others achieve their needs, as well as achieving their own – creating 'win–win' situations that benefit all concerned. Assertive people seek co-operation, show tact, and are genuine, open and enthusiastic.

Less common among managers, although elements of it are often present, is non-assertive behaviour. Non-assertive managers tend to be placid and sometimes vague and obscure, imparting messages of inferiority or lack of self-confidence. Such managers can be hesitant, defensive and subtly dishonest, being at the same time disrespectful to subordinates but deferential to their seniors.

Personal skills and interaction

Construction firms are realising more and more that their managers and other employees need good personal skills to carry out their jobs effectively. This realisation has not only dawned on the construction industry; in recent years, many other industries and professions have started to give much more attention to training in this field.

For instance, the Metropolitan Police Force included in its complete policing skills programme: (1) self-awareness; (2) interpersonal skills; (3) group awareness. This means that along with the training they receive in the more 'glamorous' side of their work – driving, detective work, firearms and so on – police officers learn such skills as how to assess their own behaviour, how to compare themselves with their peers, positive and negative aspects of verbal and non-verbal communication and how to control and change people's attitudes and behaviour, whether colleagues or the public (Mitchell, 1989).

Even very senior managers often value personal skills very highly. For instance, in a recent UK study of 45 managing directors, most of them mentioned *people skills* in one form or another as 'equally important or a very close second' to decision-making skills (Cox and Cooper, 1988). Managers in UK construction firms often rank their interpersonal skills higher than all other management skills, regardless of whether they are from a trade or technical background and irrespective of their age (Fryer, 1994b).

One reason why such skills are rated so highly is that managers realise that to get things done and to elicit co-operation from people, they have to establish rapport with them, persuade them to accept goals and motivate them. This involves creating feelings of satisfaction, approval and respect in a range of situations, such as when discussing a work problem, interviewing someone, explaining a new method, counselling or bargaining.

Establishing a good rapport is an important starting point in exercising personal skills and is achieved in a number of ways. Argyle (1983) summed up the ways in which rapport can be created:

- Adopting a warm, friendly manner; smiling; using eye-contact.
- Treating the other person as an equal.
- Creating a smooth and easy pattern of interaction.
- Finding a common interest or experience.
- Showing a keen interest in the other; listening carefully.
- Meeting the other person on his or her own ground.

Clearly, establishing good rapport with people requires skill. It involves good communication, trust and acceptance, and creating relationships in which people feel comfortable with one another. It brings into play a number of human skills which have not been taken seriously enough by most managers in the past. And these skills must mostly be practised face-to-face; not through memos and telephone calls, but through personal communication.

Effective personal communication

Even though most managers and professionals *appear* to understand the value of good communications, somehow the message often fails to get through. Managers seem clear enough that an important purpose of communication is to involve employees, so that they are committed to the business and therefore contribute effectively to its work, but little seems to be done to apply communication to make this happen.

Drennan (1989) gives an interesting case study of a large firm which wanted to 'beef up' its internal communications. This is how it did it. First, senior management redefined the firm's key goals so that they would be simple and understandable to all employees, relatively stable over the next five years and couched in such a way that every department and employee could do something to contribute to them.

Next, senior managers were asked to consider what, in practical terms, they were going to do to achieve these goals. A series of conferences were held at various levels, so that ideas and proposals about how the goals could be achieved and how to measure and communicate progress flowed back and forth among employees throughout the organisation. Each working team put together its practical programme and presented it to the next level of management for approval. The work teams set new performance targets for themselves and soon charts and graphs started to appear showing how well teams were doing.

The message is clear – if people know what they are striving for, they will largely manage themselves. But they cannot find out what they are striving for without good two-way communication and this will only happen if people – managers and other employees – want to talk to one another and know how to do so effectively.

Summary

Communication breaks down in organisations because people's interests, perceptions and viewpoints differ. People fail to see how their work affects others and their communication skills are often weak. Senior managers have the job of developing a communication network to suit the size of the firm, the projects it undertakes and the people involved.

Managers must help employees to improve their communication skills and encourage two-way communication within their groups, making time to listen to, and understand, what people say. The time will be well spent.

Informal communication channels are important but are sometimes suppressed. They must be encouraged. They supplement rather than replace formal channels, which can be inadequate on their own. However, managers must use judgement where channels are contractually prescribed.

Communication should be as direct as possible, without too many links in between the sender and the person who must act on the message. This is especially important in large organisations, where neglect of lateral relationships between people of similar rank creates problems. In construction, the site manager is largely isolated from other site managers who have similar problems and from the specialists who provide expertise. Opportunities for the exchange of ideas and information are restricted.

Good communication and willing co-operation are inseparable. Managers who stress the technical side of their jobs often fail to recognise that people may be suspicious of their motives and may misunderstand or distort what they say. Sensitivity and positive attitudes to people are vital to successful communication.

Revolutionary changes have taken place in organisational communication, with the development of the technologies associated with microelectronics and tele-communications. Information is now available to managers and other employees faster, more reliably and in larger quantities than ever before. Information now has to be systematically managed and information networks carefully designed and monitored. Communications can pass at lightning speed around and among organisations and between individuals anywhere in the world.

Group communication exercise

Attempt to observe, study, record and analyse group interaction using the quanti-tative analysis distribution technique. The technique records each communication act and the direction (person it was aimed at).

The communication QuAD (Quantitative Analysis and Direction) tick sheet shown in Figure 5.4 can be used in two ways. One way is to sit and observe the group, simply ticking the sheet as people speak and address others in the group, another way is to use a video camera and video record the group and complete the sheet by watching the recording. As long as the video camera is positioned so that all of the participants can be seen, the recording is much more accurate as it can be watched and reviewed to complete the sheet. Some observers may use more than one video camera, although this is not necessary if participants are seated in a horseshoe arrangement. When using a video, the validity of the observation is easy to check as more than one person can observe the interaction, allowing observations to be cross-checked between observers.

When observing the interaction in real time (without the aid of a video camera), it is very difficult to record consistently every communication act observed. However, it is still possible to catch most of the interaction in a reliable way. General rules for recording the interaction include:

- *Record the most obvious acts of interaction.* If a communication act has caught your attention as an observer, other members of the group are likely to be aware of the communication act.
- *You can only record one act at a time.* If two acts overlap, record the most obvious and influential (the one that won the floor).
- *Record systematically.* To make the observation reliable, the speed of recording must be constant. As a rule, assuming interaction is constantly taking place, give one tick per communication act at a rate of one tick per second.
- *When communication acts are coming too fast, stick to the one second rule.* This will mean that you miss some of the interaction, however, if people are active in the group other comments will be observed and the percentage distribution should, over the course of the observation, be representative of the interaction that took place.

	Date	Communication – Quantitative Analysis and Direction (QuAD) Tick Sheet													Sheet No. ____
	No speaker	Sender						Receiver							Group
		a	b	c	d	e	f	a	b	c	d	e	f	
1															
2															
3															
4															
5															
6															
7															
8															
9															
10															
11															
12															
13															
14															
15															
16															
17															
18															
19															
20															
21															
22															
23															
24															
25															
26															
27															
28															
29															
30															
31															
32															
33															
34															
35															
36															
37															
38															
39															
40															

Figure 5.4 QuAD – Quantitative Analysis and Direction tick sheet.

- *If nobody speaks tick the 'No speaker' box at a rate of one tick per two seconds.* (This allows for the two ticks that are normally required in each row.)
- *Record the most obvious direction.* When people speak in a group, all members of the group have the opportunity to receive the message (in most cases). However, communication is often aimed at one of the members more than the others. This can be observed by eye contact or use of a name. If it is not clear who the message is aimed at then record the direction of the communication act as 'the group'.

A considerable amount of data can be collected in a very short time when using the QuAD sheet, so a short observation can provide a quick indication of some inter-action trends that are taking place within the group.

Once the data are collected, the following information can be produced:

- Total number of group communication acts observed.
- Number of communication acts sent by each participant.
- Number of communication acts overtly directed at participant.
- Number of communication acts directed at the group.

A more detailed analysis of individuals can reveal:

- The number of times a person addresses each individual member.
- The number of times a person received a message from each individual member.
- The number of times an individual sends messages to the group or directs their communication acts towards individuals.

Through the use of the QuAD sheet the following issues can be investigated:

Group participation and:

- leadership
- friendships and alliances
- non participation or reluctant participation
- communication dominance.

Often observation sheets are criticised because they do not capture the experiences, feelings or perceptions of the group members. To make the whole exercise more interesting, members who participated in the group can be asked the following questions immediately after the discussion.

Who in the discussion do you think was most influential?
Who in the group do you think was the leader?
Which group member was least active in the group?
Whose contributions did you prefer the most?
Whose contributions did you disagree with the most?

Can you identify the members of the group who you have strong alliances with or close friendships (these can be listed in order of strongest to weakest alliance)?

Select the questions you wish to investigate and present them to the group members in a questionnaire format. Be careful, some of the questions can be sensitive if not handled correctly.

Chapter 6
Conflict and Conflict Management

Avoiding negative emotional encounters is often more comfortable than engaging in conflict (Belbin, 1993). Unfortunately both engaging in conflict and avoiding it can put a strain on professional relationships and induce stressful experiences for the individual. Conflict within organisations is to be expected, construction is no exception, and managers must develop strategies for dealing with it, not avoiding it.

Construction projects have an inherent level of technological complexity that requires co-operation and interdependence between participants from all disciplines. However, conflicts often emerge as the stakeholders in the process have different organisational objectives and conflicting interests. All of the specialists need to collaborate to realise the construction project. As problems develop, the key members of the client organisation, design team and contracting parties need to use effective communication to co-ordinate information and maintain relationships. Conflict between the professionals must be managed so that their relationships do not become volatile and adversarial. Relationships between group members are often very fragile and slight incidents during bargaining can result in members being excluded from group interaction (Ostmann, 1992).

The following model (Figure 6.1) shows how different perspectives can emerge as conflict, resulting in tension that can threaten relationships. If disagreements are not managed they can develop into a full blown dispute. In order to prevent the dispute, interaction between parties should be managed. The key professionals in construction must be aware of situations in which conflict is likely to occur, recognise when conflict emerges and develop strategies that help parties manage conflict and sustain effective working relationships.

Definition: functional and dysfunctional conflict

Conflict within construction is not only inevitable, it is often desirable. Gardiner and Simons (1992, p. 460) define conflict as:

> any divergence of interest, objectives or priorities between individuals, groups or organisations, or non-conformance to requirements of a task, activity or process.

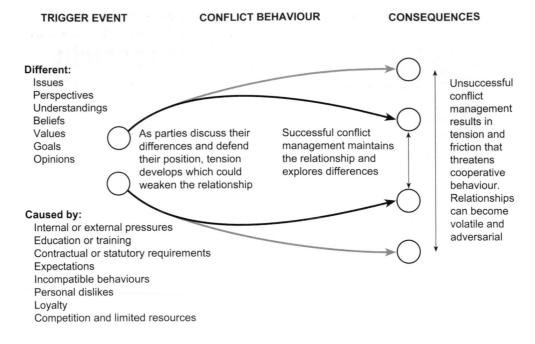

Figure 6.1 Model of conflict development (adapted from Emmitt and Gorse, 2003, p. 166).

Conflict can be natural, functional and constructive or unnatural, dysfunctional, destructive and unproductive (Gorse, 2003). Functional conflict results from challenges, disagreements and arguments relating to tasks, roles, processes and functions. This type of conflict often involves detailed discussion of key issues. Functional conflict is often beneficial, helping to:

- expose problems
- reduce risks
- integrate ideas
- produce a wider range of solutions
- develop better understanding
- evaluate alternatives
- make improvements
- develop better solutions
- improve relationships.

If functional conflict is amicably resolved, the success achieved from working through the conflict can strengthen relationships. Although conflict is often functional, with some episodes of conflict it is difficult to identify a rational purpose. Unnatural conflict is where a participant enters into an encounter intending the destruction or disablement of the other. Personal insults and criticisms that boost self-ego and put others down is often described as dysfunctional conflict. Such conflict is not aimed at improving task performance. However, the occurrence of functional conflict also results in tension, and if this tension is not defused it can build up and threaten relationships (Gorse, 2002).

Indicators that conflict has become unproductive include (O'Neil, 2002):

- Conflict deteriorates into personal conflict.
- Conflict increases with each meeting rather than reduces.
- Communications become one way.
- Parties become entrenched and will not accommodate alternative views.
- The conflict becomes a major issue – incurs costs and delays activities.

Ultimately, the completion of construction projects relies on co-operation between multidisciplinary teams and, as conflict is to be expected, managers must attempt to identify where conflict is most likely to occur and develop strategies for managing the conflict.

Conflict emergence and development

Conflict has been found to develop in multidisciplinary building design teams as the group members discover their team objectives and then attempt to enforce them on others (Wallace 1987). Conflict often emerges as people attempt to change other people's ideas, beliefs or actions.

Handy (1993) proposed five different situations in which conflict can emerge:

- *Overlap in formal objectives.* Although each party has a contractual obligation for delivering the project, they have different responsibilities resulting in different priorities when considering the same works. For example, architects concentrate on design and contractors insist on practical building methods. Each person or group has a formal goal; clashes occur when others interfere with process and activities used to achieve the goal.
- *Conflicting objectives resulting from overlap in role definition.* Two parties may both have responsibility for providing parts of the same service. Each party makes assumptions on work that is theirs and the work that is the other party's work. Where the work package is not clearly specified, gaps occur where neither party believes it is their responsibility to undertake certain works. This often occurs between subcontractors where the boundaries of two or more sub-contractors meet at an interface. Alternatively, both parties may assume work is within their remit, but have very different ideas about how to do the work.
- *Unclear contractual relationship.* Contracts are unable to identify duties, responsibilities and lines of authority for every situation, thus such situations are common. When entering into a contract, the briefing process is important to help identify the full scope of the contract and clarify responsibilities.

- *Simultaneous roles*. A party may be responsible for overseeing a part of a project, but is also responsible for providing a service. Gardiner and Simmons (1992) suggest that architects are often placed in this position where they are responsible for both the management of the project (client's representative) and designing the building; thus they are often responsible for overseeing their own work.
- *Hidden objectives*. Individuals and organisations may have a reason for performing the works that is not made clear to others, e.g. a professional's performance may be motivated by politics, promotion, competition, status, ego, etc.

Client expectations and conflict

Conflict develops during construction projects for many different reasons. Conflict often occurs due to a failure to develop and manage the client's expectations. Construction clients are either experienced or inexperienced in construction. Their organisations require one-off buildings or, inherent with the nature of their business or strategic expansion programmes, they require a continual supply of buildings and structures. Inexperienced clients do not understand the construction process; they are unfamiliar with the techniques, contracts, processes, legislative requirements and the cost of making changes during the process. The lack of building knowledge means that the construction professionals should take greater care during the briefing stages to ensure that the client develops a greater understanding of the process. Unfortunately, the information offered to clients from different professionals is often inconsistent and confusing. Even though professionals may offer the same service they tend to concentrate on aspects closely associated with their profession, training and experience (Gameson 1992). Such bias would include quantity surveyors concentrating on aspects of cost and contract, architects on design, and construction contractors placing priority on management and construction techniques. Even with their different perspectives, construction professionals need to understand the client's requirements and articulate them into a design that realises the client's building. Considerable emphasis should be placed on developing a common understanding between the professionals and the client. With inexperienced clients there is a need to delve deeply into client issues, ask questions, and check that each party understands what is being discussed. Yet Gameson's (1992) research into client briefing showed that it was the experienced clients who were seeking and obtaining more information from the professionals. When dealing with inexperienced clients' construction, professionals did not provide the same level of explanation. It may be suggested that some construction professionals take advantage of a client's naivety, offering a limited perspective rather than attempting to develop understanding by exploring alternative approaches and different solutions that may benefit the client. Failure to develop a sound understanding of each party's requirements increases the potential for disagreements occurring.

The difference in practice between what should be done and what actually happens during the briefing process means that problems emerge that result in conflict. As the client realises that the services, designs and buildings provided are inconsistent with their expectations, conflict emerges. Considerable emphasis should be placed on the briefing stage to ensure that understanding between parties is developed and the client's expectations are managed through the design development.

Integrating knowledge and resolving conflicts within a fixed duration

The pressure on delivering projects within tight schedules means that the information and activities of the various contracting organisations have to be co-ordinated and integrated. As the specialists work on the development of design information and management of activities concurrently, it is difficult to ensure that conflicting information and clashes in activities do not occur. The project manager must work together with the various parties to ensure that information is developed and integrated so that conflicts are kept to a minimum. Project managers should set up and manage meetings where parties can identify problems and resolve their differences. Where contractors and specialists work in the same space, share plant and equipment where products and components meet and have to function and fit together, conflict is to be expected. Interfaces between packages of work and different work groups are prone to conflict. Designers, e.g. architects, structural and mechanical engineers, often produce their designs simultaneously. Elements of the designs often clash or are incompatible; thus, designers have to work together to minimise wasted design. If such interfaces are not managed, conflict has considerable potential to develop into major disputes.

A level of conflict in construction can be anticipated. Gardiner and Simmons' (1992) research identified a number of reasons for conflict emerging and situations where conflict tended to occur (see Table 6.1)

Conflict and change

Change in construction results in additional work. Conflict may emerge as parties attempt to avoid redesign or take on additional work (Wallace, 1987). As Loosemore (1996) points out, change in the construction process often leads to conflict; however, Gardiner and Simmons' (1992) research found that conflict could also lead to change. Conflict can lead to change and change can lead to conflict, thus often the conflict/change process may become cyclic (Figure 6.2).

In successful conflict management processes it is to be expected that conflict and change cycle will occur a number of times. For the process to be successful, all of the parties need to be sufficiently satisfied with the resulting changes so that the working relationship can continue. With both change and conflict there is a cost; at a bare minimum, conflict incurs a resource cost of those engaged in the disagreement and its resolution. When dealing with change there is almost always a management cost of reallocating activities and resources.

Table 6.1 Reasons for conflict and situations that are prone to conflict (adapted from Gardiner and Simmons, 1992).

Inception/briefing/tendering	Design
Problems during briefing procedure	Design error
Difficulty in co-ordinating information	Design omission
Client and/or users' lack of experience	Design not meeting specification
Difficulty arriving at a consensus	Difficulty in obtaining written approval
Low recognition of users	from client or users causes conflict
Often end users have a lack of power and authority during the design stage	Clients have difficulty in interpreting drawings
Construction operations	**Project management**
Failure of the construction to meet the design requirements	Internal politics, e.g. planning and approval
Quality of work on site less than expected	Lack of agreement between users and client project manager
Exceeding project budget	By-passing single point of contact
Exceeding scheduled duration	Conflict of loyalty (e.g. clerk of works)
Functional or operational faults	Different levels of change control depending on the nature of change

Summary

Many of the changes resulted from:

- Different emphasis on proposed project.
- Difficulties in maintaining interfaces between various professionals to serve the client's needs.
- The use and misuse of quality systems and modification to standard contract conditions.

Change in construction has an impact on resources resulting in increased expenditure for one or more of the stakeholders. As the construction project progresses from inception to construction the cost of changes increases. Changes made at the design stage will involve redesign, this may be limited to the architect's details, but may also affect structural engineers' and mechanical and electrical engineers' drawings and specifications. However, changes made during the construction stage may result in redesign, modification to, or removal and rebuilding of, completed works. All abortive work results in a cost to one or more of the parties involved.

Gardiner and Simmons' (1992) research showed that a high proportion of conflict led to a change. Conflict was found at all stages in the process studied: briefing, design and construction. Although the occurrence of conflict was highest in the design stage, it was also prominent during construction. Almost half of all conflict episodes uncovered resulted in change. The potential financial impact of change on a project increases as the project progresses from inception to completion (Figure 6.3).

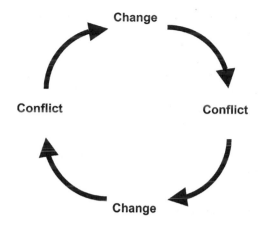

- Change can result in conflict
- Conflict can result in change
- Change may induce conflict that results in further change – the process is often cyclic

Figure 6.2 Change and conflict cycle.

Managing conflict

Most models of conflict management work on the principle that during management people perform their duties with a level of concern for themselves and a concern for their product, for example, Blake and Moulton's (1964) managerial grid model. The two dimensional approach, first considered by Blake and Moulton has been adapted to concern for self and concern for others (Pruitt and Rubin, 1986).

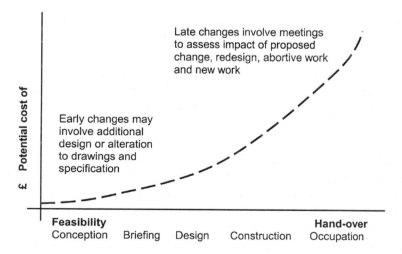

Figure 6.3 Impact of change on cost of project.

Using these two simple dimensions, various approaches to conflict management have been uncovered (Figure 6.4).

There is an argument that each of the conflict management styles has a useful purpose (e.g. O'Neil, 2002), although parties must consider how the outcome will affect the working relationship. In short-term one-off ventures, where the risk of having to work with another person is minimal, the most profitable strategy may be to dominate discussions ensuring that everything falls in an individual's favour. Such strategies may result in obtaining a bad reputation, exclusion from further projects, or legal disputes. The use of the different conflict management strategies with a goal of finding a solution that benefits all offers the optimum strategy. There is a time and place for each of the processes but, to stay in business and maintain relationships with clients and suppliers, managers must balance concern for self and others, pushing negotiations towards the ideal position.

Studies of contractors' representatives during business negotiations found that those who were most successful within their organisation made greater use of both supportive and confrontational interaction (Gorse, 2002). Although the representatives were not afraid to disagree with others they were also prepared to offer support and praise.

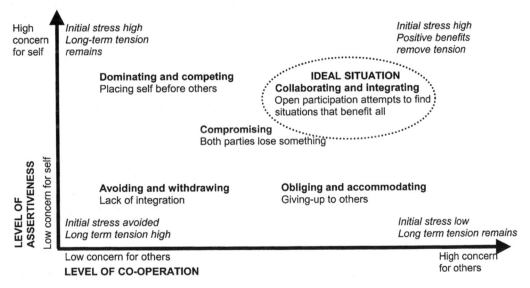

Figure 6.4 Conflict handling styles (adapted from Bales, 1950; Kilmann and Thomas, 1975; Thomas, 1976; Rahim, 1983; Gorse, 2002; and Emmitt and Gorse, 2003).

Dominating and competing (high assertiveness, low co-operation)

Those who use the dominating style are very assertive, uncooperative, have little concern for others, are selfish in attitude, forceful and unwilling to consider other viewpoints, they believe that others cannot be trusted. The style has a competitive nature; individuals pursue their own objectives at others' expense, little consideration is given to others. When two competing styles engage each other, there is a high probability that the conflict develops into a major dispute. The dominating style results in high levels of tension. During the initial discussion the level of assertive confrontation results in emotional tension for the aggressor and those being dominated. The dominant party knows that most people would rather avoid tense confrontational situations and uses this to his/her advantage. As the dominant party increasingly raises the level of negative emotion others tend to compromise, withdraw or accommodate, enabling the dominant party to win. Even though the initial tension is defused, most parties on the receiving end of the dominant person may worry about the outcome, inducing long term stress that can threaten relations. Even when the dominant style seems the preferred option, it is better to insist on exploring outcomes that will benefit all rather than asserting self gain. In business if outcomes offer little gain to other parties relationships will soon break down.

Avoiding and withdrawing (low assertiveness, low co-operation)

Those who withdraw or suppress conflict often believe that such issues will disappear if ignored; people who use this method do not attempt to co-operate and demonstrate low levels of assertiveness. Beliefs that encourage such behaviour include the belief that the opponent is too powerful or that there are no other alternatives available. Avoiding conflict may reduce tension at the outset, however, the long-term effects of suppression may result in even greater emotional stress as the problems develop and become worse through lack of attention.

If parties become too emotionally engaged in discussions, it may be wise to withdraw temporarily from discussions, but not at the expense of ignoring the issue. As soon as parties calm down, attempts should be made to continue the discussion, possibly using an independent expert, facilitator or mutual friend. A short break can often be productive, allowing parties the opportunity to reflect on the situation. People who avoid situations may have a low level of enthusiasm for the project and are not sufficiently motivated to engage in conflict. In multidisciplinary projects, parties who are not sufficiently motivated to engage in difficult discussions may also have little desire to contribute to other aspects of the project.

Compromise – partial giving-up objective (holds the middle ground on both assertiveness and level of co-operation)

Those who compromise show low to moderate levels of assertiveness and are co-operative in nature. Rather than attempting to find a solution for both parties (a win–win situation), the emphasis is on making sacrifices as one or both parties compromise their initial position. Although this approach can have positive effects on the group, helping to progress matters, it is not as effective as integrating type approaches. In integrative approaches both parties look for mutual gains rather than looking for what they can give up. Compromise has an element of self-sacrifice.

Rather than compromise a key objective, parties should look for things which they could do for their opponent, activities that would benefit each party, rephrasing of position so others can also consider alternatives. Take a break before agreeing to compromise, a short period to relax or continue discussions informally can help free up the mind for other alternatives.

Obliging, accommodating and giving up (low assertiveness, high level of co-operation)

While one party gives up something, they receive nothing in return. Thoughts that motivate such action include 'there is little positive to be gained from engaging in the conflict' and 'the encounter will be stressful, emotional and uncomfortable'. Although such action may help short-term group relations, there is little benefit for the obliging party. Such behaviour results in increased pressure on the individual. Repeatedly failing to gain successful outcomes from conflicts may threaten resources and job security. Such action should only be used where something is gained in return.

Collaborating and integrating (high level of assertiveness and co-operation)

This style is characterised by assertive interaction that is used to increased partici-pation, helping to ensure that individual and group objectives are maintained. Integrating strategies have a greater potential for win–win type outcomes. Through interaction, conflict, challenges and evaluation, parties explore a range of sugges-tions, other than just those initially proposed. The members search for solutions that have benefits for all. The process means that parities must explore the issues to identify underlying concerns and attempt to find mutually acceptable solutions. Such behaviour challenges immediate incompatible proposals. In the long-term, the parties learn from each other and have a better understanding of each other's per-spectives and insights. Although the initial levels of emotional debate tend to be high, the process avoids stress that results from long-term disputes. The benefit of the solution tends to result in member satisfaction that reduces long-term tension and strengthens relationships.

Three dimensions of conflict management

While the grid developed by Kilmann and Thomas (1975) is useful, Gorse (2003) found that when conflict emerged and was managed in task groups members did not only balance concern for self and others, they also had to consider the task. Gorse proposed that there are three main dimensions to conflict that need to be balanced when attempting to manage conflict. Long lasting, strong working relationships would be developed by parties who have a high concern for others, high concern for the task and high concern for themselves (Figure 6.5). The group task needs to be achieved, people need to undertake their own activities and co-operate with others so that they can achieve their individual goals. By working together the individual goals of all of the members and the group goal is achieved.

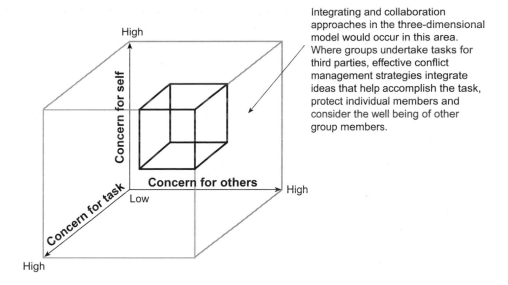

High

Concern for self

Concern for task

Concern for others

Low

High

High

Integrating and collaboration approaches in the three-dimensional model would occur in this area. Where groups undertake tasks for third parties, effective conflict management strategies integrate ideas that help accomplish the task, protect individual members and consider the well being of other group members.

Figure 6.5 The three-dimensions approach of conflict management (Gorse, 2003, p. 180).

Conflict management and construction

Loosemore *et al.* (2000) claim that construction professionals should be encouraged to engage in functional conflict and that an indiscriminate attempt to reduce conflict in construction incurs an opportunity-cost for clients and professionals. Their research, which focused on the construction phase and conflict between the contractor and architect, and contractor and subcontractor, found that the contractors' attitudes were receptive to functional conflict, although not as strongly as they originally thought. The findings suggest that most conflicts were managed by exploring alternative solutions, different perspectives, and encouraging all participants to engage in discussions and co-operate. Such behaviours were believed to be most likely to result in win–win solutions. However, a considerable proportion of the conflict-handling style used by the construction managers was not considered to offer such positive benefits. Too much emphasis on compromising, and obliging by the site manager, restricted the potential development of mutually beneficial solutions. While site managers are often considered to be unco-operative and lacking in concern for others, the style of conflict management that was least used was the dominating style, which places concern for self before others.

The conflict management behaviour observed by Loosemore *et al.* is not uncommon in work groups. Work by Farmer and Roth (1998) who examined a range of student and work-groups found that the most common conflict management strategy used was collaboration (integration), followed by accommodation (obliging), compromise, competing and the least used style was avoiding. While many anecdotal reports suggest that the construction industry is adversarial, much of the research suggests that the way conflict is handled is rather typical; however, the research also shows that there is considerable room for improving the way that conflict is managed.

Conflict and disputes

A distinction is often drawn between conflict and dispute. Conflict exists where parties have different understandings and perspectives, or differences of interest and opinion. In this context conflict can be managed, possibly preventing it from becoming a dispute. Disputes occur as a result of conflict escalation. Where differences are unresolved, matters often become adversarial, formal and often involve additional third parties (legal representation). In situations where parties are unable to resolve their differences they enter into a dispute. During disputes parties tend to communicate in a defensive manner, attempting to protect their organisation's interests. Disputes rarely result in a satisfactory outcome; in major disputes parties invest considerable time and resources defending their case, and legal and technical experts may be employed in an attempt to show who is right and who is wrong. Little time is devoted to moving practical matters forward or rebuilding relationships, and at the end of the dispute one or both of the parties will have to pay for any additional dispute services and methods used.

Emmitt and Gorse's (2003) model of conflict management strategies (Figure 6.6) illustrates the problem of not dealing with conflict early. As those engaged in conflict move from conflict management to the more extreme litigious methods of dispute resolution, the involvement of third parties increases and the potential cost of the original conflict escalates. Projects are delayed and costs increase as the dispute becomes more serious. The involvement of third parties does little for the performance of the project.

Figure 6.6 shows a number of methods for resolving disputes. The complex multidisciplinary nature of the construction industry has a reputation for being adversarial. A considerable number of cases are taken to Court every year. In recent years there has been increased emphasis on alternative dispute resolution (ADR) methods such as mediation and adjudication. With the introduction of the Housing Grants, Construction and Regeneration Act (1996) a process to refer disputes to adjudication should be written into all construction contracts. This has considerably increased the number of disputes being settled by adjudication rather than litigation.

The following list provides a brief outline of processes that can be used to resolve disputes.

- *Informal discussion*. An informal meeting can be useful to resolve many disputes. The lack of formality can encourage parties to talk openly.
- *Formal recorded meetings*. Formal discussions may be used to bring matters to a head. Holding a formal meeting lets the parties know that the situation is being

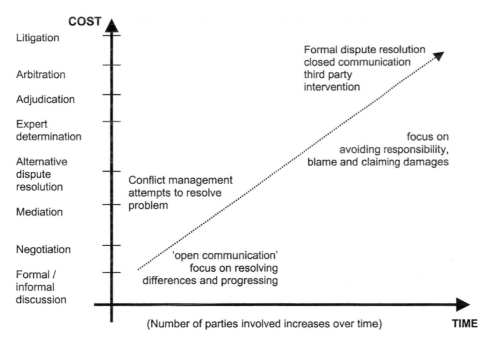

Figure 6.6 Conflict management strategies against time and cost (Emmitt and Gorse, 2003, p. 167).

taken seriously. However, when matters become formal and are recorded, parties are much more guarded about what they say.

- *Negotiation.* Rather than adopting a formal meeting, a forum may be set up so that the parties can attempt to find a way out of the problems, look for mutual benefits that can be gained from resolving the problem in a different way, or the parties may look for compromises in order to overcome the problem.
- *Mediation.* Where a relationship has broken down or experiences difficulties, a mediator can be brought in to help discussions. By listening to each side a mediator attempts to find a way out of the problems, looks for mutual benefits or helps the parties to rebuild the relationship. Having a third party to listen to one party, then transfer information to another party, helps to convey logical and rational information without negative emotion.
- *Expert determination.* If a dispute has arisen over a specialised or technical matter an independent expert can be appointed to make a decision or advise the parties on the matter.
- *Adjudication.* Under the Housing Grants, Construction and Regeneration Act (1996) all construction contracts have a clause that refers disputes to adjudicators. If a construction contract does not have an adjudication clause the Scheme for Construction Contracts applies, which means that disputes are referred to adjudication. Adjudicators are appointed under the terms of a contract to listen to the parties and make a decision within a short time (28 days in construction contracts). Generally, the costs of adjudication are considerably less than arbi-

tration or litigation. The decision of the adjudicator is enforceable under the terms of the contract. The process is private and information about the dispute is not reported to the general public. Adjudicators' decisions are binding and enforceable, at least until they are referred to Court.

- *Arbitration.* Arbitration is governed by the Arbitration Act 1996. Where disputes are referred to arbitration the processes governed by the Act are legally enforceable. During the arbitration process, which can last a considerable time, experts and lawyers can represent the parties. Arbitration can be an expensive process, but it is a private process and information about the dispute is not in the public domain. One advantage of arbitration over litigation is that the arbitrator can be appointed from a technical profession (e.g. engineering or construction) rather than a legal profession.
- *Litigation.* Referring matters to Court can be very costly and time consuming. Judges adopt strict procedures and ensure each party has an opportunity to present and defend their case. Lawyers, solicitors and barristers will represent most parties and experts may be called and cross-examined. Third party involvement is considerable and the costs associated with litigation are often disproportionate to the problem first presented; however, the decisions made are final, subject to appeal.

Summary

Within highly complex industries such as the construction industry, conflict needs to be embraced and managed. Conflicts should be exploited so that benefits are gained, but must also be managed so that relationships do not suffer. In the event that conflict results in a dispute, consideration should be given to the type of dispute resolution process that is most appropriate. In order to obtain a satisfactory resolution the parties involved in the process should be selected with care. Issues such as legal and technical expertise required and the extent of third party involvement must be properly considered before entering any such process.

Conflict management exercise

There are a number of factors that should be considered when engaging in conflict management. Different situations require different approaches. Using the conflict management profiles (Figure 6.8) each member of a team or group can complete the charts to provide an indication of how they believe they and other members behave in groups when dealing with conflict. The charts cover such things as aggression, assertiveness, intuition, support, relationships, etc. The charts can be used to provide two different perspectives of a person's conflict management techniques: a self perception and perceptions from other members of the group. Once the charts are completed, the individual self perceptions can be compared

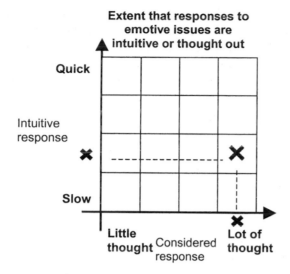

Figure 6.7 Example of completed conflict management profile.

and cross-examined against the perceptions that other members provided for the same person.

The charts do not provide an answer to how conflict should be managed – they provide an insight into different approaches that can be used. Through group discussion and evaluation the charts can be used to uncover the perceptions and observations of conflict management behaviour. Group discussions that follow the completion of the charts can be used to enquire as to the reasons why people behave the way they do and what they believe they will achieve by their actions.

In addition to the charts that look at conflict, there are also some charts which collect information about the nature of the group and the relationships. These factors should also be used to promote discussion. Conflict and conflict management may be context dependent and a cross-examination of behaviour and context may reveal that different techniques are more or less likely to work in different situations.

The charts can be used following a particular group experience where a level of conflict or disagreements emerged. It may also be useful to consider groups where conflict did not emerge as members of the group may believe that personal and group management techniques were used to such effect that conflict did not manifest.

An example of how to complete the charts when doing a self perception (how you think you behave) or perception of the behaviour of others.

1. Read the statement at the top of the chart
2. Read the statement on each axis (Fig. 6.7).
3. Place a mark in the position that relates to your behaviour or the person being observed (see crosses in Fig. 6.7).

Group member name or ID _____

Self perception Yes / No.

Charts completed for group member (Name or ID)_____

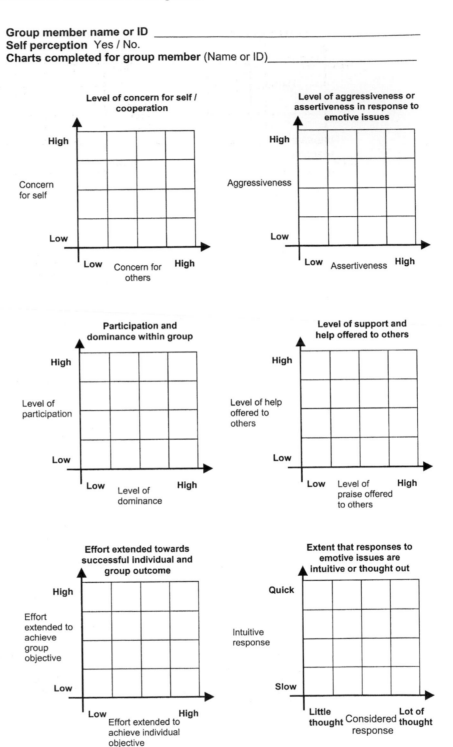

Figure 6.8 Conflict management profiles.

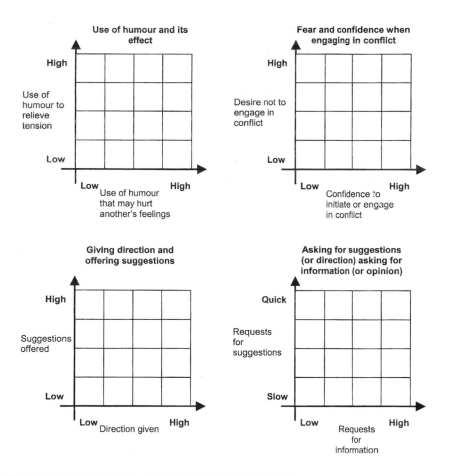

CHARTS THAT RELATE TO CONTEXT AND RELATIONSHIP

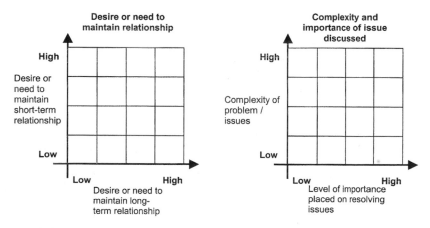

Figure 6.8 Conflict management profiles (*continued*).

4. Each person should complete the charts based on their perceptions of themselves and other members in the group.
5. It is expected that perceptions may vary.
6. Once all the charts have been completed, collect them together.
7. The differences between self perception and other members' perceptions can be discussed.
8. Then the positive and negative aspects that each person may bring to engaging and managing a conflict episode should be discussed.
9. The discussion should then consider whether the effectiveness of the techniques uncovered would vary depending on context.

Chapter 7

Individual/Group Behaviour and Teamwork

Personality and individual behaviour

The manager needs to understand what affects people's behaviour and performance at work: why people are sometimes hard working, lazy, trusting, miserable or content.

This behaviour is partly determined by *personality*, the set of characteristics by which we recognise a person's uniqueness. These characteristics are relatively enduring, but aspects of an individual's personality may change as a result of experience or circumstances.

We tend to label people as types – friendly, hostile, domineering, shy, etc. – because they have a general disposition towards certain kinds of behaviour. But in fact they behave quite differently in different situations. This is because their behaviour is affected by interaction with others and by the setting. Interestingly, people tend to describe their own behaviour in situational terms, but label others in terms of their personality.

Although some people are more rigid than others, most psychologists now agree that personality is not a fixed set of attributes. They see personality as dynamic rather than static. It depends partly on inherited factors, but is heavily influenced by the individual's experiences.

Individuals and groups

There are also *organisational* factors which influence the individual's behaviour – the structure of jobs, the tasks performed and the group to which the individual belongs.

Individual behaviour is strongly affected by the workings of groups, and managers must understand group processes if they are to manage effectively. Organisations use groups to fulfil most of their purposes. Through gangs, project teams, committees and meetings, organisations distribute tasks and responsibilities, organise and monitor work, reach decisions, solve problems, gather and exchange information, test out ideas, negotiate and settle conflicts, and agree terms and conditions.

It is now widely accepted that individuals will behave differently, and even change their ideas and beliefs, if they are members of a cohesive group. Groups can therefore exert more influence over an organisation than individuals can. Yet firms pay much more attention to individuals than groups – through career plans, staff appraisals and the like. In construction, project groups are difficult to monitor, because their composition changes so quickly.

Primary and secondary groups

One of the distinctions used by those who study groups is between small groups in which the members have some common bond or direct relationship, and larger groups in which the link is more tenuous or indirect.

The term primary group is used to describe those groups in which there is intimate, face-to-face association and co-operation. There is a certain fusion of individualities so that, in many ways, the group shares a common life and purpose.

A primary group is relatively small and its members all have close contact with one another. This can be said of a bricklaying gang or a group of buyers sharing an office. It cannot be said of a building firm or a trade union. The latter are secondary groups. Members of these larger groups are aware of a bond between them, but the link is weaker.

Every group has ways of dealing with differences among its members. In secondary groups it is through rules laid down, often in writing, and modified as conditions change. In primary groups, it is largely through unwritten rules or norms. These are also modified through time. Primary groups and their norms are very important to the manager who wants to be effective.

Developing group performance

It takes time for a work group to become efficient. The group's task has to be defined; responsibilities have to be shared out; conflicts between people and between goals have to be resolved; norms have to be worked out. Group members have to resolve two major problems simultaneously:

● How to handle the *tasks* they have been given.
● How to come to terms with one another as *people*.

In a semi-permanent group, this process can take many months (even years), depending on the group's size, membership and task. In construction, project groups have to gel very quickly – within days or weeks. Moreover, these groups alter their composition as the project moves through its lifespan. Every time people leave and others join, there is a period of adjustment, both for new and existing staff.

A further period of adjustment occurs just prior to project completion. Output falls away sharply, feelings of uncertainty develop and there is some nostalgia. The manager's aim in construction must be to get temporary groups into their cohesive, *performing* stage as quickly as possible and keep them performing.

Group cohesiveness

Group cohesiveness is the degree of solidarity and positive feelings held by individuals towards their group (Stoner *et al.*, 1995). One of the earliest researchers of group behaviour, Michael Argyle, described a cohesive group as one in which the members like each other, enjoy being part of the group and co-operate over group tasks. He noted that there was more conformity among members of cohesive groups and that members tended to spend more time with the group.

Mullins (2002) has summarised the factors which appear to influence group cohesiveness:

- *Membership* – the size and permanence of the group; the compatibility of members.
- *Work environment* – the kind of task; the group's physical setting; communications and technology.
- *Organisational factors* – management and leadership; personnel policies and procedures; success; external threats.
- *Group development and maturity* – the stage the group has reached in the development of task performance and group relations (agreement on norms, for instance).

Argyle argued that the amount of interaction among group members and the length of time they stay in the group are also important for cohesiveness to develop. This is an interesting point, because project organisations are made up of teams which change their composition and either meet infrequently or spend a lot of time working apart, whereas the project team members – architect, QS, engineers and so on – spend relatively little time together.

Group cohesiveness is important. It leads to greater interaction among members, creates a climate of satisfaction and co-operation, and can result in lower absenteeism and labour turnover (Argyle, 1989). These factors can lead to high productivity. But there can be problems if group members are cohesive simply because they are all alike. Belbin (1993, 2000) and others have argued that groups need to contain a mix of different types of people, because effective group work requires a range of different skills and behaviours.

Satisfaction, group cohesion and performance

While groups must build and establish networks and relationships, they need to deal with the task in hand. Many studies have shown that effective groups need to balance task and social interaction. Discussing tasks and proposed solutions in a critical way will have an emotional effect on others. This may weaken relationships. Where criticism is raised, supportive communication may need to be used to maintain the professional relationships. However, it is also difficult to criticise close friends. Where relationships become too close, the functional conflict that is necessary in problem solving may be reduced. Groups that are often too cohesive can suffer a reduced productivity due to the amount of social interaction that takes place. Cohesive groups may suffer from groupthink, being reluctant to challenge others when they disagree with their ideas, proposals and suggestions. A balance needs to be struck between the degree of social and task based interaction (see Figure 7.1).

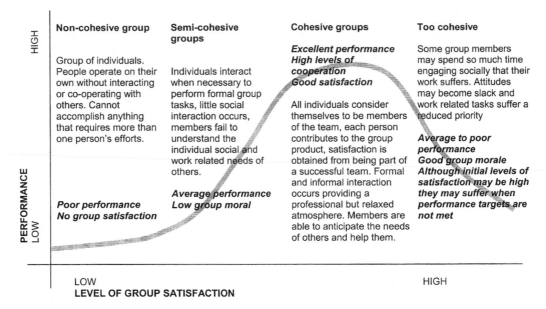

Figure 7.1 Satisfaction, group cohesion and performance.

Norms

Norms are shared attitudes, shared ways of behaving, shared beliefs and feelings within a group. They encourage the conformity and predictability which are important in any task needing co-operation. For any group to be effective, there must be some measure of agreement about what has to be done, and how.

There are norms about:

- *The work*. The best methods, how fast to work and what standards to aim for.
- *Attitudes and beliefs*. Management, unions, the importance of the group and how satisfying the work is.
- *Behaviour*. Co-operation, sharing things, what jokes to tell and where to go for lunch.
- *Clothes and appearance*. How to dress at work and outside work; the wearing of safety helmets and protective clothing.
- *Language*. Use of technical jargon, slang and bad language.

Norms develop as the group tries to solve its social and work problems. The more dominant members have most influence over norms. People new to the group have little impact and tend to shift their behaviour towards the norm.

This shift towards group norms can occur because:

- individuals are under pressure from other group members to conform; if they don't, they may be ignored or rejected;
- individuals think the majority view must be right, especially if the others are more experienced or have been found to be right in the past.

Some people conform more readily than others. People who have a strong need for belonging conform more readily than those who are independent. When people's goals or beliefs are far removed from those of the group, they are more likely to deviate from its norms. Authoritarian people tend to conform more than less rigid types.

Norms take time to evolve and must be reasonably well settled before the group can perform effectively. This is important in construction. With new projects, norm-building often starts from scratch and the norms may change as the composition of groups alters over the project's life-cycle. Many contractors try to keep certain key staff together when they move them from project to project, to speed up the process of getting groups working effectively.

Group norms represent the standards of conduct of the group. It is difficult for the members to operate effectively unless they can be reasonably sure what responses they can expect from their colleagues. Of course, groups vary in the 'tightness' of their standards; some are much more free and easy than others. Some members are tolerated even though they don't always conform. But without some agreement between members about what they will and will not tolerate, the group would find it difficult to continue.

Managers must appreciate this and have some understanding of how norms develop and how they can influence this process. They should realise that individuals may depart from their own judgement because of group pressure. Indeed the manager may sometimes be that individual!

Solomon Asch demonstrated this phenomenon in a well-known experiment in which a series of groups were asked to make a judgement about the length of a line. All but one of the group members had, however, been briefed by the experimenter to agree on an answer which was clearly wrong. The naive members were therefore faced with a group whose judgements contradicted the evidence of their own eyes.

About two-thirds of the naive subjects followed their own judgement, many showing acute embarrassment at doing so. But the rest gave in to group pressure and gave the answer they thought was wrong. They mainly did so either because they thought they must be mistaken, or they thought they were right, but didn't like to contradict the rest of the group.

Clearly, the manager must be aware of such group pressures, which may cause individuals not only to give opinions that conflict with their true feelings, but also change statements of fact.

Encouraging interaction

The work of social psychologists on small groups has given managers some useful insights. For instance, it has been shown that arranging group members in a circle, rather than sitting them in rows, often produces more interaction, and more members join in. Also, people tend to speak in response to those sitting opposite them, except when an authority figure is present. Then they tend to speak to the people sitting beside them.

Although the findings are scattered and incomplete, it is clear that *seating arrangements* for groups, chosen by architects, interior designers and managers, have a marked effect on the social structures which emerge. If people cannot easily talk face-to-face, social interaction is seriously impaired. They rely on visual, as well as verbal, feedback from others. Given a free choice, group members often seat themselves at a distance from their leader, but sit where they can see and be seen by the leader. Those who can most easily make eye-contact with the leader often do the most talking. A circular pattern, often favoured for informal discussions, maximises the eye-contact between group members. The arrangement of seats and other furniture (as in a contractor's office) facilitates or inhibits eye-contact and thus affects communication.

Roles

To understand how groups work, we have to look at how people behave towards one another. Every member of an organisation occupies a position, such as engineer, buyer, chargehand or clerk. For every position there is a *role* – the activities and patterns of behaviour typical of people in that job.

People's behaviour depends on many factors, defined by their and other people's expectations.

The different roles in an organisation interlock, like the roles of doctor and patient. One role cannot be performed without the other. Each role includes the tasks performed, ways of behaving towards people, attitudes and beliefs, the clothes worn and even aspects of the individual's lifestyle.

Managers tend to behave like other managers, architects like other architects. A bricklayer is more likely to behave like other bricklayers, than like a quantity surveyor. Organisations rely on this conformity in behaviour to ensure that work is done effectively. The task itself generates role behaviour if it can only be performed in certain ways. Each person is expected to play a part, not only by the boss, but by colleagues and subordinates. The attitudes and beliefs held by members of a group about what a particular job-holder should do are called *role expectations*. Group members often have differing expectations of how each of the other group members should behave.

Selection helps to perpetuate role behaviour. People are selected for management jobs because they look the part – or may be turned down because they don't! Self-selection operates too. If people don't like the look of managers they have seen, they probably won't apply for a management job. Several other processes encourage new job-holders to behave like the established ones, including training, imitation and coaching.

People only conform up to a point. Their individuality still shows through. This is because most people belong to several groups – family, work and leisure – and therefore have various roles to perform.

The other people in an individual's group are that person's *role-set*, the people with whom he or she has regular dealings – colleagues, bosses and, where applicable, subordinates. Managers can have quite large role-sets.

People experience *role conflict* when members of their role-set put different pressures on them. The site supervisor who has to choose between a course of action which will please the manager and another which will suit the workers, is suffering from role conflict.

Role ambiguity occurs when people cannot agree about what a role should be. For the individual this often means a lack of clarity about the scope of his or her job. The writer found that poor role definition was quite common among construction managers (Fryer, 1979). Variations in the type and size of projects, the people involved, company rules and contractual procedures, accounted for much of this uncertainty.

When people experience role conflict or role ambiguity, they may become tense and unhappy, dissatisfied with their jobs, less effective in their work or even withdraw from contact with those exerting pressure on them. They will try to resolve the conflict or ambiguity in various ways – by giving some demands priority, by seeking a ruling from their seniors or by bargaining with the people involved. Conflict can also result from differences between people's individual needs and job demands. Many site managers like to be out and about, close to the work, and resent sitting behind a desk full of papers.

The organisation can help minimise this kind of conflict by taking more care over selecting people for jobs. This may mean putting more emphasis on individual needs and interests in staff selection. Interestingly, during the early years of people's careers, there is a tendency for their goals to change to fit their roles better. This is especially true of vocations like architecture, where the training period is long. Moreover, Argyle points out that as people become more influential in their jobs, they may change organisational goals to be more compatible with their own.

Observing and analysing group behaviour

Managers perform most of their duties by leading, or taking part in, groups (project teams, departments, etc.) and therefore need to understand group behaviour. They particularly need to know how to get specific responses – how to persuade a sub-contractor to speed up progress, or the architect to provide some design information quickly. Management trainers have responded to this need by including in their courses techniques for studying group behaviour, known as *interaction analysis*. These help those taking part to:

- understand their own behaviour better;
- improve their social skills;
- analyse, understand and respond more positively to other people's behaviour.

One way to study group behaviour is to watch a group at work. The observer can either participate in the group or observe it from the outside. Psychologists have developed a number of systems for recording and analysing group interaction. These can be used, for example, to assess the roles people play (role analysis), who speaks to whom (interaction flow analysis), the social relationships within groups (sociometry) and what people say to one another (content analysis, behaviour analysis, and so on).

R. F. Bales developed one of the earliest methods of content analysis which used twelve categories, six related to the group's task and six to the social relations between group members (Figures 7.2 and 7.3).

Bales' method is still used today in studies of group interaction (Gorse 2002). In behaviour analysis, observers normally use a chart listing the behaviour categories and the names of the group members. The observer has to record, for each contribution to the discussion, the speaker's identity and the kind of behaviour used. When the group has finished, the observer totals the contributions of each individual in each category. The analysis can be extended to include non-verbal behaviour, such as gaze, facial expression, posture and so on. Analysis is not always easy. It can be difficult, for instance, to tell whether a remark is an *opinion* or a *suggestion*.

A system developed by Rackham, Honey and Colbert, uses different sets of behaviour categories to suit the situation (Rackham, 1977). One of these sets is:

- *Proposing* – putting forward a new idea, suggestion or course of action.
- *Building* – extending or developing a proposal made by another person.
- *Supporting* – deliberately agreeing with another person's ideas.
- *Disagreeing* – declaring a difference of opinion, or criticising another's ideas.
- *Defending/attacking* – attacking another or defensively strengthening one's own position.
- *Blocking/difficulty stating* – placing an obstacle in the path of a proposal or idea without offering an alternative or a reasoned argument. This kind of behaviour tends to be rather bald, 'It won't work' or 'We couldn't possibly do that'.
- *Open* – the opposite of defending/attacking. The speaker exposes him/herself to the risk of ridicule or loss of status. This would include admitting a mistake.
- *Testing understanding* – trying to find out if earlier contributions have been understood.
- *Summarising* – restating concisely the content of earlier discussion.
- *Seeking information* – seeking facts, opinions or clarifications.
- *Giving information* – offering facts, opinions or clarifications.
- *Shutting out* – excluding, or trying to exclude, another member of the group.
- *Bringing in* – a direct and positive attempt to involve another member.

		CATEGORY DESCRIPTION Types of communication act and response	
1	F	SHOWS SOLIDARITY – raises others status, gives help, offers praise and reward for effort and contribution, offers assistance, shows support	**Social emotional acts:** Positive emotional reactions Builds relationships
2	E	SHOWS TENSION RELEASE – friendly jokes, expresses light humour, laughs with others, smiles, shows satisfaction with other peoples contributions	
3	D	AGREES – shows passive acceptance, expresses understands, concurs, complies	
4	C	GIVES SUGGESTION – gives firm statement of direction, makes proposal, implying autonomy for others	**Task acts:** Gives info', direction or attempts to answer questions
5	B	GIVES OPINION – evaluates, analyses, expresses feelings, wishes, opinion or perspective	
6	A	GIVES ORIENTATION – gives information and facts, repeats statements to show and confirm understanding, clarifies, confirms	
7	A	ASKS FOR ORIENTATION – asks for information, repetition, asks for confirmation, requests further information and facts	**Task acts:** Seeks information or asks questions
8	B	ASKS FOR OPINION – asks for evaluation, opinion and analysis, asks for expression of feeling	
9	C	ASKS FOR SUGGESTION – asks for direction, possible ways of action, firm proposal. Asks person to make a suggestion	
10	D	DISAGREES – factual or formal disagreement, shows passive rejection, does not express any frustration or emotion, becomes formal, withholds help	**Social emotional acts:** Negative emotional reactions Threatens relationships
11	E	SHOWS TENSION – not only disagrees, but expresses concern and tension, withdraws out of field, expresses negative emotion	
12	F	SHOWS ANTAGONISM – deflates others status, defends or asserts self, becomes aggressive, becomes personal or overly critical	

Legend

A. Problems of orientation

B. Problems of evaluation

C. Problems of control

D. Problems of decision

E. Problems of tension management

F. Problems of integration

Figure 7.2 Bales interaction categories (adapted from Bales, 1950).

EXAMPLE OF CODED INTERACTION FROM SITE MEETING

Project: West London 1. Contractor: Fast Build   **Date of observation 8/6/04**

Structural engineers section

The speaker is identified at the start of each statement e.g. <speaker>. Each sentence, or speech act, is categorised in brackets at the end of the sentence e.g. (3) agreement. Notes regarding emotion stated in square brackets e.g. [2] shows support.

<Project manager> If we can return to the agenda **(4)**, item 5.1 roof purlins **(6)**. Ian **(6)** I sent you a letter regarding the deflection in the zed purlins last week and I still haven't had a reply to this **(6)**.
<Structural engineer> I thought I mentioned that at the last meeting **(5)**; the design is correct **(6)**.
<Project manager> The deflection does seem excessive **(5)**, are you sure it's ok **(8)**
<Structural engineer> I've checked through my calculations **(6)**.
<Project manager> Are you sure you have allowed for the air handling ducting? **(7)**
<Structural engineer> Yes **(6)**
<Project manager> The deflection must be outside the design limits **(7)**, it doesn't look safe **(10)**. Can we have another look at it after this meeting **(5) (also)** **[11]** and think about putting some extra supports in **(4)**.
<Structural engineer> OK **[3]**
<Project manager> Item 5.2 **(6)**

Figure 7.3 Sample of speech coded using Bales IPA method.

These categories are fairly easy to understand, but they can sometimes be difficult to separate. For example, if a group member states an opinion which conflicts with someone else's, it may be hard to decide whether he or she is giving information, disagreeing, attacking or making a proposal. Accurate analysis depends on training and practice, but the framework is good enough to be useful for trainers in team building exercises (Clark, 1994).

The technique often reveals that group members spend a lot of time exchanging information and the remaining contributions are often rather self-centred and negative – disagreeing, defending, attacking and blocking. Members spend more time putting forward their own ideas, than supporting or building on others' ideas. If group members are shown a video-recording of their performance and are asked to make more effort to support and build on one another's ideas, they often achieve better results the next time round.

Bales' SYMLOG (Systems for the Multiple Level Observation of Groups)

Following the development of the IPA technique, Bales *et al.* (1979) developed a different system for studying groups, known as SYMLOG. This method is an extension of the IPA system, and although it is considered to be theoretically more complex, it is believed to be more flexible. Rather than limiting observation to an independent researcher, the system is based on participant observation of other group members and a self-study of internal feelings. The advantage of participant study is that observations are not just limited to observations of overt interaction, but they also capture the participants' own feelings and values of themselves and other individuals within the group. While the method can enquire into areas of group interaction that the IPA system cannot, it requires the participants to complete a number of forms. The time required for each individual member to understand and complete the SYMLOG self-study is about three to four hours for a group of five; larger group sizes require more time (Bales 1980). The SYMLOG group self study system involves three steps:

(1)   A period of informal observation by each member of the group.
(2)   A period during which each group member, working alone, completes an observation form which describes the behaviour observed.
(3)   A meeting of the group during which the observations are compared.

The system uses three-dimensional diagrams that show the relationships of group members to each other. While the SYMLOG method has not received as much attention as the Bales (1950) Interaction Process Analysis Method, its use in studying small groups is increasing.

Simple Multiple level Observation Techniques (SMOT)

Rather than opting for the complicated SYMLOG system offered by Bales, it is possible to create a simple observation method that still captures events from a number of different perspectives.

The following steps can be useful in devising simple multiple observation methods and tools.

(1)   *Focus of data to be collected.* Decide on the topic or areas of interest that will be observed, e.g. amount of interaction and participation (how often people speak, who they speak to and how often they talk to that member), identify aspects of interaction that are to be observed, e.g. issues that develop through interaction (how conflicts emerge, how leadership develops, how decisions are made, etc.). Do not choose too many issues (one or two is fine).
(2)   *Context and group situation.* Select a group to study, and the context or situation in which they will be studied (e.g. a group or team solving a problem, or working through a task, or negotiating, or meeting).
(3)   *Perspective 1. Observation.* Ask a person to observe the group. Ask them not to engage in the group discussion, but to sit and observe the process – simple data collection sheets can be devised so that the observer can collect the data easily, e.g. number of times a person speaks, who they talk to and when the topic of research emerges.
(4)   *Perspective 2. Collect perceptions from group members.* Following the group interaction give all group members a questionnaire. The questionnaire should ask how all the other members of the group interacted, reacted and behaved. Focus the questionnaire on the research topic.

(5) *Perspective 3. Ask individuals to state how they interacted.* Ask members to state how they interacted with others, how they sent messages and how they responded, prompt the participant to focus on the aspect of interest.

(6) *Perception 4. Ask individuals to reflect and state the feelings that emerged during interaction.* Ask members to state how they felt when interacting with others and when certain situations or events manifested themselves.

(7) *Perception 5. Video or record the interaction (optional).* Use a video recorder to monitor the group. This is not always necessary and can be considered too invasive on the group and may change their behaviour. If a video recording is taken of the group's activities both the observer and the group members can use the video to review the data captured.

(8) *Analyse the data.* An immense amount of data can be collected from one situation. Ask members and the observer to analyse the data individually, then ask them to analyse the data as a group. Finally, ask the group to reflect and review the analysis. It is normal to find that differences are found between the perceptions of what happened and what the video data showed or the observers found. Differences of opinion may be quite diverse; however, participants develop greater understanding of what is happening during group activities.

Such tools can be used to research interaction, leadership, power, authority, conflict, negotiation, teams and sub-teams, decision-making, dominant and reluctant communicators, and many other aspects of group behaviour. While it may be possible to study more than one of these aspects at any one time, researchers should be careful to focus their research so that the data collected are manageable and meaningful.

Summarising individual and group behaviour

In a labour-intensive business like construction, managers need a good understanding of human behaviour. Many factors influence the way an employee behaves in a given setting. Personality is one of them, but not necessarily the most important. Indeed, an individual's characteristic way of behaving often alters in response to different people and problems. Moreover, employees' behaviour is affected by the work they do and the groups they work with.

A work group can exercise considerable power over its members. Just as organisations have rules which govern what people can and cannot do, so groups have norms, which dictate what behaviour is acceptable and unacceptable within the group. Individuals often conform to group norms, even when this conflicts with their personal preferences. Some people deviate more from norms than others; they are often the independent thinkers.

People perform various roles in the organisation. These roles cause them to shift their behaviour towards that of other people doing similar jobs. The way people behave may not be the way they want to behave, but how they think others expect them to behave.

A number of techniques have been developed for analysing group behaviour. When these are used in training sessions, they can help group members to evaluate their behaviour, leading to improved group performance.

Teamwork

The mid 1980s saw an upsurge of interest in teamwork in the construction industry. This focused both on teams working in individual firms or practices and, more importantly, multi-disciplinary project teams, where ineffective teamwork had led to mistrust, communication breakdown and faulty management.

Since the energy a team can devote to its work is finite, it follows that time spent dealing with the shortcomings of the team and its workings is time lost to *real* work. So, it is important for the industry's managers to know about teamwork and, in particular, about how to build a team fast and maintain its performance throughout a project.

But, first, why has teamwork become so important? In the past, it was possible for an individual to have a good command of most aspects of construction management. Such a person, if reasonably competent, could be relied on to provide an adequate and comprehensive service to the client. However, the last decade or so has placed huge demands on organisations and managers to rethink what they are doing and how they are doing it. Projects have grown more complex – technically, organisationally and contractually – and it has become increasingly difficult for an individual to possess *all* the know-how to manage a project from inception to completion. We are having to acknowledge that large-scale modern building requires a team effort, simply to share out the total project into manageable tasks, to keep customers and society happy and to maximise the chances of a project's success in what is often a fiercely competitive environment.

Such success is embodied in project goals, which have traditionally been expressed in terms of the design, quality, cost and speed of erection of a building, but other criteria have increasingly been introduced, such as energy efficiency, flexibility for future adaptation and costs-in-use. Moreover, major new issues about *sustainable* development and environmental protection are triggering an urgent review of business objectives and the emergence of new disciplines, such as environmental accounting. The inescapable 'greening' of the economy and business practice will cause the industry's managers to further reappraise the priorities for teamwork among the professions.

In the construction industry, creating good project teams isn't always easy. This is because project teams are temporary, *task-force* groups, whose members are brought together only for the duration of the project (and some for only part of the duration) and then disbanded on completion. At the best of times, such teams can be difficult and frustrating to manage.

Of course, teamwork isn't always the answer. Some professional tasks don't need it – and to use it would be inefficient. Many of the routine tasks that professionals carry out are best discharged independently and only require intermittent co-ordination. There is a well-established maxim which says that if an individual can do a job perfectly well, don't give it to a team.

Nevertheless, even the fairly disparate tasks which can be performed by different professionals working in isolation, need to be co-ordinated – and teamwork is essential for *integrating* specialist work into the total scheme of things.

Moreover, as Baden Hellard (1988) points out, the network of human relationships in a project team becomes a network of *contractual* relationships that is at the root of many disputes. Disputes are far more likely to arise from deficiencies in organisation and communication between the different groups than from failure of technology, materials or problems arising from unforeseen circumstances.

There are several other reasons for giving special attention to teamwork in construction:

- *Location*. The specialists who make up the project team are not located in the same place and do much of their work away from the site and away from one another. They only meet intermittently to exchange information and to solve problems and co-ordinate their actions.
- *Different firms*. Team members work for different parent businesses, each of which has its own values, goals, strategies, ways of working and so on. Team members may experience conflict of loyalty between the project and the firm.
- *Individual differences*. Each profession tends to attract different types of people to its ranks; they are likely to have different interests, skills, backgrounds and personalities. These differences can be reinforced by the pattern and focus of education and training adopted by each profession.
- *Late involvement*. Team members are often appointed at different stages, sometimes after key decisions have been made. This can make it difficult to create commitment to the project and, if meetings are infrequent, it can take a long time before the team can function fully.
- *Teambuilding*. Project teams are not usually put together in a systematic way – but rather in an idiosyncratic way, depending on who is available (and when), who has the necessary experience for this particular type of building, who recommends whom and so on. Moreover, many of the participants are 'part-time', in the sense that they are also contributing to other projects or they are not involved in the project for its full duration. All this makes teambuilding difficult.
- *Delegation*. Project managers and senior managers in parent companies who employ members of the project team aren't always good at delegating. Thus, some members may feel their hands are tied and that they lack the responsibility to commit themselves to major decisions, without consulting their bosses. When this happens, team effectiveness can be drastically reduced.

These issues are set against a backcloth of inter-professional problems, including differing perceptions of status, power and role among the professions; lack of mutual respect, understanding and trust; and a reluctance on the part of some professions to adapt to new roles and relationships to meet clients' changing demands and expectations (Fryer and Douglas, 1989).

However, it is possible to identify some basic requirements for good teamwork, all of which can be achieved with good management.

- Managers should receive training in teamwork skills.
- Teams should be much more carefully selected.
- Clear goals need to be set for the team, so that they develop a common purpose.
- Adequate resources should be made available to the team.
- Good communication between team members needs to be established from the outset.
- The team members need to develop mutual trust and understanding.
- Simple but effective procedures should underpin the actions of a team.

Features of a good team

A number of studies of teamwork, notably by Hastings, Bixby and Chaudhry-Lawton (1986) at Ashridge Management College, have helped unravel many of the secrets of good teamwork. Some of the qualities they have observed in highly effective teams include:

- *Persistence*. The team perseveres in its efforts and is obsessive in pursuing its goals, but it is also creatively flexible in getting there.
- *Tenacity*. The team is very tenacious and is inventive in removing obstacles – whether people or situations – which lie in its path.
- *Commitment to quality*. Team members are committed to quality performance and excellence in teamwork, with high expectations of themselves and of other people.
- *Inspiration*. The team has strong vision and sense of purpose; it knows where it is going and has a realistic strategy for achieving its aims.
- *Action-orientation*. The team makes things happen, responding rapidly and positively to problems and opportunities. Team members are optimistic even when the going gets tough.
- *Strong leadership*. The team has a really effective leader who fights for support and resources for the team.
- *Excitement and energy*. Members are lively and thrive on success and the recognition it brings.
- *Accessibility and communication*. Members proclaim strongly what they stand for, but welcome outside help and advice.
- *Commitment to success*. Team members are committed to the success of their organisation and thrive on the responsibility and authority delegated to them.
- *Drive*. The team is never complacent; members are continually striving for ways of doing things better.

- *Flexibility*. The team likes to work within guidelines and principles, rather than rigid rules, thus maintaining the important quality of being adaptable.
- *Prioritising*. Team members can distinguish between what is important and what is urgent.
- *Creativity*. The team prides itself on being innovative and will take risks to achieve significant results.
- *Influence*. The team has a significant impact on parent organisations because of its credibility.
- *Co-operation*. Teams always try to work with others, rather than for or against them.
- *Keeping things going*. Team members are able to maintain momentum and communication even when they are working apart.
- *Values*. The team values people not for their position or status but for their contribution, competence and knowledge.

For comparison, two other lists of desirable team characteristics are summarised in Table 7.1, although Adair stresses that some of these are often missing, even in good teams. Indeed, to expect a building project team or a contractor's site team to exhibit all these characteristics would not be realistic. But the lists help to pinpoint problem areas and aspects of teamwork on which managers can focus when trying to make practical changes in a team's performance.

Table 7.1 Some characteristics of effective teams.

• People smile, genuinely and naturally	• People care for each other
• There is plenty of relaxed laughter	• People are open and truthful
• People are confident – a 'can do' group	• There is a high level of trust
• They are loyal to the team and to one another	• There is strong team commitment
• They are relaxed and friendly, not tense and hostile	• Feelings are expressed freely
• They are open to outsiders and interested in the world around them	• Conflict is faced up to and worked through
• They are energetic, lively and active	• Decisions are made by consensus
• They are enterprising and use their initiative – proactive not reactive	• Process issues (task and feelings) are dealt with
• They listen to one another and do not interrupt	• People really listen to ideas and to feelings
(Adapted from Nolan, 1987)	(Adapted from Adair, 1986)

Teamwork roles

Meredith Belbin has also provided valuable insights into teamwork by looking at the roles people perform in teams. His research shows that people play a *team role* in their work groups as well as a technical or functional role. The team role defines a person's contribution to the team's internal functioning. Belbin argues that most people have a preferred role, which will to some extent reflect their personalities, values and attitudes – and their roles are not static. People often carry out a number of roles or a kind of composite role which includes several of them.

Belbin (1993) identifies nine team roles which are consistently found in work groups. They are summarised in Table 7.2.

Team leadership

Whilst leadership has been discussed in Chapter 4, the comments which follow draw attention to aspects of leadership specific to managing teams. In the most effective teams, the leader is likely to be straightforward, honest, trusting, considerate and respected – and not dominant or power-orientated. Although team leadership, like all management, must be flexible to suit the situation, in most team settings the leader must show integrity, enthusiasm and consistency; and lead by example.

For managers in construction, the physical separation of team members can pose problems. For example, contracts managers controlling wide areas or site agents running extensive engineering projects may find that their teams lose their sense of identity and cohesion. Such team leaders must be especially sensitive to the needs of their subordinates, who may feel out of touch and neglected. Ways must be found of keeping these people updated and involved. Telecommunications can help, but will not provide the whole answer. Part of the solution rests with the leader's own behaviour.

Adair (1986) summarises a useful analysis by J. R. Gibb and L. M. Gibb, who suggested five broad classes of leadership functions within teams (or groups as they called them):

- *Initiating* – getting the team going or the action moving, by identifying a goal, suggesting a way ahead, recommending a procedure, etc. This function mainly applies at an early stage in the team's activities.
- *Regulating* – influencing the pace and direction of the team's work, by indicating time constraints, summarising what has happened so far, etc. This becomes an important function as the team gets into its stride.
- *Informing* – providing the team with helpful information or opinions. Like regulating, this will mainly apply when the team is established in its work.
- *Supporting* – creating a climate which holds the team together and helps members to contribute effectively, by giving encouragement, showing trust, relieving tensions in the team, etc. This function is needed all the time.

Table 7.2 Belbin's nine team roles (adapted from Belbin, 1993).

Role title	Description and team contribution	Allowable weaknesses
Plant	Creative, imaginative, unorthodox. Solves difficult problems.	Ignores details. Too pre-occupied to communicate effectively.
Resource investigator	Extrovert, enthusiastic, communicative. Explores opportunities. Develops contacts.	Over-optimistic. Loses interest quickly.
Co-ordinator	Mature, confident, good chairperson. Clarifies goals, promotes decision-taking, delegates effectively.	Can be seen as manipulative. Delegates personal work.
Shaper	Challenging, dynamic, thrives on pressure. Has drive and courage.	Can provoke. Hurts people's feelings.
Monitor evaluator	Sober, strategic, discerning. Sees all options. Judges accurately.	Lacks drive and ability to inspire others. Overly critical.
Teamworker	Co-operative, mild, perceptive, diplomatic. Listens, builds, averts friction, calms waters.	Indecisive in crunch situations. Can be easily influenced.
Implementer	Disciplined, reliable, conservative, efficient. Turns ideas into practical actions.	Somewhat inflexible. Slow to respond to new possibilities.
Completer	Painstaking, conscientious, anxious. Searches out errors. Delivers on time.	Inclined to worry unduly. Reluctant to delegate. Can nit-pick.
Specialist	Single-minded, self-starting, dedicated. Provides knowledge and skills in rare supply.	Contributes on narrow front. Dwells on technicalities. Misses the big picture.

- *Evaluating* – helping the team to monitor the effectiveness of its actions and decisions, by testing for consensus, taking note of team processes, etc. This function will become more important as the team approaches completion of a task.

Hastings *et al.* (1986) identify four primary team leadership functions:

- *Looking forwards* – giving the team a vision and a sense of direction, being able to anticipate events and obstacles and creating an environment that encourages high performance.

- *Managing team members' performance* – defining success criteria for the team, showing interest, keeping performance on course and rewarding significant achievements.
- *Looking inwards* – continually analysing how the team is working and how it can be improved, taking an objective view of what is happening and what is likely to happen.
- *Looking outwards* – creating links with other parts of the organisation and the outside world, ensuring a two-way flow of information, resources and support between the team and others.

Hastings and his colleagues emphasise the need for team leaders to create the right *climate* for effective teamwork, by being more aware of their own behaviour and attitudes, demonstrating their values and expectations and putting forward an exciting vision of what the team can achieve. They also suggest that, wherever possible, leaders should influence the composition of their teams and should spend plenty of time with team members discussing the kind of climate and ways of working which could best contribute to *joint* success. The values and qualities associated with such team leaders include: unshakeable confidence and trust in the team; persistence and positiveness; optimism tempered by toughness and realism; a sense of urgency; accessibility and an openness to ideas.

Day (1994) points out that an important leadership role of project managers is that of *integrator*, pulling together the efforts of the organisations and people contributing to a project. This involves unifying a group of diverse specialists, who may have different ideas about priorities – and, perhaps, tunnel vision.

Team leadership and the self-managed team

The leadership qualities identified above are reasonably consonant with the ideas of empowerment and self-managed teams, but they imply certain features of hierarchical structure and the kind of organisational culture which goes with it. Stewart (1994) questions whether existing organisation structures and cultures provide a suitable basis for empowered, self-managed teams. She discusses the need for an *empowerment culture* within organisations. Pointing out that even modern organisations, with their flatter pyramids and shorter chains of command, are still hierarchical, Stewart suggests that team leaders can create new cultures and structures by inventing their own team hierarchies and their own roles – so that they play a supporting, rather than figurehead, role.

One implication of such action would be the reappraisal of some of the leadership functions stated earlier. Leaders might spend less time initiating, regulating and evaluating their teams; they would spend more time looking forwards and outwards and less time looking inwards and managing team performance. In their inverted pyramid structure, these managers give up their top role of command and see themselves at the bottom, providing a firm foundation for their teams. In this kind of structure, Stewart argues, leaders use the experience and skills of their front line team and deploy important new management skills, which include:

- *Enabling* – ensuring the team has all the resources it needs to be fully empowered.
- *Facilitating* – removing blocks and delays which prevent staff from doing their best work.
- *Consulting* – with staff to harness their knowledge and experience and use it in both operational and strategic ways.
- *Collaborating* – going beyond consultation to collaborate fully, freely and openly with team members, harnessing all their expertise towards the organisation's goals. This requires seeing the staff as full partners, not just junior members, and is the ultimate test of the leader's skill in empowerment and the *will* to implement it.
- *Mentoring* – helping team members develop and play a fuller role.
- *Supporting* – giving help when it is needed and being especially supportive when someone makes a mistake (Stewart, 1994).

These abilities suggest a radical departure from conventional management thinking – and they are just that. They may not be appropriate in all situations or suit all team members. But for the leader who really believes in empowering a team, they represent the kind of shift that is needed. This is no abdication of leadership, for it leaves many important tasks for the leader to perform. Stewart calls these the eight Es of empowerment:

- *Envision*. Ensure the staff have a shared vision of the goals.
- *Educate*. Train staff to use their own judgement, make decisions, develop understanding and special skills.
- *Eliminate*. The barriers to empowerment.
- *Express*. What empowerment is, what it can achieve, what needs to be achieved, what is going wrong.
- *Enthuse*. Generate excitement, encourage enjoyment; be energetic.
- *Equip*. Devolve resource power/budget control; ensure training happens.
- *Evaluate*. Including self-evaluation; monitor what happens, appraise and give feedback; receive feedback from staff.
- *Expect*. Resistance to change, errors, teething problems; plan to avoid them or overcome them. Also expect success.

Training in teamwork and team leadership

Construction organisations are increasingly recognising the value of training their managers and other employees in teamwork skills. There are many approaches to this, but an interesting example is where team members are brought together, away from the normal pressures of their work, and given the opportunity to analyse their teamworking methods and ponder on how they might improve them. Nolan (1987) argues that teams benefit from a regular workshop or teamwork course of this kind, because it provides them with some commonly agreed processes and structures and a common language, which the team can subsequently use.

Nolan describes the Synectics' Innovative Teamwork Programme (ITP), an example of a training technique developed by his organisation, which specialises in teamwork training. Participants bring tasks from their workplace and these are used as vehicles for learning in group and individual training sessions. This makes learning more relevant and helps people tackle real problems back in their jobs. The emphasis is on creative problem-solving and on developing in individuals responsibility for their own actions.

Group sessions are video-taped and replayed, giving participants a chance to see themselves in action – and the group is able to analyse each person's behaviour and contributions. Used in this way, video is a powerful learning tool; a tape can be replayed again and again to pick up subtle nuances – and the action can even be re-recorded, with team members doing things a different way and comparing this with earlier versions.

The role of the trainers is quite a humble one, because their main responsibility is to set up a relaxed, non-threatening environment conducive to learning; to be positive and encourage risk-taking, and to be good listeners – open-minded and responsive. Good team-workers, in fact!

In this way, participants on Synectics' courses learn firstly from the problem-solving sessions, secondly from video feedback, next from each other and *lastly* from the trainers. As in many of the modern approaches to training, these trainers don't impose their views on the participants, don't even do most of the talking, but basically set up a learning event and allow it to happen. If feedback is given by the trainer, it is constructive and positive and it accepts the ideas of participants as 'true or valid for them'. The trainer at all times respects the autonomy, experience, competence and self-respect of course members.

George Prince, co-founder of Synectics, has concluded from his studies of thousands of meetings, that when participants act destructively, this is 'grounded in their need to *apparently* win'. He says he deliberately uses the term 'apparently' since 'no-one really wins anything' in a meeting, except that too early a criticism of an idea often results in it being dropped. He identifies other discouraging behaviour, such as pulling rank, acting distant, insisting on early precision or proof, being impatient, making fun of the person who puts forward the idea, or not listening. On the other hand, groups which treat every suggestion as a starting point and try to build on them are often much more productive and creative (Prince, 1995).

Other teambuilding exercises

Adair (1986) classes teambuilding activities as either *substitute* team tasks – business games or outdoor activities – or *real* tasks, such as a weekend conference devising a major business plan. He argues that such activities are crucial in making a group into a high performance team, particularly because they reinforce valuable informal relationships and mutual understanding among team members.

Since the 1950s, a variety of courses have been devised under the banner of Outward Bound or Adventure Training, with the common feature of using outdoor activities to focus on developing individual and teamwork skills. Laings are among the construction firms which have used this kind of activity. It is very popular with many managers, but some don't like it at all. Among the strengths of outdoor activities are the bonding effect of shared experiences, the 'Hawthorne' effect of concentrating on teamwork matters and, of course, the potential for learning which is present in any new experience (Nolan, 1987). However, Nolan questions the relevance of such training for managers and others working in a commercial and creative environment. Adair points out that business games and outdoor activities are only simulations of corporate teamwork and have the advantage of being risk-free; however, there is also the possibility that they may be seen as irrelevant diversions and not taken seriously.

Nolan quotes Reginald Revans' argument that business games and outdoor activities have more to do with solving puzzles than problems. A puzzle entails finding an already known solution, whereas a problem involves finding a solution where none yet exists. Since management is about creating the future of organisations, training ought to be about solving problems, not puzzles.

The concept of a business planning conference, where the task is real, has the benefit of providing an immediate and relevant task. However, participants may become so immersed in the reality of the problem that the training value of the exercise is overlooked.

Also, if the conference is poorly organised or participants are unable to become actively involved, it can lead to increased discontent and scepticism among team members.

Evaluation of teamwork training

As with most kinds of management training, it is difficult to evaluate teamwork training, because the results aren't easy to quantify. Ideally, one would try to measure progress in the team's performance, but this is difficult because improvements aren't always easy to assess and many other factors are at work in influencing the team's achievements. Moreover, as with many kinds of management development, the benefits of training may not be seen immediately and may only show up in the long-term performance of the team.

The most effective teams will tend to regularly engage in self-evaluation and select their own criteria for evaluation. This is probably the best form of teamwork evaluation. Careful reflection and discussion can lead to major insights into the complexities of teamworking and the team's situation; they can also create a better understanding of the values, attitudes and concerns prevalent among the team members.

Summary

Not until the 1980s did the merit of good teamwork become firmly established in management thinking and in most other areas of human activity. In the construction industry, sound teamwork is now widely regarded as crucial for the achievement of increasingly complex and interrelated social and economic goals, not only within departments and organisations, but on widely dispersed sites and (most importantly) within multi-professional teams, which perform major aspects of project management.

Most of the characteristics of effective teamwork are now well understood, as are the conditions under which teams are likely to succeed. Teams which benefit from good organisational support and competent leadership are, for example, more likely to be highly motivated, cohesive, flexible, tenacious and committed to success and quality. Good team leaders create the right climate for teamwork, lead by good example and spend time with their teams negotiating ways of working which contribute to joint success.

Some leaders are keen to empower their groups and help them become self-managing teams. To do this, the leader must undertake a thorough reappraisal of his or her own roles, skills and attitudes, team members' roles, the group culture and the implications of empowered teams for the organisation.

Teamwork training is now taken very seriously and many approaches have been tried. The most effective techniques seem to be those which involve participants working in groups on realistic and relevant problems, sometimes in the workplace but often away from the job, where the day-to-day pressures and interruptions can be temporarily forgotten. The very best teams learn to evaluate their own performance and to choose the criteria on which to judge their own achievements. Such teamwork is likely to be increasingly valued in an industry which finds itself under growing pressure from its customers to deliver a better co-ordinated service.

Exercise

In a group, look at the different skills and attributes of each member and identify the roles that each member would be best suited to when running a complex construction project. Consider who would be the best members to plan, start-up, monitor, check progress, control, give instructions, chair meetings, ensure relationships are maintained, supervise (safety and quality), handle disputes, bring the project to an end and close, plus any other aspects of project management.

Chapter 8
Motivation and Human Performance

People and work

People have mixed feelings about work. To some it is liberation, to others slavery. In industrial societies, much of the work consists of ready-made jobs. Many of them don't offer much scope for individual expression or fulfilment. Yet, work is undeniably important. Robert Kahn and his colleagues asked nationwide samples of American workers the same question over a period of more than 25 years:

> If you were to get enough money to live as comfortably as you'd like for the rest of your life, would you continue to work?

The answer did not change very much. About three-quarters of employed men and the majority of employed women said they would carry on working even if they didn't need a wage. Seventy per cent of all workers surveyed said they have met some of their best friends at work. Even the small number of people who would give up work if they could afford to, mentioned their co-workers when asked what they would miss most. The majority who would carry on working pointed out that having a job keeps them from being bored and gives direction to their lives (Kahn, 1981).

Kahn defines work as human activity that produces something of recognised value. All elements of this definition are important to the worker's well-being:

- The activity itself.
- The experience of making something.
- The fact that the activity or product is valued by the worker or by others.

One of the problems of industrial work is that one or more of these elements is often poorly provided for.

For many people in an industrial society, there is no alternative to paid employment; nothing to replace it for providing activity, meaning, reward and social status. The industrialisation of society has reduced many people's jobs to merely making a living. For them, work is just a means to an end. But for most people, having a job means much more than just earning a wage. They want to work, even if they don't need to. They might not do the job they do now, but they want work of some sort.

Many features of work are important. It can create dependence or autonomy, danger or safety, isolation or belonging, monotony or variety. The social reformer, Gandhi, argued that the object of work is less the making of *things* than the making of *people*. Work brings people together to co-operate, in direct contact with materials, giving them knowledge of those materials, engaging the whole person, mind and body. Work gives people a sense of belonging to society, of having something positive to do, of having a purpose in life.

Of course, work is not the only way in which people satisfy needs. And too much work can be as unsatisfactory as too little. The way in which work meets people's needs varies. In particular, different occupations satisfy different needs. Senior contracts managers may achieve status and power through their jobs, steelfixers may not. However, steelfixers may get satisfaction from making something with their hands, whilst contracts managers sit at their desks worrying about the piles of paperwork. Some people satisfy most of their needs through work. Their jobs become a main life interest. Others mainly satisfy their needs outside the workplace. For them, paid employment is a means to an end.

Employee performance

Within any group of people performing the same job, some will do it better than others. This applies to all employees, whether operatives or managers, engineers or clerks. One reason is that the better workers are more skilled or more experienced. They have more *ability*. Another explanation is that the high performers are willing to work harder. They have more *motivation*.

Other factors affect job performance too (see Fig. 8.1). Employees must have a clear idea of what the job requirements are – *role clarity*. Misunderstandings about what they should or should not be doing can lead to wasted effort and poor performance, even if the employees are able and highly motivated.

Employees' *personalities* can also have a bearing on performance. If their characters are ill-suited to their jobs, they will not be so successful. Managers whose jobs involve co-operating with people and influencing their behaviour, are unlikely to be successful if they are arrogant, intolerant or poor listeners.

Performance can suffer if any factor is weak. The most able employees will not work well if their motivation is low. The most highly motivated workers will not be a success, if they lack the skills or personality needed for the job.

A great deal has been written about improving workers' motivation with a view to finding out how to get the best out of employees – including managers themselves! But human performance depends on many other variables, including the task and the individual's level of alertness, anxiety and fatigue.

In most tasks, people set themselves standards which they are content to achieve. Often they don't exceed these targets, even though they are capable of doing so. The level individuals set for acceptable performance depends on the situation, and on their past successes and failures. It is not always possible to predict how successes or

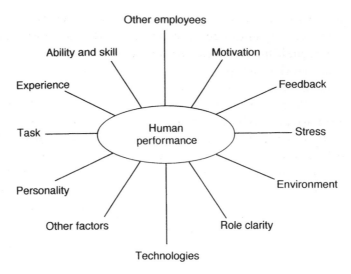

Figure 8.1 Factors affecting human performance.

failures will affect people's future performance, but psychological experiments have suggested that successful performance leads to an increase in the standards employees set themselves, whilst failure leads to a decrease. However, there are exceptions to this. For instance, continued success may eventually lead to boredom and an unwillingness to expend further effort.

Many psychological experiments have shown that performance is influenced by people's expectations. For instance, in one well-known experiment, C. A. Mace improved subjects' performance at an aiming task by adding more concentric rings around a bulls-eye, making a previously good score look mediocre.

People work closer to their capabilities when given *feedback* comparing their performance with other people's, or with their own earlier achievements. Objective criteria for measuring performance are needed to achieve this.

For many operative and clerical tasks in construction, measurement is fairly straightforward and is relied on in management control, estimating and bonusing. The performance of technical and managerial work is less easy to measure because there are so many variables. We can say that one task is harder than another, but we cannot always say how much harder. We can see that a manager's performance has improved, but cannot say by how much.

Human performance partly depends on *skills*. People develop hundreds of skills during their lives, including highly developed skills for listening, observing, understanding and dealing with social situations. Some skills are used so often and so naturally that people don't recognise them as skills at all. People are very versatile at developing skills for coping with life, but their capacities are not unlimited. The rate at which skills are learned and the level of performance finally achieved depend on the body's musculature and nervous system, as well as on the tasks themselves (Fitts and Posner, 1973).

Skilled performance depends on organisation, awareness of a goal, and feedback. But even a well-organised sequence of activities directed towards a specific objective is not enough, if the individual receives no feedback.

Feedback

There are two kinds of feedback. *Intrinsic* feedback comes from the individual's own senses. *Extrinsic* feedback comes from other people. Both provide the individual with information and, if used properly, can enhance motivation. Feedback can serve as a reward, providing strong motivation to continue a task, because it gives information about progress towards a goal. For this reason, feedback is important for both effective performance and learning. Its importance may not be recognised until it is missing and performance has declined as a result.

Intrinsic feedback

Normally, intrinsic feedback is automatically present and is immediate. In construction, the operative receives constant feedback from sensations like pressure, vibration, noise and movement. Seeing is an important source of feedback in many tasks.

Sometimes part of the feedback is missing. For example, operatives working in noisy surroundings cannot hear the sounds made by their tools. This can disrupt their performance. Similarly, if operatives working in cramped, poorly-lit conditions cannot see what they are doing, their speed and efficiency will be impaired.

Gould (1965) conducted an experiment in which subjects were able to watch themselves on a monitor as they performed a task. Selectively blocking out parts of the intrinsic feedback (by excluding them from the picture on the screen) always impaired performance, although the subjects did slowly adjust to the lack of feedback. Other experimenters have reached similar conclusions. It seems that:

- performance is disrupted when any part of feedback is eliminated or distorted;
- when feedback is missing, performance improves with practice, but only up to a point;
- people carrying out a task without proper feedback seldom perform as well as people receiving adequate feedback.

Managers should be aware of the importance of intrinsic feedback, the lack of which may seriously disrupt speed and quality of people's work. People adjust to lack of feedback in the same way that they learn new skills, but they rarely achieve their full potential.

Extrinsic feedback

Feedback from others is very important and can provide strong motivation, leading to high performance levels. If this feedback is augmented from another source, its value is increased. For instance, Smode (1958) gave two groups the same task. One group was given a feedback report after each trial. The other received a report *and* a display of their cumulative score. From the outset, the group receiving the extra feedback performed much better and continued to do so, even when conditions returned to normal.

One of the problems with extrinsic feedback is that it is often delayed; the busy manager forgets to tell an employee how well he/she is doing; bonus payments are received a week after the work was done. This feedback 'lag' is most serious when the individual moves on from one task to another, so that feedback is received when it is too late to influence behaviour at that task. If the individual is carrying out similar tasks over a long period, the feedback lag is less serious.

Ability and skills

It is quite commonly believed that some people are inherently more able than others. Psychologists increasingly think that ability depends more on matching people to tasks and giving them proper training, than on any inherent factor. An individual's ability in a particular task is affected by many factors, of which one of the most important is the level of skill attained.

Understanding the stages in the acquisition of skills can help the manager to:

● devise suitable training programmes and job experiences for new recruits and less skilled employees; and
● monitor their progress as they acquire skills.

The stages of skill development, although not necessarily sequential, are summarised by Taylor, Sluckin *et al.* (1982) as:

● *Plan formation.* New skills are built on to existing skills, which are numerous. Before people can modify or extend their existing skills, they need a plan of action. They need to understand the task they are learning and its purpose.
● *Perceptual organisation.* The learner begins to sort out the important information from the less important, recognising patterns in incoming information, e.g. A is usually followed by B, rarely by C and never by D.
● *Economy of action.* The unskilled operator has to work harder than the skilled one. The apparent effortlessness of the skilled worker comes from knowing when to act or respond.
● *Timing.* This is an important feature of skilled behaviour and often the last to be learned. Skilled workers become expert at timing their actions and this is the first aspect of skill to be lost under stressful conditions.
● *Automatic execution.* The elements of the skilled behaviour become so automatic that many of them are performed unconsciously and the operator can work whilst thinking about other things.

Learning does not stop here. Fitts and Posner and others have shown how skill continues to improve until limited by the age of the operator or the constraints of the task. To maintain automatic performance, especially in complex tasks, regular practice is needed. This would apply to driving a large crane or excavator. Operators may think their performance remains at peak, but what deteriorates is their ability to cope with incoming information. The less practised operator is less able to respond effectively when the demands of the task suddenly increase, as in an emergency.

In considering human performance, there are those who would argue that this should also be looked at in the context of the project and the wider organisation(s) involved in the project. At this point, it is important to mention that the UK construction industry has developed what are called Key Performance Indicators (KPI) (DETR, 2000).

A key performance indicator is the measure of performance of an activity that is critical to the success of an organisation. Its purpose is primarily to enable measurement of project and organisational performance throughout the construction industry. The KPI framework consists of seven main groups:

- Time
- Cost
- Quality
- Client satisfaction
- Client changes
- Business performance
- Health and safety.

In order to define the KPIs throughout the lifetime of a project, it is suggested that five key stages need to be addressed. These are:

- *Commit to invest.* Client decides in principle to invest in a project, sets out the requirements in business terms, and authorises the project team to proceed with conceptual design
- *Commit to construct.* Client authorises the project team to start the construction of the project
- *Available for use.* Project is available for substantial occupancy of use. This may be in advance of the completion of the project
- *End of defect liability period.* The period within the construction contract during which the contractor is obliged to rectify defects.
- *End of lifetime of project.* The period over which the project is employed in its original, or near original, purpose ends. This is a theoretical point over which concepts of full life costs can be applied.

Details of the KPI groups and their associated indicators (and definitions for these indicators) can be obtained from the KPI–DETR document (DETR, 2000) entitled *KPI Report for the Minister of Construction.*

Performance and stress

Stress can be defined as the demands that a task and the environment make on an individual. The structural engineer uses the term in a similar way to describe the demands made on materials.

It has been found that people perform best under *intermediate* stress. If all the demands of the task and environment are removed, the individual becomes bored, less alert and may even fall asleep! Hopefully, this will not happen too often on site. When the work is too demanding or working conditions are very unfavourable, people also perform poorly.

People can cope with a range of physical conditions and can tolerate wide variations in temperature, lighting, noise, ventilation and humidity. But extreme physical and social conditions are stressful.

When the job demands and working conditions are reasonable, the employee is most likely to find the work stimulating and challenging, and will put in maximum effort.

Job stress

Unlike physical and chemical hazards, job stressors respect no occupational boundaries. Therefore, the potential for exposure to this class of health risk is ubiquitous. There are many reasons why organisations should take account of stress and do something about it. These include:

- Organisations have the social responsibility to provide good quality of working life.
- Excessive stress causes illness.
- Stress can result in an inability to cope with the demands of the job which, of course, creates more stress.
- Excessive stress can reduce employee effectiveness and therefore organisational performance.

A model of job stress and health is depicted in Fig. 8.2. The model shows that job stressors can produce acute reactions and strains, which can lead to chronic illness. Although job stressors are listed as a single category, usually they are grouped into several broad categories such as: factors intrinsic to the job; role in the organisation; relationships at work; career development; and organisational structure/climate.

From the model presented, it can be seen that the factors that influence (moderating factors) job stressors include personal characteristics (e.g. personal traits), non-work factors (e.g. family matters, financial issues, social relationships), and buffer factors (e.g. social support, coping skills, physical exercise). The moderating factors operate to strengthen or weaken the relationship between job stressors and health outcomes.

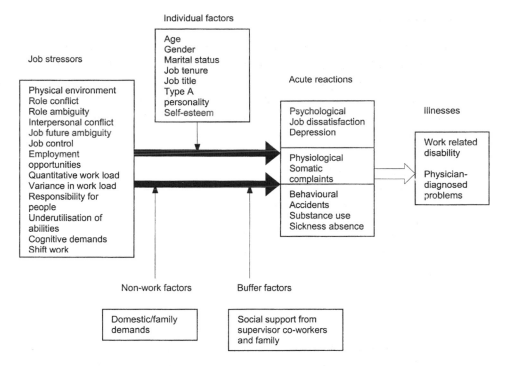

Figure 8.2 Model of stress/health relationship (from Murphy, 1995, pp. 41–50).

The model also highlights the complexity of the problem of stress, as it cuts across work and non-work domains. These cross-cutting effects suggest that the study of job stress, and the design of stress management interventions, should be approached from a multidisciplinary perspective. It is only through this that an accurate picture of the nature of stress and how it should be managed could be produced.

People react to excessive stress in various ways:

- *They work faster*. They act without weighing up all the available information and allow more errors to happen.
- *They work out priorities*. They filter incoming information, discard some or set it aside for later attention, delegate some tasks to subordinates and concentrate on the important ones themselves. Many people work fairly effectively in this way. Managers often have to.
- *They put all work in a queue*. Jobs, important or trivial, just wait in line. This evens out the individual's workload, but causes delays. Some delays lead to costly mistakes, but others may be productive. In some tasks, human performance improves if information has been absorbed before being acted on.
- *They stop working*. Under extreme pressure, people cannot carry on. Taking a break may seem undesirable in the short-term, but can lead to better performance later on.

The causes of stress do not necessarily have a cumulative effect. Stress involves many factors which interact in various ways. For example, if someone is doing a job which involves *reading*, a small amount of extraneous *conversation* will be more disruptive than a loud mechanical noise. Lack of sleep produces a low level of arousal, but loud noise increases it. So, a noisy workplace would offset tiredness.

Optimal stress

It is not easy to specify an optimum level of stress. Its effects can change as a task progresses. Normally, moderate levels of stress produce the best performance, so a demanding task should be counterbalanced by favourable environmental conditions, and vice versa. Talking to fellow workers or listening to the radio can make the performance of a routine task more efficient, but would hamper a demanding task which needed concentration (Fitts and Posner).

Alertness and fatigue

Alertness drops when an employee has been doing a job for too long under low stress conditions. Tasks which need vigilance, as in checking a bill of quantities for errors, usually result in a steady fall in alertness and performance. Tasks of this kind are becoming more common, as routine operations become automated.

A typical task in which fatigue occurs is mechanical excavation, where the driver is continually adjusting controls in the cab. At first, performance improves as the operator adapts to site conditions. But efficiency declines if the task goes on for too long. The continual demands of this 'tracking' task cause fatigue and loss of attention. The longer the task goes on without a break, the more the worker makes mistakes. Performance can be improved by adding variety to the work, using frequent rest pauses and giving feedback on performance.

Anxiety

People respond to excessive stress in different ways. It has been cited as a contributory factor in heart disease, cancer and stomach ulcers. Often, however, the reaction is simply anxiety or anger. As with stress, a *moderate* amount of anxiety can improve performance, but too little or too much is usually counterproductive.

People become too anxious if the job they are given is too hard. Employees need to be given goals which are challenging but attainable, and should be encouraged to organise their work more efficiently. Anxiety caused by personal problems can interfere with an employee's ability to concentrate. This may be harder to remedy.

Stress and its management

Many factors beyond the task and working conditions also act as stressors. What an employee finds stressful depends on his or her characteristics, the situation and the interaction between the two (Payne *et al.*, 1982). Research by Arsenault and Dolan (1983) also suggests a contingency theory of job stress. For an introduction to potential sources of stress, see Cooper (1978, 1984).

Much effort has gone into suggesting ways of managing stress at work. Methods of stress management training include muscle relaxation, biofeedback and meditation. Murphy (1984) suggests that these can be cost-effective, but must take account of sources of stress at the organisational, ergonomic, group and individual levels. He points out that while stress factors cannot be designed out of some jobs, work environment and organisational factors *can* be modified, through organisational change, job enrichment and job redesign. Murphy sees stress management methods as a useful support to these techniques, but not as a substitute. The organisation must still tackle the causes of excessive stress. Whilst some people doubt whether techniques like job enrichment can increase productivity, it seems that they can help to reduce stress.

In summary, the following can be offered as ways in which stress can be managed by an organisation.

- *Job design*. It is important to clarify roles, reducing the danger of role ambiguity and conflict and giving people more autonomy within a defined structure to manage their responsibilities.
- *Targets and performance standards*. It is important to set reasonable and achievable targets that may stress individual workers but do not place impossible burdens on them.
- *The placement of people in jobs*. Organisations need to take care in placing people in jobs that are within their capabilities.
- *Performance management processes*. There is a need to have performance management processes that allow meaningful dialogue to take place between managers/supervisors and individuals about the latter's work, problems and ambitions.
- *Counselling*. It is important to give individuals the opportunity to talk about their problems with a member of the HR/personnel department, or through an employee assistance programme if one exists.
- *Management training*. It is important to provide management training in performance review and counselling techniques and in what managers can do in order to alleviate their own stress and reduce it in others.
- *Work–life balance policies and strategies*. Having work–life balance policies is important. This could be operationalised to take account of the pressures on employees who have responsibilities as parents, partners or carers, and which can include such provisions as special leave and flexible working hours.

Information technology

One of the technologies which is having an increasing effect on employee performance is information technology. IT is changing work practices and hence employees' attitudes towards work. It is changing the nature and content of individual jobs and work groups; the tasks of supervision and management; and the hierarchical structure of work roles.

On the positive side, this 'new work' (as it was sometimes dubbed in the 1990s) may offer more employees work which is less repetitive, less boring; tasks which empower them and involve them in constant learning; roles which require them to do more problem-solving, decision-making, innovation and other higher-order thinking.

Motivation

Miller (1966) and others point out that what constitutes motivated behaviour is very diverse. Its study encompasses biochemistry, sociology, psychology and anthropology, to name just a few.

There have been numerous attempts to explain motivation and the boundaries between approaches are not clear cut. Few theories embrace the full complexity of motivation. Instead, they provide partial explanations of motivated behaviour and offer the manager sketchy advice about how to influence the process.

Even in psychology, many kinds of explanations have been put forward, some more plausible and useful than others. A contingency approach seems most appropriate – what motivates one person will not necessarily motivate another. For each individual, what motivates depends on the circumstances and how the person perceives them.

Although some theories are more soundly constructed than others, there is no single theory which explains all motivated behaviour. Most introductory psychology books expand on some or all the ideas mentioned below, but vary in the way they categorise them.

Approaches to motivation

Needs and drives

Drive-reduction theory

Hull's theory (1943) was built on by Mowrer (1950). People have a range of primary biological needs, e.g. hunger, thirst. These activate primary drives such as searching for food. *Anxiety*, caused by the fear of being unable to satisfy primary needs, is also motivating. This anxiety drives people to strive for success, power, social approval and money. Mowrer maintains that the need for security keeps most people in their jobs.

For many reasons, most psychologists are disenchanted with drive-reduction theory.

Self theory

Snyder and Williams (1982) claim that people have a basic need to maintain or enhance their self-image. They suggest that this theory could provide a unifying theme for a range of cognitive theories of motivation and be a useful addition to operant conditioning theory (see below).

Other needs theories

There are numerous needs theories. Some, such as Murray's (1938), list about 20 needs. Others list a few arranged in a hierarchy. Maslow's hierarchy (1954) is perhaps the best known to managers.

Maslow claims that needs operate in a kind of hierarchy, where *reasonable* gratification of one level triggers the next level to operate. These needs include basic necessities like food, clothing and shelter, which lead to physical well-being and security. Then there are higher needs, like affection, respect and self-fulfilment, which are said to be triggered when the basic needs are reasonably well catered for.

At first glance, the idea seems plausible. People will not be interested in gaining one another's respect, if they are starving to death. They will be motivated by the need to obtain food.

Managers should certainly be aware that for any individual, some needs will be *prepotent* at certain times. Despite its appeal to many managers, Maslow's theory has been heavily criticised.

Some needs theories mention only single items, such as the need for affiliation (Schachter, 1959), achievement (McClelland, 1961), and competence (White, 1959).

Cognitive theories

These relate motivation to cognitive processes like thinking, perception and memory. They include the following.

Cognitive consistency theories

Motivation depends on how the individual perceives the world. An example is Korman's theory (1974), which argues that an individual is motivated to behave in ways consistent with his or her self-image. Festinger's cognitive dissonance theory (1957) is another well-known example.

Expectancy theories

Examples are Vroom (1964); Porter and Lawler (1968). Motivation is seen as a joint function of *expectancy* – a belief regarding the probability that a particular course of action will lead to a particular outcome – and *valence* – the value an individual attaches to each probable outcome. If the most probable outcome is highly valued, motivation will be high; if the likelihood of achieving the most valued goal is low or if the most probable outcome is not highly valued, motivation will be low.

Expectancy theories assume that people always make rational choices, but the evidence throws doubt on this (Wason, 1978 and others).

Instrumental conditioning

The theory of instrumental conditioning looks at the relationship between performance and rewards. The general assumption is that people will work harder if rewarded for their efforts (B. F. Skinner, 1953). The reward, and hence the motivation, is *extrinsic* to the task. Skinner found that if a reward is given when people behave in a certain way, they are more likely to repeat that action. There is solid evidence to support this.

To establish a desired level of behaviour, managers must reward improvements in performance, until eventually they reward only behaviour which closely approximates to the desired behaviour, and finally only the behaviour itself. Rewards should be given regularly until the desired behaviour is well-established. Ideally, the reward should follow the desired behaviour fairly soon.

A reward is anything valued by the individual. Some people prefer tangible rewards, others a word of praise or hint of promotion (Higgins and Archer, 1968). Moreover, what the individual regards as rewarding varies over time.

Even if employees don't want a monetary reward, money can be motivating if it helps them *buy* the rewards they do want. In this role, money acts as a *secondary reinforcer*.

What Skinner found out about the repetition of rewards is not widely known among managers. First, he noted that if the performing/rewarding sequence is repeated regularly after the desired behaviour is established, there is a gradual decline in performance. But rewards given unpredictably lead to continued motivation to repeat the task. This is how gamblers are rewarded; they do not know when they are going to win and they do not always win.

Second, Skinner noted that when a person's behaviour is ignored, there is a tendency for it not to be repeated. Thus, the absence of a suitable reward can cause performance to decline, unless something else takes its place.

Reward systems often have unpleasant connotations, implying that some controlling person offers rewards to some less influential person. In practice, extrinsic motivation operates continuously in all aspects of human relations and is a two-way process. Construction managers and workers reward, or fail to reward, one another all the time (Fryer and Fryer, 1980).

Intrinsic motivation

Even when rewards are absent, people often work at a task for no other reason than the pleasure of doing it. Motivation is derived from the task itself. Such a task is said to be *intrinsically* motivating. Bruner (1966) identifies three reasons for this.

Curiosity

We become curious about a task when it is unclear, uncertain or unfinished. Our attention is maintained until the problem is solved. Operatives are motivated in this way when they have an unusual construction detail to work out. But motivation of this sort will only be sustained if employees are given tasks which are slightly different from, or a little harder than, those they have done before. If a task is too easy, employees become bored. If it is too difficult, they become frustrated. Either way, motivation suffers.

Setting challenging tasks for subordinates demands ingenuity and imagination on the part of the manager. Fortunately, construction work is often more varied and interesting than mass production and processing work, because projects are quite challenging and diverse. The manager must look for opportunities to restructure tasks to provide people with a challenge. The scope for this may be limited by outside constraints, such as rigid job specifications and job demarcations agreed with the unions. Overcoming such problems may require help from senior managers in the firm.

Sense of competence

It seems that most employees are motivated by a need to become more competent. Bruner argues that unless people become competent at a job, they will find it difficult to stay interested in it. To achieve this sense of competence, employees must have some measure of how well they are doing. This relies on having a clear target and some feedback on performance. A vague task, stretching far into the future, offers little scope for measuring progress. Its effect on motivation will be small.

The need for competence may vary with age, sex and background, and managers must be sensitive to individual differences if they are to provide opportunities for competence needs to be met. Construction managers recognise the need to treat their subordinates as individuals (Fryer, 1979), but may not recognise how much their needs differ.

Employees often have *competence models* – individuals with whom they work, whose respect they seek and whose standards they wish to make their own. They identify with such models even when the latter are not in positions of authority – hence the success of many informal group leaders. People are very loyal to their competence models.

The need to co-operate

Many people need to respond to others and work together towards common objectives. They satisfy this need in different ways. Some are natural leaders whilst others contribute to the group by offering helpful suggestions, by evaluating ideas or simply by doing what is asked of them.

Managers must be tolerant and flexible; it is in cultivating these varied but interlocking roles that they help their subordinates to get a sense of working together. If individuals can see how they contribute to their team's effectiveness, they are likely to become more motivated. Construction work is often organised so that it is carried out by small gangs. Tasks like bricklaying, which often depend on co-operation, will encourage motivation, if properly managed.

Goal-setting theory

Developed initially by Edwin Locke in the 1960s, goal theory argues that performance in almost any work activity can be improved if clear goals or targets are set in relation to specific tasks, the employee accepts those goals, and performance of the task can be measured and controlled. Cooper (1995) says that extensive research into the effect of goal-setting on performance shows that:

- Difficult goals lead to higher performance than moderate or easy goals.
- Specific, difficult goals are more effective than vague, broad goals.
- Feedback about the person's goal-directed behaviour is necessary if goal-setting is to work.
- Employees need to be committed to achieving the goals.

Bruner argued that if tasks are *too* difficult, there is an adverse effect on motivation. Locke made a similar point; employees will not be motivated if they don't possess – and know they don't possess – the skills needed to achieve a goal. But of course lack of skills is not the only reason a task may be too hard. Some of the obstacles may be beyond the employee's control. The manager must ensure that such obstacles can be resolved and that targets are therefore feasible.

Locke has suggested that goal-setting should be viewed as a motivational technique rather than a motivation theory.

Motivation and job satisfaction

There is a widely held view that if people are satisfied with their jobs, they will be motivated to work harder. However, it is difficult to draw a distinction between satisfaction, rewards and needs. Many rewards are sources of satisfaction, in the sense that they satisfy needs.

Researchers have put a lot of effort into showing that job satisfaction and motivation are correlated, but the evidence does not support this. Moreover, it has been difficult to separate cause and effect. Effective performance could lead to job satisfaction, rather than be the result of it. Also, removing the causes of dissatisfaction does not automatically lead to satisfaction!

It does seem that unconditional rewards – fringe benefits not directly related to performance – do help the firm attract and hold employees. They can therefore contribute to productivity by reducing absenteeism and labour turnover. But it seems very doubtful whether they lead to increased motivation.

Financial incentives

Financial rewards are based on instrumental conditioning. The construction industry introduced financial incentives after World War II to improve productivity. They have generally not worked well, because of the complexity of motivation and even of reinforcement (see instrumental conditioning, above). They are still widespread because most managers have not appreciated how complex motivation really is. Moreover, bonus schemes are 'visible' and are relatively easy to operate. Agreements about financial incentives have been reached by employers and unions over many years and these are formally written into working rule agreements.

For an incentive scheme to work even reasonably well, the following conditions must be observed:

- *Simplicity*. Operatives must be able to calculate or check their bonus earnings.
- *Honesty*. The scheme must be seen to be fair.
- *Agreement*. The terms must be fully accepted by workers and management.
- *Targets*. These must be reasonably attainable.
- *Size of task*. Bonus should be based on small parcels of work, enabling the operative to assess progress and bonus earned.
- *Group size*. Bonus should be related to individual effort (although, in practice, rewarding a small group can be effective and can encourage co-operation).
- *Availability of work*. There must be adequate bonusable work available, so that operatives are not prevented from earning bonus.
- *Payments*. These should be regular and prompt (although this is more likely to keep operatives happy than maintain their performance).
- *Scope of scheme*. As many tasks as possible should be bonusable.

For a discussion of the types of incentive scheme and their implementation, see Harris and McCaffer (2001).

Sub-contractors

Site managers are concerned with the motivation of two distinct groups – direct labour and sub-contract labour. Whilst they have direct influence and control over their own labour, sub-contractors pose a different problem.

The manager, having no direct authority over sub-contract operatives, must identify ways of supporting the sub-contractor's own efforts to get high performance levels.

A great deal can be done to support sub-contractors' site personnel. Although managers do not dictate the rewards to sub-contractors, the attention they give to target setting, planning and co-ordination can create better prospects for sub-contract staff to achieve their goals. Construction managers can do much to provide favourable conditions for sub-contract performance, but their efforts will be of little value unless the sub-contractor's own management is making an effort too.

Job design

Job design is about improving motivation and performance at work. Some people thought machines would help, by automating dull, monotonous work out of existence. But the replacement of human energy by machines has created problems too. It has encouraged specialisation and the deskilling of many jobs, making it hard to provide varied, interesting (and hence intrinsically motivating) work. We have paid a price for our technical progress. Some would argue that the price has been too high. What can be done to make work more meaningful? Several techniques have been tried.

Job enlargement

This involves making a job more interesting or challenging by widening the range of tasks. Normally the extra work is no more difficult. The challenge comes from the greater variety of tasks the worker handles. Monotony is reduced because each task is repeated less often.

Work can be restructured in this way for humanitarian reasons, but the underlying purpose is to improve performance. Individuals may see it as a management ploy to get more work out of them for the same pay.

In construction, job enlargement would certainly involve removing some trade demarcations. This could create problems, but would lead to more flexible use of employees. The gain in terms of human satisfaction might not be great. Many of the more skilled construction jobs already offer variety and interest.

Job enrichment

Sometimes called vertical job enlargement, this allows workers to take more responsibility for their work. This could include quality control and decisions about work methods and sequencing. The *autonomous work group* or *self-managed team* is an extension of this idea.

Whether job enrichment succeeds is hard to evaluate. There is some evidence that workers are more satisfied, but production levels are not always higher. The aim is to promote productivity by providing challenging jobs. One difficulty is that if people are given more responsibility, their bosses will have less. The effects of job enrichment on higher management levels must be considered, although there is often scope for senior staff to turn their attention to strategic problems, which might otherwise be neglected.

Some people may not want their jobs enriched or may not be able to cope with the responsibility. Extra training may be needed and this cost must be set against the benefits. Quality of work may improve when jobs are enriched and this must also be taken into account.

Ergonomics

This involves an interdisciplinary approach to work, using knowledge of anatomy, physiology and psychology. It is used to design better workplaces and to improve machine layouts and controls. Its main aim is to improve efficiency rather than job satisfaction, but designing jobs to suit people can help to increase satisfaction and reduce frustration. This will not be achieved if the purpose of the exercise is only to achieve efficient and cost-effective production. To maximise productivity *and* satisfaction, there must be a trade-off between technical efficiency and the employee's well-being. In ergonomics, human satisfaction should be a key consideration.

Job rotation

A rather more straightforward way of enlarging people's jobs is to move them around the business. This can widen the range of their work, creating interest and motivation, but its value may be short-lived if they see it as moving from one boring job to another. It is also costly to move people from jobs they are good at to jobs they are unfamiliar with. It can lead to union demarcation problems and, unless the tasks involved are well-designed and interesting, it is unlikely to lead to a substantial change in workers' attitudes to their jobs.

There is scope for varying the length and timing of job rotation and making it voluntary rather than compulsory. This may make it more attractive, but too much rotation can cause confusion, breaking down the task and social bonds which exist within the organisation.

Job rotation has another function – in staff development. In a study by the writer, construction managers said they valued job rotation as a way of developing managers (Fryer, 1977).

Time management

Put simply, time management refers to the development of processes and tools that increase efficiency and productivity. Time management is often thought of, or presented as, a set of time management skills; the theory is that once we master these skills, we will be more organised and efficient.

Over the years, practitioners and students of time management have attempted to analyse and understand the time use of those persons who want to become more efficient on the job, in their home lives and in the other activities that they undertake. Some sets of common precepts have emerged. These include the need for prioritisation, the creation and use of lists, and the assigning of activities to particular time slots on an individual's calendar (Macan, 1994). All these approaches are based on the premise that activities can be organised longitudinally and completed in manageable bits, allowing a person to work through the obligations of the day to achieve desired goals. However, there have been studies that examined the relationship between traditional time management behaviours and the concept of polychronicity (Slocombe and Bluedorn, 1999). Polychronicity is the extent to which people prefer to engage in two or more tasks or events simultaneously. Thus, polychronic behaviour, at first glance, appears not to fit the more traditional step-by-step, one-thing at a time suggestions that characterise efficient time management. Rather than prioritising and ordering activities one by one, polychronic time use is characterised by overlaps of activities, interruptions, and the dovetailing of tasks.

In business, time management has morphed into everything from methodologies such as Enterprise Resource Planning through consultants' services such as Professional Organisers. There are people who find management tools, such as Project Information Management (PIM) software and PDAs (Personal Digital Assistants), useful for managing their time more effectively. As an example, PDAs can make it easier to schedule and keep track of events and appointments.

There are useful personal time management skills. These include:

● Goal setting
● Planning
● Scheduling
● Prioritising
● Decision-making
● Delegating.

The typical manager's day is full of activities, some of which are planned beforehand and some of which are new and need to be addressed as they arise. This makes time management difficult. However it is important that managers are able to make the relevant effort to consider some of the following issues:

● *Planning and organisation.* Organise your same-day appointments geographically. Arrange any errands to coincide with any outings. Also, organise materials for a meeting the day before to gain more time on the day of the appointment. Resolve to do today what you wanted to do tomorrow. Make a catch-up list of tasks you have avoided or put on the 'back-burner' and rank them from the most important to the least important. Resolve to do at least one task from this list each day.
● *Set priorities and focus on goals.* Determine what is most important – prioritising is the key to mastering the use of time.
● *Work breakdown.* Break down a major project into smaller component tasks that can be done in a short time. Target a date to complete each of these.
● *Avoid interruptions.* Be able to set aside a period of the day as 'off-limits' during which time you will not be interrupted. Use this time to work on your top priorities. Time your calls strategically. Make your contacts with colleagues, staff and clients productive, but crisp. Be able to schedule free time. Find at least

one day each week during which you do no work or business at all. You will get a better perspective on your work after you have time away from it.
● *Delegation.* Be able to assign tasks to another or others to give you more time to handle the tasks which only you can do. Effectively multiplying your time is the ultimate time management technique.
● *Establish work habits.* There is a need to have a routine to take care of the work activities. If not, it is more difficult to focus, to get things done, to schedule activities, and to control distractions. With an effective routine, one can increase efficiency and get more things done.

Summary

Human performance is complex and difficult to control. Whether or not an individual works hard or works effectively depends on many diverse factors, such as skills, age, personality, past experience and motivation. The type of task, developments in technology, the job design, the feedback given and the organisational setting are also important. Moreover, all these variables interact with one another.

Stress affects performance, but not necessarily in a negative way. People often work best under moderate levels of stress.

Motivation is extremely complex. It seems that an individual's motivation depends on factors intrinsic to the task and on extrinsic rewards that the individual values. However, people are different and what motivates one person will not necessarily motivate another.

There is no simple relationship between job satisfaction, motivation and performance. Indeed, some evidence shows that poorly motivated people performing badly can be more satisfied than highly motivated workers doing good work.

To improve performance, managers must be aware of all the variables involved and take a contingency approach, recognising that what will work in one situation may not work in another.

Discussion

Discuss the important aspects of organisational and project strategies that are likely to contribute positively to the reduction in stress levels of employees.

Discuss the strategies that a construction manager might employ as part of effective time management.

Chapter 9

Problem-Solving and Decision-Making

Most managers, including construction managers, regard decision-making as a key aspect of their work. Studies have shown that managers do not always spend a lot of time on decisions, but making a good decision is often the result of much careful information gathering and analysis, involving discussions with a range of people, scrutiny of recorded information and, for some decisions, manipulation of data using computer programs.

So a decision reached in minutes may be preceded by many hours of collating and analysing information. Even a key business decision may be reached quickly, but only after prolonged consideration of information, a process which may have been spread over weeks or months and involved other staff.

Problem-solving has not enjoyed the same status in management thinking as decision-making. Problem-solving occurs all the time as people try to achieve their goals, find they cannot do so directly and search for ways round the problem. Much problem-solving, though quite elaborate, is performed without the individual's awareness of the process.

Some problems do not involve a decision, because there is only one course the manager can take. A decision almost always involves choosing between several courses of action. If the choices are well-defined, the problem can be described as routine. There may already be procedures for dealing with it. If the choices are unclear, the problem is non-routine and the manager may spend a lot of time looking at the options before reaching a decision.

The decision will be more difficult if the number of choices is large or the outcomes are hard to compare. If the manager lacks information about the problem or about the options available, the decision can become very difficult indeed.

Most decisions are routine. They may not take up a lot of the manager's time, but they interrupt other work. They distract the manager from more critical decisions, which are less structured but have long-term consequences. The manager must guard against this and get priorities right. But routine problems cannot be ignored. They can be urgent too!

The conflict between short-term and long-term decisions is a real one, as this site manager lucidly describes:

The long-term decision is to some extent a stab in the dark, an attempt to decide policy some distance in the future based on today's standards and events. Immediate, operational decisions require no crystal-ball gazing. A problem presents itself and the manager makes a decision with most of the facts available. It is perhaps unfair to say that long-term decisions are 'neglected' but rather that they are 'shelved' – until today's long-term decision becomes tomorrow's immediate decision.

To completely neglect the long-term is to court disaster, but equally if the short-term is neglected then, as fast as the organisation thinks it is making money in the future, it is definitely losing money in the present.

As a site manager with too few supervisors under me, I have had to make the decision to spend an afternoon in the office, scheduling and programming, when I know full well that labour is idle or not fully employed on site. I have had to balance the effects of 50 per cent production against the possibility of a total lack of direction, or no materials on site with which to work. But in all honesty, I invariably end up scheduling and planning at home! That is, these items take second place to the immediate decisions.

Consider too my contracts director. He feels it his duty to oversee existing contracts and seek out new work at the same time. But if one of his contracts is doing badly, he will feel that time spent in finding new work will be to the detriment of the existing contract.

The site manager will be judged on the performance of the contract to date, rather than on the final result, and is therefore unlikely to plan too far ahead if the problems of today are pressing.

Our industry is subject to change at short notice, often negating weeks of preparation and planning. The factors influencing a long-term decision may have changed before it has been implemented.

Clearly, managers face a dilemma, but the situation can be eased. The same manager had these suggestions for striking a better balance between long- and short-term decisions:

If I had more intermediate supervisors, I would delegate more and my time could be spent more effectively on long-term tasks. Managers must do as much as possible to control the changing environment. Good long-term policies can ensure that many immediate operational decisions have already been made as part of a longer-term view.

Types of problem and decision

Management problems come in all shapes and sizes. They vary with the type of work, the rate of external change, the levels of management involved and so on. Some problems are easily resolved; others need a long and difficult period of creative thinking and decision-making. Igor Ansoff developed one of the best known analyses of decision types or categories (see Ansoff, 1987):

- *Operating decisions* relate to the firm's day-to-day activities and to making current operations profitable. They absorb a lot of time and energy and include decisions about allocating resources and people, planning and monitoring projects, scheduling routine tasks and co-ordinating sub-contractors.
- *Strategic decisions* are about long-term problems, risks and uncertainties. Senior

managers have to decide about markets and clients. They must review objectives and consider new techniques, to guarantee the firm's long-range survival. They must have a policy about sub-contracting work and employing direct labour.

- *Administrative decisions* bridge the gap between operating and strategic decisions and deal with how the firm functions effectively. Some of these decisions are about organising the business: what decision-making to centralise and decentralise; how to structure responsibilities, work flow, information flow, and location of facilities. Others are about obtaining and developing people and resources, and the financing of operations and capital assets.

Important decisions come mixed up with trivial but time-consuming demands. Somehow the manager must strike a *balance* between them. On a single day, a senior manager may have to make a decision about the firm's future, reconcile a conflict between two members of staff and advise on a host of operating problems. The strategic decisions are the ones most likely to remain hidden, or be pushed aside. The manager must actively pursue them.

Some managers write down their problems and arrange them in order of priority. However, importance and urgency do not always coincide. Managers must try to delegate routine decisions to give themselves more time for important ones. These are not always obvious and managers may have to search for the opportunities and threats looming up. Many contractors try to solve their trading problems using operating decisions, like cost-reduction exercises, when what is needed is a complete rethink of the business.

Site managers work mainly at the operating level, leaving the main strategic problems to their seniors. However, viewing the site as a separate organisation, some of the manager's decisions are strategic in relation to the project goals. The site manager must strive to balance the immediate and long-range issues, albeit within the narrower timespan of the contract.

H. A. Simon suggests another way of classifying decisions:

- *Programmed decisions.* These are repetitive and can be dealt with using tried procedures. If a problem occurs often – how much spot bonus to pay for sweeping up, how soon to call up a delivery of timber – a routine will be worked out for dealing with it.
- *Non-programmed decisions.* These are the difficult ones. They relate to problems which are novel and unstructured. There is no obvious method for dealing with them because they haven't happened before, or their structure is complex. To solve them, the manager must rely not only on techniques, but on judgement, intuition and generative thinking.

Many decisions are taken under pressure. The manager hasn't time to think them through and may seem to behave irrationally. Thorough planning, thinking ahead and the use of some decision rules can help the manager to cope. Decision rules evolve when a problem occurs regularly. Once a problem has been solved, the manager knows, more or less, what to do if it happens again.

Stages in problem-solving and decision-making

Problems and decisions vary so much in complexity and importance that the manager needs to be flexible to cope with them. On site, some of the problems are technical and can be quantified. Others, like some sub-contract problems, are organisational or contractual and demand judgement and compromise. The manager may have to decide what is *reasonable* rather than right.

Managers must know when a problem should be tackled alone and when to involve others who have some special knowledge or skill. They must be able to judge when others want a firm directive and when they expect consultation.

There have been many analyses of problem-solving and decision-making processes, but for many simple problems the steps are passed over quickly and without much conscious thought. A decision can emerge without anyone being sure when it was made or who made it; indeed, without anybody realising a decision was reached at all.

More complex problems must be approached systematically.

Deciding priorities

Problems rarely crop up one at a time, but come in thick and fast, important ones mixed up with trivial ones. The first step is to decide which problems need to be tackled first. This is not easy. Information will be incomplete and it will be difficult to judge priorities objectively.

Defining the problem

For straightforward problems, this stage can be passed over quickly, but classifying a problem too soon can limit one's thinking. Many problems need clearer definition before a solution can be sought. It often helps to write the problem down in simple language, identifying causes and the desirable outcome (although defining an acceptable solution is not always easy). If the problem is complex, it can be helpful to break it down into a series of 'problem statements' (Parnes, 1992). This makes the task more manageable and is more likely to lead to novel solutions.

Collecting information

Information is gathered, often from many sources. Opinions must be separated from facts and accuracy of data checked. Some of the information can be converted into numbers, graphs and diagrams, which make the problem more visual (but perhaps more abstract).

Major or complex problems may have to be tackled piecemeal to make them manageable. For instance, with materials wastage it may be necessary to tackle one cause of waste at a time (say, multiple handling) or one material (presumably one causing high wastage costs).

Generating choices

Possible solutions must be identified, but there may not be an ideal one. Choices emerge as information is analysed, evaluated and synthesised. However, the information is often incomplete and the validity of each possible solution can rarely be accurately assessed. The manager may have to be content with a course of action which is acceptable rather than correct. Most books on decision-making stress the importance of considering alternatives, but there are times when only one course of action is open.

Drucker has pointed out that one choice is to do nothing. Even this requires a decision, for it will produce an effect, just like any other course of action.

Reaching a decision and acting on it

Choosing between the alternatives is not easy. The full facts are seldom available, so the manager simply doesn't know which decision is best and has to fall back on experience and judgement.

Some decisions need two kinds of knowledge: that which comes from knowing the local situation and that which comes from knowing where the local situation fits into the wider picture. The person on the spot – the site manager, for instance – understands the local situation better than the senior manager, who may be very experienced, but is distant at head office. However, the senior manager is better able to judge the effect of a local decision on the whole firm and must decide when a decision needs this wider perspective.

Once implemented, the effects of a decision should be monitored to ensure that the solution is working.

Problem-solving and decision-making demand a mixture of experience, intellectual ability, skill in rearranging the problem, and insight. Previous habits play an important part in the process. Skills and principles previously learned can be used in solving problems, but people may persist in using solutions that worked in the past, but which are no longer appropriate.

Lack of skills for dealing adequately with any part of problem-solving can lead to poor results. Managers who are good at generating ideas will not necessarily be able to solve a problem if they cannot diagnose it properly in the first place.

Human reasoning and problem-solving

Before the development of experimental psychology, philosophers thought that all human thinking followed the laws of logic. We now know this is not the case. People are not always logical or rational. Instead, they often solve problems intuitively. They don't always know how they arrived at a solution, but are fairly sure it is correct.

H. A. Simon contrasted these two views of people as decision-makers. In the *rational* view, the manager has perfect knowledge of the problem and a clear idea of the alternatives and the kind of solution wanted. The other view is that the manager solves problems in a much more *intuitive* way. The manager rarely has perfect knowledge and cannot operate entirely rationally.

Some interesting research on human reasoning has been carried out by psychologists such as P. C. Wason and J. St. B. T. Evans. Logical reasoning involves two processes – deduction and induction. *Deduction* involves drawing specific inferences from a general set of statements, as in:

> All construction workers are mortal.
> Alex is a construction worker.
> Therefore, Alex is mortal.

However, consider:

> All construction workers wear safety helmets.
> Alex is wearing a safety helmet.
> Therefore, Alex is a construction worker.

The reasoning in the second example is faulty. Wason (1978) argues that people are often poor at reasoning; this can be counter-productive, but at times invaluable. Faulty reasoning forms the basis of much prejudiced thinking, but it becomes invaluable when it allows people to base useful conclusions on hunches. Strictly logical reasoning cannot be used in this way.

Induction involves generating a rule based on some specific instances. An example is 'All construction workers wear safety helmets' based on seeing a number of operatives wearing hard hats. Inductive inferences can always be disproved, for instance, by the appearance of a construction operative *not* wearing a hard hat.

Much of the time people do not reason logically, unless they have had special training. Wason argues that instead of looking at people's abilities to reason logically, it is more productive to study how they perform when given *closed* tasks, where they must choose among fixed alternatives, and *generative* tasks, where they have to think up their own hypotheses and examples. Wason evolved a number of experimental tasks which mirror the processes involved in everyday problem-solving. The results have been surprising. They show that people faced with difficult problems may regress to a simplistic approach. Having reached a solution, often latching on to information given in the problem, they strongly resist attempts to persuade them to change their minds.

In the mid-1970s, Wason and Evans published their *dual-process hypothesis* of reasoning. They argue that for simple problems, people can state accurately how they reached a solution. For more difficult problems, they cannot always do so. How people say they solved a problem may bear little resemblance to how they really did. They cannot recall how they reached the solution and therefore tailor their explanations to fit the result. Wason says this is like the intuition of mathematicians who 'know' when a solution is correct and work out a proof afterwards. It is interesting that research on logical reasoning should result in statements about intuition and irrational thought.

Don Norman made some interesting comparisons between the reasoning capabilities of human beings and computers. People make elementary errors in perception, have poor memories and make mistakes in their reasoning. Computers handle vast amounts of data quickly and accurately and make logical inferences from data given. On the other hand, people play violins, paint masterpieces and understand language. Almost all the things computers are good at, people do badly and vice versa. Ironically, the aspects of human behaviour that we understand best are the things we do most poorly (Norman). One reason for this is that errors give clues to how people think. Indeed, psychologists have devised many experiments intended to cause people to make mistakes so that they can study human thought processes.

Some of the strengths of people and computers are compared in Table 9.1. We are lucky to have the best of both worlds. We can use computers for tasks that people are poor at or find boring. This gives people more time to concentrate on those creative tasks which they do better than computers.

Table 9.1 Comparison of human and computer attributes (adapted from Norman, 1978).

Computers	People
Fast at computation	Flexible
Accurate	Have vast stores of information and learned
Good at computation	strategies
Good at storing and manipulating	Good at applying things already known
information	Good at exploiting new situations
Good at storing abstract data (codes and	Capable of insight
figures)	Good at intuition
Good at making logical inferences	Can tackle novel problems
Known memory capacity	
Rigid	Slow at computation
	Prone to errors of perception, logical
	reasoning and recall

Artificial intelligence (AI)

AI uses computational approaches to simulate the characteristics of intelligent human thought and behaviour. One of the applications of AI is in the development of expert systems, which can emulate human judgement and expertise – diagnosing problems, recommending alternative solutions, identifying possible strategies, and so on. It is likely that computers and hence robots will eventually be able to do many of the things humans can do.

It appears that human skills and computer programs are arranged in the same, hierarchical way. That is to say, human behaviour is governed by programs or sequences of instructions, similar to those used by digital computers. Parts of these programs, like the actions of a scaffolder tightening a coupling, or a crane driver slewing right, are repeated again and again. These *sub-routines* are under the control of higher-level instructions called *executive programs*, which decide the overall plan of action and call upon the various sub-routines at the right points in the process.

Critics of this computer analogy stress that AI programs have not yet come anywhere near modelling the complexity of human cognitive processes. The human mind is so elaborate that even the most advanced AI models leave major aspects of thinking unexplained. But one should remember that a model does not have to be complete or correct to be useful. Its value lies in its ability to foster new understanding and stimulate research that extends our knowledge (Smith, 1993).

Group decision-making

A lot of interest has centred on whether groups make better decisions than individuals. The conclusion is that there are benefits and drawbacks. Some groups are very creative and produce consistently good decisions. Others never get things together.

Managers should avoid judging the value of group decision-making solely on the quality of decisions reached. The very process of co-operating to solve a problem can have a powerful effect on employees' satisfaction and motivation, and this may outweigh the disadvantages of a few poor decisions.

Hunt (1992) and other writers have summarised the advantages and disadvantages of group decision-making.

Advantages

- More skills and experiences are brought to bear on the problem.
- Groups can generate more ideas and information than individuals.
- Members can spot one another's mistakes.
- The task can be divided up between members.
- Group involvement can increase commitment, motivation and satisfaction.
- Groups sometimes average their answers and eliminate extreme positions (but see 'Group polarisation' below).

Disadvantages

- Members may be too alike.
- Members may be so different that they cannot communicate with, or understand, one another.
- Averaged answers may end up as ineffective compromises.
- Decisions often take much longer to reach.

- Members may not identify one another's skills and experiences, so that their contributions are wasted.
- Discussions go off at a tangent, wasting time and effort and creating frustration and annoyance.
- Some members don't understand the problem as well as others.
- Time is lost dealing with personal clashes and social issues.
- Some members dominate the others and do not listen to their ideas. The more passive members stop making suggestions.
- If there are too many in the group, some will not get a chance to express their views.

Despite these problems, the trend towards group decision-making, sometimes called *management by committee*, has continued. For one thing, organisations have become larger and more complex, making it increasingly difficult for one person, or even one department, to reach a decision without consulting others who have relevant information or are affected by the outcome. Moreover, people want to be involved in decision-making about matters that affect them.

Hunt suggests that groups are more effective at decision-making if certain guidelines are followed:

- Give the group a clearly defined, 'concrete' task, with a clear objective.
- Give the group autonomy to carry out the task, and feedback on its decisions.
- Reward the group as a whole, not as individuals.
- Give the group a task which needs a variety of skills and experiences.
- Teach group members about group processes.
- Appoint a good leader who will co-ordinate the group and keep it on course.
- Restrict the size of the group. Five or six is often about right.
- Don't give a group a decision which only justifies one person's attention. Decisions should be assigned to groups only when there is a clear benefit to the members or the organisation.

The problem-solving abilities of a group depend to a large extent on the interaction within the group (for further discussion, see Chapter 7).

When unanimous agreement is a problem in decision-making: groupthink

Cline (1994) is one of the few researchers who has investigated what happens when group members seem to agree, but probably do not. The findings show that, during difficult tasks and stressful situations, members of the group are more inclined to pursue relationship goals supporting each other than to deal with the problem and enquire about the risks involved. Also, individuals may avoid disagreeing or asking for further information as they feel that their view is inferior, or they believe that if they offer a different point of view they may offend or upset other members of the group. When those considered to hold more senior or authoritarian positions in the group put forward a view, some members of the group may be so fearful that they will not put forward their true opinion, even if asked to. Some members within a group may have such a desire to be part of the team included in future projects, and so worried about rejection or conflict, that they may withhold potentially important information so that they don't offend influential group members. Pressure to agree may be so strong that group members may continue to agree blandly whilst unwittingly consenting to their own destruction.

'Groupthink' is an expression used to describe a group that feels that they are moving forward agreeing on issues, when privately some members of the group are not in agreement, but do not express this. Cline (1994) suggested a few ways of avoiding 'groupthink'. This included asking questions, noting an absence of disagreement (which serves as a warning to group members to reassess alternatives) and being aware that the risk of illusory agreement heightens as external stress increases. Hartley (1997) also points out that a seemingly unanimous agreement by the group may disguise a silent minority.

Group polarisation and 'the risky shift'

Groups often make more risky decisions than individuals. This was discovered in the late 1950s. It came as a surprise because it had been assumed that committees and other groups tend to stifle individual boldness and produce cautious, unimaginative decisions (Taylor *et al.* 1982).

Various explanations were put forward for this *risky shift*. The most successful was by Brown (1965; 1986). He suggested that risk is valued in Western culture. We admire risk-takers more than people who are timid and cautious. People discover that there are others in their group who are prepared to take higher risks, so to maintain their self-image, they shift their level of risk-taking towards the higher level.

However, the shift is sometimes towards caution. A group may make a safer decision than the individual members would have done. Brown's explanation is that caution is also valued in some situations (as in investment decisions).

In the late 1960s, it became apparent that an effect called *group polarisation* was at the root of the risky and cautious shifts. An important French study by Moscovici and Zavalloni (1969) showed that individuals are drawn towards the predominant attitude in their group. When group members' individual judgements tend towards the risky pole, a risky shift occurs. When individuals are tending towards the cautious pole, there is a cautious shift. This effect has been amply shown by experiments.

Brown identified another factor which contributes to the polarisation effect. He argued that individuals tend to support the dominant group opinion because they want to remain popular. This reinforces the tendency towards a risky or cautious shift.

Managers should be aware of this polarisation effect, as it will affect the choice of individual and group decision-making for specific problems. For instance, it has been found that trade union mass meetings produce more militant decisions than ballots among members. Similarly, when members of bargaining teams set high targets, discussion results in even higher targets.

Complexities of group decision-making

Owing to the many variables associated with group decision-making, it is a difficult topic to understand. Because people seem to make effective decisions every day, writers often oversimplify the process, neglecting many of the factors that influence choices (Hirokawa and Poole, 1996). Some treat decisions as discrete events that are distinguishable from social behaviour, overlooking interaction, influences, constrained choices, assumptions, power, dominance, norms, rituals, practices, support, relationships, conflict and other group phenomena that may affect the decision-making process.

Many aspects of group behaviour affect group decisions (Fig. 9.1). Group practices can develop into norms or ritual type behaviour that eventually provide a backdrop against which decisions are made (Hackman, 1992). Many groups operate almost automatically and fail to consider what they are doing and the possible effects and limitations of the procedures that they use.

While decisions can be made by an individual alone or within a group context, the difference between making a decision alone and making a group decision is

Figure 9.1 Factors affecting group interaction and decision making potential.

considerable. Group chemistry has the potential to release the best that every member has to offer, generating ideas and synthesising view points, or it can stifle contribution contaminating the group product (Hirokawa and Poole, 1996). To assist the group's effectiveness, an examination of the nature of the task, perceptual barriers, procedures and methods used during decision-making is needed. Decisions made by groups where members are prevented from participating fully do not make use of the group's decision-making potential and these decisions can result in adverse consequences (Hartley, 1997). Failing to consider issues of group interaction and behaviour when making decisions can lead to disasters. There are a few notable examples of institutional group behaviour that can help contextualise poor decision-making processes, for example Capers and Lipton's (1993) research into the behaviours of engineers involved in the development of the Hubble Space Telescope, and the many studies and theories now emerging from the Challenger space shuttle (Hirokawa *et al.*, 1988; Cline, 1994; Hirokawa *et al.*, 1996). In all of the examples reported, members within the teams knew the problems that caused the disasters before the projects were launched. Probably more interesting is that the specialists within the teams had the knowledge to solve the problems, yet for some reason this information was not used. The behaviours reported in such disasters are often common in everyday groups, although the consequence of ignoring information, blocking contributions and not considering others does not always lead to such major disasters.

The phenomena of 'groupthink or risky shift' are closely related to such issues and should also be considered.

Groupthink is where individuals within the group believe that all other group members are in agreement and, even though the individual does not agree with the opinions expressed, they choose not to disagree or put forward their alternative view. The individual feels that their opinion is inferior to those of the other members of the group, or that it will not receive a favourable response. Often when groups seem to be in unanimous agreement individual members may be suppressing alternative views, and may even agree when they actually disagree.

Risky shift often occurs in group situations. Research has shown that groups are more likely to decide on courses of action that involve greater risks than an individual would. The more flamboyant, dominant and courageous members of the group may appear more interesting and exert greater influence on other group members, encouraging others to take increased risks. Also, it is often perceived that the individual is protected in a group and if things go wrong they will not be held personally accountable. The lack of individual responsibility may contribute to groups accepting greater risks.

Video observation and reflection on group decision-making

Although a vast range of what-to-do and what-not-to-do decision-making models do exist, they all have their limitations (Jarboe, 1996). Rather than overwhelming practitioners with the vast array of decision-making models, attempts should be made to develop a better understanding of the group decision-making process.

No amount of theoretical discussion seems to have quite the same impact as a study of one's self in a decision-making and problem-solving context. Possibly one of the best methods for evaluating a group decision-making process is to video-record group meetings and allow the members to investigate and evaluate their own and others' performance (Gorse and Whitehead, 2002). Previous research on video feedback has found that it can also affect future group behaviour. Weber's (1971) research on team building and the effects of video feedback on group interaction found that simply exposing groups to video footage of earlier meetings moderated subsequent behaviour. Over a series of meetings, control groups that did not receive feedback during the experiment increasingly engaged in less co-operative behaviours and became more negative. Members in groups who watched footage of their previous interaction reduced their communication dominance and encouraged others to participate in the group. Thus, when individuals observe their own behaviour they become much more aware of the group's behaviour and how they influence and affect other members of the group.

During the discussions prior to examination of video footage of group decision-making people are prone to making very simple generalisations about specific features that they believe have the greatest effects on the group and decision-making (Gorse and Whitehead, 2002). As the participants take part in group decision-making (which is video recorded) and then later observe themselves engaging in the group decision-making process, they become more aware of their own action and behaviours that affect other members of the group. Early observations by the participants tend to concentrate on the interaction behaviour of themselves, others with whom they engage, and those who affect their individual contributions. This self-interest results in criticism, defence and general statements regarding mannerisms and behaviours used by themselves. Some participants suggest that their behaviour is different from what they imagined. Participants have noted characteristics that they were satisfied and uncomfortable with. The first important discovery is that perceptions of how one behaved can be quite different from that observed.

As participants are allowed to analyse the video data using various quantitative methods (e.g. counting the number of times a person contributed, the number of times a person spoke, offered a suggestion or disagreed with another) and qualitative methods (e.g. making notes of how conflict developed and how problems were resolved) they discover factors that affect the group's discussion, and ultimately the group's decisions. Examinations of interaction often reveal that participation within the group is not distributed evenly. There are dominant group members, those who participate less or are reluctant to communicate, members to whom most of the group interaction is addressed. Thus, the skewed interaction is used to inform the group's initial decisions. Participants may realise that some members are more likely to offer direction, others show emotion and disagree more than other members.

Once data on the number of times a person speaks to or interrupts another are collected, group members often re-examine the video data attempting to determine how members dominated discussions. Those less willing to contribute also become subjects of interest. Attempts are made to disclose why individuals did not contribute; whether they were blocked, suppressed, chose not to contribute, became free-riders (or loafers). Observations often uncover different styles of interaction behaviour that seem to be more effective in encouraging others to engage in interaction.

Following the identification of unequal participation, discussions develop over the strengths and weaknesses when compared with a more balanced level of contribution. Proposals then emerge on how greater contribution from less dominant members can be achieved. The group's investigations are often iterative, with members reviewing the text, examining the video and discussing observations a number of times and changing behaviour between meetings. As well as observing others both during the process and using the video data, participants can also reflect on the internal feelings experienced, providing an extra dimension to the exercise.

The learning process equips the participants with a greater awareness of factors that will affect their own and others' ability to influence and contribute to decision-making processes.

The video exercise has potential to be used as an in-house company training tool. With relatively minor alterations, practitioners can engage in a decision-making process that focuses on an aspect of decisions at work. Using video and audio recording techniques described above, Conflict Management Profiles (Fig. 6.8) and the Quantitative Analysis and Direction (QuAD) tick sheet (Fig. 5.4) participants can be encouraged to evaluate their own and other colleagues' behaviour. With the aid of a facilitator to control monitoring equipment and subsequent discussions, the exercise develops greater awareness of oneself and others during the group decision-making. Such training has the potential to improve understanding of multi-disciplinary decision-making processes.

Suspending judgement in problem-solving

When managers think about a problem, their purpose is not to be right, but to be effective. The education system instils in us the idea that we should be right all the time, but the manager only needs to be right *in the end*. The danger of trying to be right all the time is that it puts the manager's thinking in a straitjacket. It shuts out ideas that are not right in themselves, yet could trigger an original approach to the problem. An effective solution could depend on identifying this fresh angle.

Approaches to thinking stressed by people like Liam Hudson and Edward de Bono rely on the premise that we may need to be wrong on the way to a solution if we are to come up with a good one.

Lateral thinking – a term introduced by de Bono – is not concerned with the *logical arrangement* of information, but with where it will *lead*. De Bono stresses that we have been taught to reject silly or impractical ideas; we judge ideas as useful or useless almost as quickly as we think of them. The impractical ideas are pushed aside so quickly that further thinking which they might have generated is cut off. Instead, we immediately channel our thinking into well-trodden paths that often end with unimaginative solutions.

If managers suspend judgement, ideas survive longer and may breed further ideas. If a manager resists the urge to label an idea as good or bad, subordinates may feel safer in making suggestions – suggestions the manager might find very helpful. Ideas which don't fit into the manager's current framework of ideas may survive long enough to show that the framework itself needs modifying.

In lateral thinking, the manager suspends judgement because exploring an idea is much more productive than evaluating it. The longer the idea survives, the more likely it is that it will lead to a fresh insight.

Creative problem-solving (CPS)

There are a large number of techniques which can be used by managers and others, individually or in groups, to generate and evaluate original and imaginative solutions to problems. VanGundy (1988, 1992) provides an excellent summary of some of them.

CPS techniques help people to break away from entrenched thinking habits and generate truly original ideas. Participants are required to suspend judgement on their own and others' ideas, so that they can explore very unusual, even fantastic, ideas without fear of ridicule.

Brainstorming, also known as information showering, is one of the better known techniques and uses spontaneous group discussion to generate more ideas and better solutions to problems.

For brainstorming to be effective:

- the problem must be stated clearly and simply;
- participants should not criticise one another's ideas;
- self-criticism is discouraged;
- all ideas are recorded, preferably in a way which allows everyone to see them;
- free association of ideas is encouraged;
- quantity of ideas is important – they should come thick and fast;
- building on and relating to previous ideas is encouraged.

Suspending judgement is vital in creative problem-solving. If members criticise one another, this inhibits thinking and discourages people from sharing their ideas. Fear of looking a fool or being proved wrong is probably the biggest barrier to creative thinking. The barrier is heightened when people of varying seniority work together on a problem. Juniors are afraid to put forward unusual ideas for fear it will damage their prospects. Seniors are reluctant to make wild suggestions which might damage their image or credibility.

It is vital to get rid of such barriers in CPS sessions. Participants must *free-wheel*, letting go of inhibitions and allowing themselves to think freely about the problem. Even a wild idea may quickly be modified by someone else, exposing a fresh insight into the problem. Usually, ideas are not evaluated until later, often some time after the session.

Participants should actively develop one another's ideas, allowing one idea to spark off another. In normal meetings, this rarely happens; people are so busy deciding what they want to say next, that they ignore other people's suggestions.

Before a group CPS session, the leader should remind the group of the rules and perhaps start with a warm-up on an unrelated theme. This helps overcome initial anxiety and lack of self-confidence. Ideas often dry up after half an hour. It may help to have a break, before returning to the problem.

Study of CPS groups in business suggests that members learn to show greater empathy and tolerance for their colleagues' ideas. It can also lead to improved morale because there is more interaction and everybody feels they are making a contribution.

Alex Osborn is credited with founding the technique of brainstorming. His process was elaborated by Parnes and is now known as the Osborn–Parnes creative problem-solving process. The CPS process can be used by individuals and teams and, in essence, involves a systematic approach to idea generation carried through to implementation of the chosen solution. At every stage, problem solvers are required to first think broadly (as in brainstorming) and then analytically, before finally homing in on the chosen course of action.

Another creativity development programme, *Synectics*, deliberately brings together people with different expertise to work on a problem. The techniques used involve drawing analogies which may relate to quite disparate disciplines. For example, designers and engineers often use biology as a fruitful source of ideas. They might explore how plants cope with harsh climatic conditions. They then consider what analogies can be drawn with their design brief. Synectics uses analogies in complex ways (Gordon, 1961). Gordon points out that often the really productive ideas result from noticing the points of similarity between otherwise unrelated phenomena.

Other idea generation tools

There are many methods that are used to help groups generate ideas, evaluate alternatives and solve problems, a few of those most commonly used are identified below:

● *Attribute listing*. Very useful for tackling product or process problems. A problem and its objectives are clearly stated. Next, all the attributes of the problem are listed. Witholding all evaluation, each attribute is systematically modified until ideas for a solution emerge.

- *Reverse brainstorming*. The same as above, but generates ideas that would make the problem worse.
- *Role storming*. Each person is asked to brainstorm an issue from another person's perspective.
- *Idea writing*. This is used to explore the meaning of ideas generated. The process is normally undertaken in four stages: (1) Divide into subgroups; (2) Each member writes responses based on an idea, topic or suggestion; (3) Response forms are shared and exchanged and (4) Each member then reads their initial response and a group discussion follows.
- *Delphi method*. Experts work independently listing their individual ideas. The ideas are then reported to all group members and then ranked in order of usefulness and, finally, the ranking of ideas is reconsidered.
- *Focus group*. A group directly affected by an issue or that has a specific interest in a topic are brought together to gather responses and thoughts.
- *Buzz groups*. Large groups are divided into smaller groups to generate ideas on a topic and then bring their ideas back to the larger group forum where they are evaluated.
- *Brainwriting*. Members generate ideas silently. In one variant, each member writes down one idea on a card and passes it on to the next member. The process is used to help members to build on ideas and stimulate thought. Following brainwriting *consensus mapping* can be carried out. In consensus mapping, cards from members can be sorted into various classifications to develop more informed solutions.

Before selecting and using the above methods, which either deal with problems using groups or individuals in isolation, it is important to consider factors that affect individual and group performance.

Will the problem-solving process inhibit members' contributions?

These are the factors to be considered before selecting and using idea generation and decision-making tools:

- Individual's confidence when working alone, when working in groups.
- Members' abilities to use communication methods (reading, writing, talking, presenting skills, mathematical computations and language).
- Reluctant and dominant communicators. A socially inept member may be the sole expert in the team, yet they may find it difficult to contribute their ideas. Alternatively the dominant person may have little relevant knowledge, yet can be highly influential.
- Members' ability to identify relevant knowledge and expertise within a group. Regardless of skewed participation, groups are capable of recognising individual expertise (Littlepage and Silbiger, 1992); however, plenty of legal decisions show that some professionals do lie and exaggerate.

- Members' ability to control and deal with differences of opinion and conflict.
- Members' ability to ask questions, explore options and make detailed enquiries.
- Informal and formal leadership. Those who are socially liked or respected often assume such roles; members of groups may try to please these members by agreeing with them without fully considering proposals.
- Members' ability to build on others' ideas. Sometimes hopping on another person's idea is frowned upon, yet such behaviour can be successful, rewarding to the group and should be encouraged.
- Members' ability to think laterally and come up with ideas that are based on a very different approach. Those with minimal experience and expertise often propose unconstrained solutions.
- Risk – note that groups tend to take riskier decisions. The belief that no one person will be accountable if the decision goes wrong is considered to be a key reason for the riskier decision. Should one person assume sole responsibility for the group decision, possibly reducing such effects?
- Problems of groupthink, individuals may withhold contributions.
- The effect of taking turns in groups reduces the amount of contributions that a member can make.

In order to deal with some of these issues it may be necessary to use a combination of the methods previously described. Different approaches can be used to allow people to work alone and/or in groups, use methods that develop ideas that build on another person's idea and/or produce alternative approaches to ideas, select techniques that use facilitators or experts and/or allow groups to work unaided. There are benefits to be gained from both working in isolation and in groups. While some group members may suppress others, the combined discussion of two or more people may develop a proposal that none of the members could have individually envisaged.

Summary

Problems crop up in an endless stream, important ones mixed up with trivial ones. When a problem can be solved in various ways, a decision is needed. Decisions vary from routine and short-term, to unstructured and long-term. Managers often neglect the strategic decisions and spend too long on the operational ones. It is important to strike a balance between the two.

There are definite steps in problem-solving and making decisions, but for simpler problems they are passed over quickly and without much conscious thought. Many decisions are reached intuitively rather than logically. A common difficulty is that people classify problems too soon, failing to collect and interpret all the relevant information. Lack of skill in any stage of problem-solving will lead to poor decisions and solutions.

Group decision-making has benefits and drawbacks. It is most effective when there is a clear task with specific objectives, and when participation helps secure the group's commitment to the task. Groups may reach a riskier or safer decision than they would have as individuals.

There is a wide range of creative problem-solving techniques for generating and evaluating new ideas. They help remove the barriers to creativity, leading to more imaginative solutions to problems.

Exercise

Within a group, propose a fictional or real problem, which is related to a particular situation, and identify a number of problem-solving and decision-making techniques that would be suitable for exploring the problem and options available. Discuss why the techniques selected are more suitable than other techniques.

Chapter 10
Managing Change

In the late 1990s Sir John Egan's report, *Rethinking Construction*, issued a challenge to the construction industry to commit itself to change, change that might lead to dramatic improvements in overall performance. As such it was a wake up call to the construction industry, setting out five key drivers for change:

- Committed leadership
- A focus on the customer
- Integrated processes and teams
- A quality driven agenda
- Commitment to people.

Importantly, Egan set targets against which efficiency gains could be compared. Namely, annual reductions of 10% in construction cost and construction time, reductions in defects in projects and reportable accidents together with increased productivity and turnover, and profits of 10% per year. Whilst the report was not without its critics, these statements of intent were backed up by radical proposals to change the processes and culture that were endemic in the industry. Integrated teams and supply chains, coupled with an emphasis on delivering value rather than lowest price, provided the platform for delivering improvements, much as they were perceived to have done in other industries.

Four years later, Egan (2002) reported that change was 'already underway' and, consistent with his ethos of continuous improvement, he proposed further strategic targets that focused on the factors that underpin the successful delivery of projects. Namely, teamwork, stakeholder collaboration and education. There could be no doubting, therefore, Egan's desire to place the workforce and their concerns 'at the heart of the industry's agenda'.

To achieve cultural change within any organisation, long-term people-oriented strategies are required – a fact recognised in The Respect for People Working Group (2002) report, *A Framework for Action*. The report features key Action Themes, i.e. equality and diversity, the working environment, health, safety and career development and lifelong learning, and reiterates the business case for investing in people.

When organisations are operating in a stable, predictable environment, there is little pressure to change. But most organisations are not. They are having to face up to the need for quite dramatic change and this must be planned.

Future studies

Future studies is a developing discipline which uses a number of quantitative and qualitative methods to study change. These methods range from forecasting based on trend extrapolation, which mostly uses numerical data, to more judgemental methods, such as scenario building and the construction of forecasting models, some of which are very complex (e.g. global models).

All the methods suffer from a common problem – the future is basically unpredictable – it can only be assessed in terms of possibilities and probabilities. As the possibilities are almost endless, a whole range of futures can be posited, from highly optimistic to totally pessimistic. Trend extrapolation, in particular, can be very misleading, because past trends may be totally unreliable as a guide to what will happen in the future.

Whilst acknowledging the difficulties associated with predicting the future, May (1996) recognises that successful firms increasingly need to respond to changing markets. May suggests a method of classifying the ways of thinking about the future. Futures techniques are placed on a continuum between those that are concerned with foreseeing the future and those that aim to help managers to create the future. Techniques concerned with the management of change, e.g. impact-assessment, cost benefit analysis, scenarios and risk assessment, are centrally positioned on the continuum as these are deemed to be useful in situations where it is difficult either to foresee the future or plan a pre-determined goal with any certainty. May examines in detail the techniques associated with foreseeing, managing and creating the future.

The process of organisational change

Recognising the need for change

Part of the pressure for change originates outside the organisation, in the form of shifting market structure, technological development and government measures. Other pressures come from within the organisation. They include new attitudes to work and industrial conflict. Any effort to change the organisation must take account of both external and internal forces.

External pressures

Companies are experiencing unprecedented pressure to change the processes, structures and functional divisions of their organisations. The forces for change include the following:

- Increasingly complex and onerous client requirements.
- Globalisation of the world economy.

- The impact of Internet-based information and communication.
- Public pressure for efficiency improvements and best value.
- The growth of public private partnerships.
- Public concern about the environmental impact of business.

Winch (2000) suggests that the internationalisation of construction clients and construction firms, together with economic imperatives, underlie the changes currently taking place in the construction industry. The British construction industry is relatively successful in international markets but Winch believes that it faces important challenges. Sustained client commitment to a best value ethos, re-definition of professional roles, greater understanding of whole-life building performance and the potential of concessions contracting and renewed focus on organisational capability and the control of the production process are advocated.

Inside pressures

The manager must take account of the organisation's *climate* or ethos, because it creates pressure for change. The norms, values and attitudes of managers and other employees are among the factors affecting its ethos.

An organisation's climate can be assessed by looking at the following.

- Workers' perceptions of whether the atmosphere at work is friendly or hostile.
- The kind of leadership style adopted by management.
- The extent to which people have to conform to rules and procedures.
- The production standards set by managers and workers.
- The ways in which employees are rewarded.

The climate can centre around power, relationships or achievement. In a power-oriented climate, the decisions are centralised, communication channels are clearly defined and authority is clearly established and frequently used. There is little room for individual discretion.

In a relationship- or affiliation-oriented climate, the firm is organised along more democratic lines. Workers participate in problem-solving and are encouraged to bring their difficulties to the manager.

The climate is said to be achievement-oriented when senior managers formulate objectives, but allow groups to work out their own procedures and rewards. Top managers expect high performance from employees and give them feedback on their achievements.

Planning organisational change

The firm's problems must be thoroughly investigated before any action can be taken. Managers must agree on the scope of the problems and the need for change.

Data must be collected and analysed with care and presented in a suitable form to employees affected by the changes or involved in putting them into action.

The goals of change should be realistic and clearly stated. Where possible, they should be quantified, so that progress can be measured. Many attempts to change organisations have failed because the purpose was not clearly stated and misunderstandings arose among employees who had to implement the change. There should be a clear statement of the timescale for change and the activities needed to achieve it. The process cannot be monitored or controlled unless there is a clear plan of action.

Implementing change

Change can be structural, technical or social. Structural changes introduce new systems of authority, work flows, rules and decision-making systems. Technical change stresses new work methods and layouts, the use of computers and so on. Social changes include such things as the modification of social skills, changes in attitudes and organisational cultures and new approaches to motivation.

Systems thinking stresses that these variables are interdependent. For example, a change leading to decentralised decision-making will affect the attitudes and skill needs of more junior managers. Similarly, the introduction of a technical change, such as a management information system, may change the structure of the organisation and alter the tasks and skills of some employees.

Different firms adopt different strategies for coping with change. Some firms lack rigid structure and rules and are therefore inherently adaptable. Bureaucratic firms find it harder. They often respond to rapid change by setting up new departments or functions, or by strengthening the formal structure. They redefine managerial roles and working relationships along conventional lines, making reference to organisation charts and manuals. Unless a new department has been set up, problems with change tend to be referred up the hierarchy and end up on the desks of senior managers. The latter become heavily loaded with decisions. If a new department is set up to deal with the demands of change, a communications problem arises between the new and existing departments. In the hierarchic firm, people are not encouraged to move freely across functional boundaries. In flexible firms, these problems hardly exist.

Once changes have been introduced, they must be closely monitored for some time to ensure that they are working properly.

Managing change

Change can be implemented at various levels. At the organisational level, it involves activities like strategic management, marketing and organisational development. At the individual level, it involves changing employees' attitudes and helping and encouraging them to develop creative and adaptive skills (Fig. 10.1). To integrate

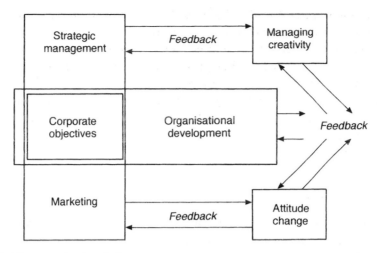

Figure 10.1 The organisational change process.

these levels of change, the composition and tasks of groups and departments may have to be altered.

Strategic management

Terms like strategic management, strategic planning and corporate strategy tend to be used interchangeably. The term strategic management is used here as the umbrella term. Grundy (1994) defines *strategy* as:

> A pattern in the decisions and behaviour of an organisation, team or individual in creating and responding to change.

Although strategic management usually focuses on the organisational level, Grundy's definition is revealing because it stresses the human elements of strategic change (decisions and behaviour), that they are organised (patterns) and that they occur at various levels (including teams). Indeed, Edgar Schein's concept of the psychological contract was built on the premise that there is a consensus between management and employees about the organisation's mission, goals and the strategies for achieving them.

Many people find this hard to accept – seeing strategy as the prerogative of top management. But this belief needs to be reviewed in the light of modern management thinking, with its emphasis on participation, empowerment and ethical and social responsibility. Most would, however, agree that strategic management is fundamental. It deals with significant change, ambiguity, complexity – everything that is non-routine. It is about where the organisation is heading, why, and how it plans to get there.

Deciding how to get there is given names like strategic planning or corporate planning. One analysis uses strategic planning as the overarching level which divides into:

- *Corporate planning* – planning that can't be delegated[1].
- *Business planning* – decisions that are critical to sustainable competitive advantage.
- *Functional planning* – to develop the organisation's core competencies, the sources of its competitive advantage (Hax and Majluf, 1994).

Strategic management establishes the mission of the organisation and its long-term goals, assesses its strengths and weaknesses and searches for the opportunities and threats on the horizon. These questions also form the starting point for marketing. Strategic planning addresses the question of how to implement the long-term goals. This also impacts on marketing.

Wheelen and Hunger (2002) break down strategic management into: environmental scanning, strategy formulation, strategy implementation, and evaluation and control. Strategy formulation includes establishing the organisation's mission and long-term goals, leading to strategies and policies. These strategies and policies are the basis of strategic planning. Figure 10.2 summarises the elements of strategic management using Wheelen and Hunger's broad framework.

Environmental scanning

Decisions about the organisation's future have to take account of both external and internal constraints. Actively searching both the external and internal environment for signs of new opportunities, strengths, trend shifts and dangers is sometimes called environmental scanning. An internal appraisal should include company performance and assets, financial standing, the organisation's structure and systems, and employee strengths and weaknesses. An external appraisal would include a study of the expected pattern of future competition and other events and trends which may influence the organisation's future success.

The aim is to collect reliable data and identify the environmental factors which will most influence the business – and how they may impact on it. The main difficulty is, of course, uncertainty – data is unavailable or unreliable; things change all the time; today's opportunities may become tomorrow's threats. The changes themselves are complex. As Wheelen and Hunger (2002) acknowledge, environmental scanning can be the source of reasonably reliable data on the present situation and current trends, but intuition and luck are needed to predict accurately if these trends will continue.

Strategic planning

The plan describes how the organisation will try to meet its strategic goals. It can be broken into a hierarchy of sub-plans using the kind of breakdown suggested by Hax

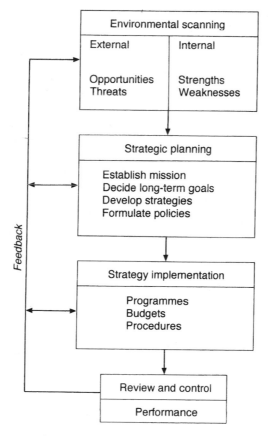

Figure 10.2 The strategic management process.

and Majluf above. This ensures that issues like markets, competitive advantage, financial constraints, resources and employee competencies are systematically examined, making the plan as realistic as possible.

However, because its timescale is long, it is impossible to exercise strategic planning in too great a detail. Too much detail would, in fact, make the plan too rigid. Like all plans, the strategic plan must be flexible enough to meet unexpected events. Much of the process will rely more heavily on the experience and judgement of managers than on quantitative techniques.

Implementing the plan

The plan must be disseminated throughout the organisation. Co-ordinated, goal-directed action cannot occur if employees don't know what the plan is. The plan is divided up and allocated to departments and groups. The information will include targets, programmes of action, a budget for each programme, and procedures for operationalising it. Clearly, this stage is important and requires close attention. Clear communication of this information is vital to effective implementation.

Review and control performance

The strategic plan, like any plan, must be continually monitored, and corrective action taken if performance varies from it. Systems must be set up to provide reliable feedback.

There are a number of broad options open to a firm when reviewing its long-term plans:

- *Containment.* Attempt to maintain present workload in existing market(s).
- *Expansion.* Try to increase workload in present market(s).
- *Contraction.* Reduce current workload.
- *Diversification.* Try to enter new market(s). These can either replace or be additional to present ones.

Benchmarking has become an increasingly popular management tool for understanding how value is delivered for customers (Winch, 2001) and is widely acknowledged as having gained momentum in the construction industry since the mid-1990s. Pickrell *et al.* (1997) define benchmarking as:

> a continuous process of establishing critical areas for improvement within an organisation, investigating the extent to which others carry out the same or similar tasks more efficiently, identifying the techniques that give rise to better performance, implementing them and measuring the outcome

As such, benchmarking has proved to be a flexible tool that can be used in isolation or to support other strategies (Garnett and Pickrell, 2000). Further details on benchmarking and Key Performance Indicators can be accessed on the KPIZone website.

Wheelen and Hunger observe that benchmarking has been found to produce best results in companies that are already well managed.

In smaller firms, strategic planning is often neglected owing to pressure from recurrent operational problems, some of which are quite important and urgent. It must be actively pursued, with time set aside for it. This may mean delegating other tasks and responsibilities. The degree of formality adopted in strategic planning will reflect the size and circumstances of each firm.

There are various tools and techniques for helping to deal with the uncertainties and risks associated with strategic management. Future studies has already been mentioned. Another is *risk management*, a discussion of which can be found in Chapter 11.

Marketing

A company's survival and success depends on its ability to satisfy its customers. Marketing identifies what clients want and how the company can most profitably meet those wants. A firm's marketing policy must be flexible, to cope with market changes.

Amongst the most notable changes in the UK construction industry is a move away from traditional contract procurement towards partnerships and alliances. Indeed, current literature is replete with case study examples of successful partnerships. Hence it comes as little surprise that a considerable number of clients and contracting organisations are now adopting a partnering strategy within their commercial relationships, as evidenced by the *Building* '50 Top Clients' review (2003).

Pheng (1999) draws comparisons between partnering and relationship marketing, which is very much focused on building strong relationships with individual customers. In this new commercial environment, Pheng recognises the need to place trust, dedication to common goals and a shared understanding of each other's individual expectations and values at the centre of a contractor's marketing strategy. Moreover, Walker (2000) suggests that a wider view of the 'customer' is needed. No longer should the client be seen as the 'paying customer', but as a 'stakeholder' alongside all those in the community who are directly affected by construction projects. He argues that an awareness of the community makes marketing and business sense.

A successful marketing strategy, therefore, must recognise the interplay between external, social, political and economic environments both at home and abroad, as these determine the opportunities for work and thereby potential profit (Yisa *et al.*, 1996). Indeed there are many examples today where UK contractors have seized upon opportunities to increase their activities in other parts of the world. As Pettinger (1998) points out, there is no such thing as a closed market. Every market is open to global competition.

The elements of marketing

Market research

This involves identifying the market structure and systematically collecting information about markets, clients, competitors, competitive pricing and general trends. Several trends are important in arriving at marketing decisions, including political, economic and social trends. Trends aren't always reliable because there can be discontinuities or step-changes, which alter the expected future dramatically.

Unless contractors undertake field research, which often requires the services of an outside agency, information will normally be collected via desk research. Data can be gathered from internal sources, e.g. accounts records, sales reports and customers records, or from external sources, e.g. government statistics, trade associations, business magazines and the business pages of the national newspapers (CBPP, 2002).

Today, much of this information is available at low cost or freely accessible via the Internet. Information about customers and competitors can quickly become out-of-date, therefore the Internet has become a valuable tool for contractors. For example, up-to-the minute financial information, in the form of annual reports and financial statements, can often be viewed on-line together with press cuttings about a firm's activities.

Marketing strategy

Pearce (1992) defines marketing strategy as that part of corporate strategy and business planning which considers the needs of customers, identifies the customers on whom the firm should concentrate its efforts, anticipates their needs and plans how to go about satisfying them. The data used in this process comes from numerous sources and much is gathered through market research.

Unlike some marketing writers who stress the importance of deciding which products or services to develop, Pearce puts a lot of emphasis on deciding who the firm's customers should be and on building good relationships with those customers. In an industry like construction, where products and services cannot be displayed in shops, this seems a thoroughly sensible approach to marketing.

A marketing strategy should ultimately answer the question 'What business are we in?' and this analysis must be based on sound research and imaginative thinking.

SWOT analysis

One of the first stages in developing a marketing strategy is normally to carry out an audit of the firm's strengths, weaknesses, opportunities and threats – a SWOT analysis, as it is usually called. Managers can exercise some control over most of the firm's strengths and weaknesses, but opportunities and threats are external and often cannot be directly influenced. However, it is essential that the organisation knows what these outside factors are and responds in an appropriate way.

Positioning

To be successful, most construction firms have to limit the scope of their activities and concentrate on particular market segments. Positioning is about choosing what products and services the organisation will offer to clients; at what price, quality and timescale; and for what type of client. This is a key market strategy decision; it identifies what business the firm is in. It is, of course, essential to let potential clients know where the firm is positioning itself in the market-place.

A *unique selling proposition* (USP) is an exceptional characteristic of a service or product which clients know separates it from services or products offered by competitors.

Building personal and corporate relationships

Contractors used to say that they could exercise very little influence over what competitive contracts they were awarded, other than by submitting a price that was so low they were unlikely to complete the project at a profit. Nowadays, firms are much more active in pursuing business opportunities and, apart from giving assiduous attention to all kinds of marketing information, one way is to build good relationships with clients at personal and corporate levels.

This makes use of the human preference for working with people we know and trust. It gives the contractor a more thorough understanding of the customer's business needs and can also give rise to earlier information about the client's building needs.

Enquiries and contracts

An enquiry can literally be an enquiry from a potential client or it can be any project a firm is interested in undertaking. An enquiry which may lead to a firm's first contract with a new client is especially important. It could lead to further business direct from the client, without competition, so such a relationship helps a contractor to keep down the costs of abortive tendering. The sooner the contractor makes contact with the customer, the more influence can be exerted over the customer's decision-making and this can benefit the contractor.

Meetings with clients

These can take several forms, including initial face-to-face contact with customers to establish personal and corporate relationships, formal interviews, tender presentations and contract negotiations. Selection interviews are now commonplace prior to awarding a contract. Contractors and professional consultants make a presentation to the client and answer questions; and their selection can depend on the quality of that meeting. Contractors take a lot of trouble preparing for such meetings, choosing interview teams carefully, rehearsing presentations, and including presentation skills training in their management development programmes.

Corporate identity

The phrase 'corporate image' has been largely superseded by 'corporate identity', which can be defined as the way a company expresses to the outside world what it is and what it stands for. A company's corporate identity reflects its mission or objectives and its products and activities. Corporate identity is demonstrated through the communications and publicity materials the firm produces, the appearance of its buildings and sites, and the behaviour and appearance of its employees. The attitudes of managers and other employees are very important. They will be visible in many communications the company has with clients and others. At the other extreme, the design of the company's letterhead will convey a distinct impression of the firm – for better or worse.

Brochures

Many construction firms now routinely use brochures, sometimes accompanied by CD Roms, to help communicate their corporate identity and give positive information to potential and existing clients. If the firm operates in several market segments, it will usually have different brochures for each, probably with an over-arching company brochure. Preece and Male (1997) recognise that different pro-motional strategies are required for established key clients, prospective clients and for those who may influence the client's decision. Personal recommendations, however, from previous satisfied clients and professional advisers are always highly influential. Accordingly, they advocate the use of testimonial, or third party endorsement, in brochures that are informative and attractively designed.

Other marketing activities

A number of other activities can support the marketing effort. They include advertising, press coverage, sponsorship, entertaining, exhibitions and other forms of publicity. These activities are expensive; they need to be carefully targeted and their cost-effectiveness assessed. An advertisement aimed at potential customers is useless if it appears in a publication that those clients don't read! Advertising in the right place can help to establish a firm's corporate identity, by giving details of successful projects and conveying information about the company's philosophy and standards. In all advertising, the critical factors include repetition, timing, careful wording and good layout of 'copy'.

Marketing audit

The purpose of the marketing audit is to review the organisation's marketing intentions, performance and methods. The audit can be carried out by internal staff or by external consultants, but either way it must be performed systematically, objectively and by people who have a thorough understanding of marketing. The outcome should be a report summarising what can be learned about the organisa-tion's marketing efforts and how they might be improved.

   The audit can be carried out annually to coincide with the preparation of the annual report, but it may be advisable to do it less frequently. After all, any audit is a dis-ruptive process, which unsettles people and interferes with the smooth running of an important business function. There is an argument for having no set frequency and only performing an audit when there are signs that it is needed (Pearce, 1992). Nevertheless, when the audit is carried out, it must be done thoroughly and may lead to major changes in marketing activities.

Organisational development

Companies affected by continuous and rapid change may draw up a formal policy for dealing with it. Organisational development (OD) is a term used to describe formal approaches to organising change. Ways of implementing OD include:

- Employing consultants to advise on change.
- Setting up a specialist department to do the work.
- Integrating the change process with the mainstream activities of the firm.

Each approach has its strengths and drawbacks. Consultants bring new ideas and expertise into the organisation, but their services are costly and they will not know the business as well as the employees do. Staff may resent outside interference, especially when the consultants start telling them what to do. Most important of all, perhaps, the people who benefit from the exercise – the ones who develop new skills and gain real insights into the organisation's problems – are the consultants themselves and not the firm's employees.

For this reason, many firms prefer to get as many as possible of their own personnel involved in change, so that they can learn from the experience, as well as becoming committed to new systems and methods.

Whether internal OD work is best done by a specialist department or spread through the organisation, depends to some extent on the firm and its problems. Line managers may be too busy to devote time to development work. Specialists will have the time but, like the consultants, may lack a detailed understanding of how the firm operates. If the specialists design all the changes, those who have to implement them may lack commitment. They won't understand the need for change and how it might benefit them.

Sometimes the firm combines the talents of its line managers (thus gaining their ideas and commitment), a specialist department (which can offer both the time and techniques for developing and implementing the changes), and outside consultants (who will see the problems more objectively, having experienced other change activities).

The difficulty here is how to co-ordinate people and get them to trust and share their ideas with one another. In some organisations, the OD specialists are isolated from other employees – the people who will have to implement and live with their ideas. To overcome these difficulties, OD specialists need to have good social skills and be prepared to network extensively.

An increasingly popular way of using consultants without losing the benefit of employee involvement, is to alter the role of the consultants, so that they 'facilitate' change rather than carry out the work themselves. The consultants act as mentors, helping the organisation's employees to learn the skills for engineering and coping with change.

It is widely believed that people resist change. Perhaps it is more accurate to say that there is a time lag between the introduction of a new idea and people's attitudes catching up with it. Certainly, the social aspects of change present special problems. To introduce a new process or alter a work method may require a major shift in the attitudes and behaviour of the people affected by it. In OD work, the biggest task is not changing the system, but changing the people. As Hannagan (2002) and others have pointed out, OD may involve changing the organisation's *culture*.

When specialists and consultants are involved, no single person is likely to have all the knowledge and skills needed to cope with the whole change programme. Many consultants who are experts in designing a computer system or setting up an automated production process, would not know where to begin to help or persuade employees to adapt to them. It is therefore essential that OD is a team effort, using people with a wide spectrum of skills. It usually means that any development must be tackled as a multidisciplinary task at three levels: organisational, group and individual.

Modifying individual and group behaviour is often the most important and time-consuming part of a change programme. It involves creating the right organisational climate, in which two-way exchange of ideas is actively encouraged. It is important that managers and workers trust one another and know that the other group will listen to them.

Employees who know their jobs well often have worthwhile ideas about what the firm ought to be doing and what changes are needed. If managers are receptive to these ideas, the firm may become more efficient and employees will feel valued and gain a sense of identity with the firm and its goals. Human factors play a major role in organisational change.

Business process re-engineering

BPR is a technique which became more widely known among UK managers in the early 1990s. It involves the organisation in a radical rethink and redesign of business processes aimed at making major improvements in key performance areas, like cost, time and quality. This process sounds similar to OD and the two processes are quite difficult to separate. Both can be approached in different ways; both can lead to major changes. BPR is perhaps a little more focused.

Leading BPR proponents Hammer and Champy (1994) claim that the outcome of re-engineering a business process can be dramatic; it can lead to narrow, task-oriented jobs becoming multidimensional, to functional departments losing their reason for existing, to workers concentrating more on customers' needs than bosses' needs and managers behaving more like coaches than supervisors. Almost every aspect of the organisation changes.

As with many new management concepts, BPR describes actions that many would see as long-established aspects of the manager's job. Love and Li (1998) doubt whether BPR approaches to change can improve the performance of projects. Whilst traditional BPR may lead to improvements within an organisation, i.e. intra-organisational processes, the fragmented nature of the construction industry is perceived to be a barrier to inter-organisational change. They suggest that improvements in project performance can only be achieved if the concepts of lean construction, concurrent engineering, continuous improvement *and* process re-engineering are embraced by multidisciplinary teams. Improvement is likely to be a continuous process rather than the radical improvements sought by BPR (see Figure 10.3).

Elsewhere BPR has focused attention on the use of electronic commerce

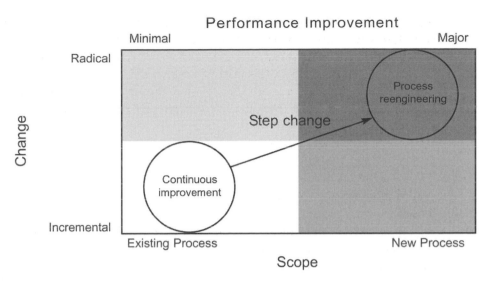

Figure 10.3 The process change continuum (adapted from Love and Li, 1998).

(e-commerce) applications to improve information flow throughout the supply chain (Elliman and Orange, 2000; Ruikar *et al.*, 2003). For example, Ruikar lists reduced project cost, reduced time wastage, reduced errors and the avoidance of disputes amongst the many benefits that e-commerce offers the construction sector. Hence the value of initiatives like BPR is that they can help focus managers' attention on the need for change and the ways to achieve it.

Changing people's attitudes

Corporate planning, marketing and organisational development will fail unless the policies they introduce are accepted by employees. People will adapt to new strategies more willingly if they have been properly consulted and involved in decision-making.

Many factors have a bearing on attitude change and there is a spectrum of ways of influencing people. *Education* and *persuasion* are the more socially acceptable methods. Education is normally the mildest form of social influence. One of the aims of education, at least in Western society, is to give people information in a reasonably unbiased way and present as many viewpoints as possible. Even so, the material chosen and the way it is presented involves value judgements. Implicit biases may be passed on, without the receiver being aware of it. The purpose of education is to develop people who will help to steer organisations into the future.

Persuasion, like education, is considered socially acceptable by most people. If managers are to do their jobs properly, they will have to persuade employees to accept new methods of working, adapt to new technologies, be more safety conscious, and so on.

Propaganda comes further along the spectrum of influence methods. It is not considered very ethical, but companies, governments and advertisers do resort to it from time to time. It can involve censoring or doctoring information, concealing its source or using emotive language. The aim is to get the message accepted and acted upon.

However, in a pluralist society, propaganda is no more likely to be effective in changing attitudes than education or persuasion, except amongst poorly educated and insecure individuals. Propagandist activities may go on within an organisation or a whole industry and this can sometimes cause difficulties for those trying to foster good labour relations.

Extreme forms of social influence include indoctrination, brain-washing and torture. Clearly, they are beyond the scope of the manager's job! The manager is left with education and persuasion.

Making persuasion work

There are two conflicting views about the effectiveness of persuasion. One is that people are very malleable and can easily be persuaded to change their attitudes and beliefs. The other is quite the reverse – that people are stubborn and resistant to change.

These opposing views partly result from early research on persuasion by social psychologists at Columbia University and experimental psychologists at Yale. They studied persuasion in different ways. The Columbia centre used surveys to monitor the effects of media campaigns on the public. The Yale psychologists concentrated on laboratory experiments with individuals.

The Columbia group found that only about one in 20 of the population were affected by persuasive communications, but the Yale group found that between a third and a half changed their attitudes. The Yale experiments showed that persuasive communications can be very powerful if there are no conflicting influences. The Columbia studies found that *personal contact* is more effective for changing opinions and behaviour than mass media campaigns. This is not a problem in small firms, but large organisations often have to use impersonal, formal communications. This may be ineffective unless supported by *opinion leaders*. These are popular or respected employees who take an interest in developments by attending meetings or actively seeking information in other ways. They influence others as they pass on new information and ideas.

The importance of personal contact in attitude change has been demonstrated in many studies. People take much more notice of those they admire or identify with, than they do of impersonal communications, however well these are formulated. For instance, researchers found that Iowa farmers took more notice of their neighbours' opinions about a new seed corn than they did of information from a government department.

Formal communications are not a waste of time, but are more likely to be successful when supported by opinion leaders and when the audience is already mildly interested. For example, a circular letter from the managing director can help awaken latent ideas or beliefs. It creates favourable conditions, in which personal persuasion by those closer to the work is more likely to succeed.

Persuasion can be more successful, especially in larger organisations, if:

- empowerment is used to include everyone in the change process, even those who are indifferent or antagonistic to change;
- the communicator considers the individuality of those at whom persuasion is aimed: differences in their goals, values, attitudes and beliefs, lifestyles;
- the goals for change, as presented, are SMART: *specific, measurable, achievable, relevant, timed*;
- the communicator arouses a moderate level of anxiety about the proposals. Too much anxiety can reduce susceptibility to persuasion; too little can lead to indifference.

Many factors are involved, such as the complexity of the message and employees' anxieties about the subject being communicated. Employees who are anxious about redundancy, may become unco-operative if put under pressure to accept new working practices which they feel would increase the likelihood of redundancies. Similarly, employees worried because they can't cope with their work are unlikely to respond positively to threats about what will happen if they don't improve. Low-threat persuasion has considerably more effect on anxious people than it does on calmer ones.

Encouraging dialogue

Attempts to change attitudes in organisations are unlikely to meet with total and instant success – and this is just as well! In a free society, a plurality of viewpoints exists and is encouraged (one hopes!). When people come into contact with one another daily, this partly offsets the effect of persuasive media communications. In an organisation, attitudes are influenced by information from many sources – the media, friends, family, opinion leaders and management. The firm is more likely to successfully engineer change if, rather than imposing its views on employees, it encourages dialogue and values a range of independent viewpoints.

It is important that differences of opinion can be aired without fear of recrimination. This encourages independent thinkers as opposed to conformers (Jahoda, 1959). The independents are likely to be the creative members of the organisation. To create something better, people must question and criticise what exists. If suggestions are welcomed and taken seriously, the firm is more likely to have an innovative and enthusiastic workforce, well equipped to face the future.

Managing creativity

The old view that creativity was mainly a matter for the arts has gone. It is now widely agreed that creativity is necessary in all fields and is central to business success. In stable conditions, creativity helps businesses to improve their vitality and competitiveness, avoiding stagnation. When rapid change is the norm, creativity is vital, helping firms to drive change, innovate boldly and meet new demands and trading conditions energetically.

Moreover, as we move into the tertiary era, core questions emerge that demand serious creative thinking. For instance, 'what should replace the management principles and practices learned in industrial society in the new organisations that prevail in our (so-called) post-industrial society?' (Davis and Scase, 2000).

Creativity is about pushing forward boundaries in ways that are *original* and *effective*. The best ideas are not just original but often *radical*. In business, creativity involves looking for archetypal ideas, keen insights, fresh perspectives and new opportunities. These translate into new products and services, unique designs, novel ventures and fresh markets. Creativity demands a rethink of current policies and practices and can lead to better use of resources (Fryer, M., 2002).

Managing this creativity is about finding ways to encourage *creative behaviour* among employees. It used to be thought that creative talent was enjoyed by only a few, but current thinking is that most people have creative potential. The management skill is in tapping this ability for the benefit of the organisation and all its stakeholders.

Although exceptional creativity may depend on special knowledge and talents, much of the creativity found in businesses stems from the actions of ordinary people, using well understood skills and know-how. Research has shown that commonplace qualities like *determination* and *perseverance*, *curiosity*, *hard work*, *independent judgement*, *willingness to take risks*, *tolerance of ambiguity* and a *readiness to challenge orthodox thinking* are central to most creative achievements. Many of these characteristics can be regularly observed in *highly motivated* employees, suggesting that management action aimed at boosting motivation may be one key way to stimulate creativity at work.

Creativity pioneer E. Paul Torrance (1995) found that highly creative people are more likely to really enjoy their work and enjoy thinking deeply about things. They also tend to have a strong sense of purpose, are not afraid to express a minority viewpoint and are prepared to challenge established wisdom. They often seek autonomy, are attracted by complexity and display self-confidence. Davis and Scase, among others, also stress the importance of autonomy, seeing it as a major feature of the creative process in organisations.

If managers can encourage or support qualities like these in their staff, they are more likely to release the kind of new thinking that leads to positive change and future business success.

It takes about a decade working in an occupation for someone to become highly innovative or 'expert' in their field (Stein, 2002). Once they have developed this expertise, they tackle complex problems and issues more effectively than do beginners. For example, they are better at recognising patterns, thinking in terms of underlying principles and seeing analogies (Weisberg, 1993). So, helping employees to develop expertise can contribute directly to their creative skills.

However, there are many barriers to creative thinking, some *personal* and some *organisational*. Innovative ideas are not likely to be produced unless both kinds of obstacle are tackled. Organisational barriers include unnecessary or outmoded customs and procedures, as well as an inappropriate organisational *ethos*. Nothing kills creative thinking faster than an atmosphere where new ideas are frowned on or ridiculed. This is especially important because some of the best ideas often involve moving way beyond current thinking; or else, going off at a tangent (as in Edward de Bono's popular concept of *lateral thinking*).

Indeed, some exploratory ideas may seem quite bizarre, even ludicrous, but if they are rejected at the outset, their potential for suggesting something really worthwhile will never be realised. Rejecting unusual ideas is highly damaging to any business that wants to foster innovation. Personal barriers to creativity include resistance to change and other unhelpful habits. Habits lead us to become more set in our ways, yet creative thinking demands challenging assumptions, breaking with old ways and thinking differently.

Developing creative thinkers

The task for the manager is to encourage more creative thinking among as many employees as possible, whilst paying particular attention to employees who show exceptional potential or whose occupations depend more heavily on fresh thinking and new ideas. Increasingly, creativity is likely to involve teamwork (M. Fryer, 2002), so management action and staff development that build team skills and flexibility also contribute to creativity training. Indeed, Taylor Woodrow Chairman, Robert Hawley, has pointed out that creative feats of engineering are routinely achieved by project teams in the construction industry (Hawley, 2003).

Many creativity training courses concentrate on *creative problem solving*, taking participants through the well-defined steps of the process – problem finding, fact finding, idea finding, solution finding and so on. Other courses deal with the so-called *creativity tools* or *techniques*, which use analogy, metaphor and other procedures to spark new thinking.

However, the courses need careful evaluation. They can be very expensive, and high cost and slick packaging don't guarantee quality. Also, few courses are comprehensive. Most tend to be limited in scope and fail to address the full complexity of creativity or the problems organisations face in implementing it. The Creativity Centre, a UK organisation based in Leeds, has tried to address these problems seriously and offer a more balanced service.

Apart from creativity courses, managers can help by paying more attention to the way that they organise the firm or construction project; and the way they treat employees. VanDemark (1991) offers some useful guidelines, stressing that organisations need to create a stimulating and challenging environment, with clearly articulated goals. Managers need to be dynamic and 'people-oriented'. This tallies with UK research, which also supports an employee-centred management style (Fryer, M., 1994). Fryer explains that the behaviour and attitudes of managers need to be similar to those of student-centred teachers, who are more successful in developing creativity in their students than are their mainly task-oriented colleagues.

VanDemark also points out that younger organisations are more conducive to creativity. They have important qualities of openness, flexibility and fluidity. There is enthusiasm and a willingness to experiment. Older organisations, especially as they grow larger, tend to get bogged down with bureaucracy and rigid management structures. In these organisations, people may become protective of their territory and unwilling to tolerate change, which they see as a threat. In construction, temporary projects are a positive feature because they are fluid and provide the scope for flexibility that creativity needs.

West, among others, has stressed the importance to creative success of a *supportive* and *challenging* work environment, where managers offer not only warm support and flexibility, but also intellectually demanding work (West, 1997). Earlier research also suggests that people respond creatively to a challenge.

So, managers need to nurture their teams, foster motivation, stretch their staff and encourage and reward independent ideas. They need to protect their most creative employees from the ridicule of others long enough for them to try out and modify their more extreme ideas. Unfortunately, managers aren't always sufficiently sensitive or knowledgeable to provide this kind of support.

There is a common belief that creativity rests on some mysterious processes and thrives in informal, unstructured settings. But the evidence suggests otherwise. Creative people use specific thinking strategies, for instance:

- *scenario building* – building mental pictures of current and alternative states of affairs in order to select the most promising option;
- *attention-directing devices* – helping to generate and evaluate fresh ideas at every stage from identifying problems to implementing solutions;
- *metaphor and analogy* – identifying useful connections between previously unrelated ideas (Fryer, M., 2002).

Creative people make use of extensive knowledge and a raft of well-established skills, and often work best when there are tough constraints on their actions.

Summary

The need for organisations to face up to change has been stressed throughout this chapter and elsewhere in this book. The impact of information technology and telecommunications and the globalisation of trade are just a few of the irreversible trends which are changing the structure of work and the shape of organisations. Information technology is altering the nature of clerical work, and office jobs are changing fast.

People's expectations and values are changing. They demand more from their jobs and are less deferential to the authority of their bosses. They don't expect a job to be 'for life'.

Organisations must manage change if they are to remain in tune with society's needs. Managers have to respond to changes which are beyond their control, but they must also shape the environment in which their firms operate – they must engineer change and involve their teams in the process. Innovation is a central process of change, over which managers and teams can exercise some control.

The management of change starts with strategic management and marketing. These strategic activities give the company a clear idea of where it is going and how to get there. If the impact of change is likely to be great, the company may introduce a programme of organisational development or business process re-engineering. E-commerce is providing managers with new tools to improve communication and so foster effective multidisciplinary working.

No programme of change will succeed unless accepted by the people who will be affected by it. Changing the organisation involves changing people. Employees may have to be persuaded to alter their work practices, learn new skills and change their attitudes. They must learn to live with change.

Creativity is central to the survival and success of all kinds of construction organisations. Creative skills are not the preserve of the few; these skills are widely distributed in the population. Today's managers need a sound understanding of how to nurture inventive behaviour, releasing employees' full potential to contribute new ideas to the business. In particular, managers need to focus on improving the organisational ethos and the motivation and expertise of staff, as these have a major impact on innovative thinking.

Exercise

Book mark the following industry specific Web sites:

- Constructing Excellence
- Strategic Forum for Construction
- Housing Forum
- IT Construction Best Practice
- CITB Construction Skills
- DTI Construction Sector Unit

Choose three Demonstration Projects in the M4I Projects database that employed best practice in working relationships and analyse the features that were common to all.

Notes

[1] As firms adopt practices such as partnering, empowering staff and allowing them significant decision-making, employees will develop a growing interest in the mission, goals and long-term plans of the business. Perhaps corporate planning can't be delegated, but it could be co-determined.

Chapter 11
Value and Risk Management

Best value

Whilst the achievement of best value in construction has long been the aim of clients and contractors alike, today it has become a platform for radical performance improvement in the public and private sector. In the public sector, the 'Best Value' initiative launched by the UK government seeks to raise standards and provide better quality provision at reasonable cost (CBPP, 2002). The Housing Corporation describe this challenge as 'thinking the unthinkable' so that new and innovative ideas can be developed (Bennett, 2002). Hence there has been a noticeable shift in government procurement policy towards value for money and not lowest price alone (OGC, 2002). In the private sector, however, the concept of best value is not used in quite the same context, as informed clients resort to new techniques to maximise investment opportunities. But in essence both the public and private sectors are seeking the same outcomes – effectiveness, efficiency and economy.

Not surprisingly, in these post-Egan days, such ideas no longer remain the preserve of clients and their professional advisers. The boundaries between the consultant and contractor are becoming increasingly blurred, largely due to the growth in partnering and collaborative relationships. Additional demands are being placed on contractors in their dealings with clients, and the techniques used in achieving the client's objectives and in managing the supply chain are becoming ever more sophisticated. Best value in construction, therefore, is relevant to all project stakeholders.

CIRIA (1998) published a report, *Selecting Contractors by Value*, in which they identified the opportunities for contractors to add value through better:

- teamwork
- programming
- design and specification
- care of the environment
- budgeting
- management of risk and value.

Two techniques that are increasingly being used to add value and improve project success are value and risk management. For some, this split between value and risk is artificial. Rather than perpetuating a cyclic process of adding things in (because of risk) and taking them out (because of value), many practitioners argue that both concepts should be combined. But it is not entirely clear from practice how this might best be achieved and opinion remains divided as to whether value constitutes a subset of risk management or vice versa.

The discussion that follows considers risk and value management separately (drawing extensively on Ellis and Wood's research presented at the RICS COBRA 2001 and 2003 conferences), prior to a discussion of the likely benefits of bringing all project stakeholders together in a collaborative workshop.

Collaborative agreements

In the wake of the Egan Report, the Movement for Innovation and the Construction Best Practice Programme, many see alliancing and partnering as the panacea for all the construction industry's problems (Standing, 2001). Certainly, such approaches are aimed at improving productivity, profitability and value and reducing confrontation and dispute. In short, the creation of win–win situations.

Collaborative agreements are said to avoid the waste of resources that traditional fragmented and adversarial relationships engender, and encourage innovation, which leads to the enhancement of value. Value management (VM) and value engineering are techniques for enhancing value within a project by defining what will deliver value in the specific project, engineering a best value solution to meet those defined value parameters, and then delivering that solution cost effectively.

VM studies take place at strategic, tactical and technical levels. The aims and objectives of the different levels are shown in Fig. 11.1

The contribution of the contractor's expertise to such activities is helpful in the enhancement of project value and long-term collaborative agreements can provide greater opportunities for such involvement. The benefit to the client of such collaboration is real and obvious. Any financial benefit to the contractor may be less easily identified and obtained (Standing, 2001). Pain/gain share arrangements within partnering are generally fixed against agreed target costs. Standing contends that in the UK there are no formal value-based incentives.

However, the trend towards partnering is the most radical shift in procurement thinking for many years and the move away from a purely cost-driven process now has the backing of the largest construction clients in both the public and private sectors.

Value management

The CBPP (1998) define value management as a structured approach that seeks to establish what 'value' means to a client in meeting a perceived need, by clearly defining and agreeing the project objectives and establishing how they can best be achieved.

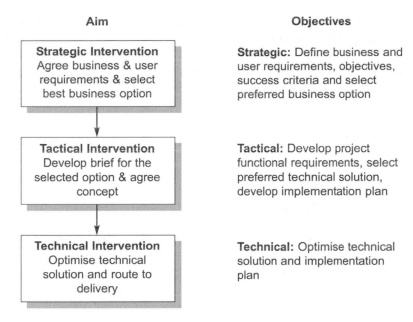

Aim

Strategic Intervention
Agree business & user
requirements & select
best business option

Tactical Intervention
Develop brief for the
selected option & agree
concept

Technical Intervention
Optimise technical
solution and route to
delivery

Objectives

Strategic: Define business and
user requirements, objectives,
success criteria and select
preferred business option

Tactical: Develop project
functional requirements, select
preferred technical solution,
develop implementation plan

Technical: Optimise technical
solution and implementation
plan

Figure 11.1 Aims and objectives of value management (Locke, 2002).

Having originated within USA manufacturing, the business technique of Value Engineering (VE) was first applied to UK construction projects during the mid-1980s. Kelly and Male (1988) concluded that VE had a place within the industry, but that its form would need to be adapted to suit UK practice – a view subsequently endorsed by the CIOB in a Technical Information Sheet (Palmer, 1990) advocating that value studies should take place in the early concept stages of the design process. Since then the technique has evolved into a much broader based approach that seeks to achieve best value and the term Value Management has become popular. Kelly and Poynter-Brown (1990) were among the first to recognise that value management was a natural progression for the QS and an opportunity to develop leading edge skills. Moreover, authors (Kelly *et al.*, 1993; Green, 1994) began to identify a range of possible value interventions and to adopt more sophisticated techniques of analysis such as SMART (Green, 1992) and FAST (Kelly and Male, 1993). Today, these services have become commonplace amongst cost consultants and contracting firms and have been found to have widespread application throughout the project lifecycle (Fig. 11.2).

Value management is applicable to all construction projects and complements Private Finance Initiative schemes, two-stage design and build, prime contracting and other partnering-style arrangements throughout the project life cycle. There is also agreement that with the growth of facilities management, the emphasis on capital spend is becoming less important than operational spend.

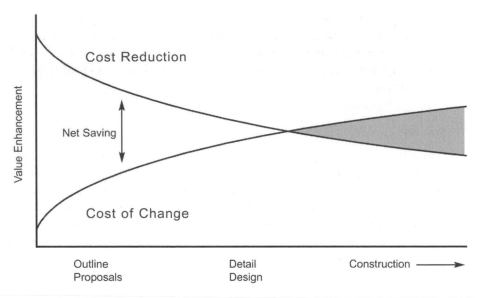

Figure 11.2 VM benefits through the project life cycle (adapted from Kelly and Male, 1993).

Value management and enhanced team-working

The shift in focus away from lowest cost to value enhancement in recent years is attributed to people looking for better ways of working, bringing about a growth in non-traditional procurement techniques. Value management has a key role to play in these new approaches. The CBPP state that value management can:

● integrate a variety of proposals from stakeholders with client's own objectives;
● identify revenue-earning potential and features that are likely to attract funding;
● accelerate the development of a project performance brief;
● achieve consensus from disparate groups of stakeholders; and
● inspire confidence that the chosen project team was focused in the requirements of the client, end-users and other stakeholders.

Moreover, the technique has applicability throughout the supply chain. Value Stream Mapping seeks to maximise cost reduction, quality and delivery, and minimise lead times and waste (CBPP, 1998). However, it is widely acknowledged that a common spin-off arising from all VM interventions is enhanced team-working and the development of closer working relationships.

Value management interventions

Much has been written about the processes and techniques involved in Value Management (ICE, 1996; Defence Estates Organisation, 1998; BRE, 2000), but whilst the length, approach and detail may vary according to the nature of the project, the time available and the timing and purpose of the specific intervention, some form of workshop is universally the norm. To gain maximum benefit from the workshop activity, pre-planning, appropriate facilitation and adequate follow-up are essential.

Pre-workshop

Two aspects of pre-planning are vital, i.e. the value manager briefing and the briefing of the workshop participants. The former is necessary to ensure that the value manager understands the client's key functionality aspects. The latter, so that the participants, who are influenced by their own professional traditions and backgrounds and who are often 'solution-driven', understand both the need for a broader, functional, analytical approach to value management and the essential multi-disciplinary nature of the activity.

Workshop

As with all group activities, clear guidance and easily understood objectives are essential. Thus construction managers, charged with facilitating value management sessions, must follow a clear agenda. The Job Plan approach, comprising information/analysis, creativity, evaluation and development is commonly used, although there is no necessity to adhere rigidly to these basic methodological stages. During the first stage, some form of functional analysis and prioritisation is required, although Ellis *et al.* (2003c) found that the approach adopted by facilitators varies according to the experience, training, understanding and openness of the workshop team, the needs of the particular project and the view of specific clients. The need to think creatively, to identify innovative alternative solutions and generally indulge in some 'out of the box' thinking is arguably the key aspect of value management. During the creativity stage, alternatives are generated that enhance value, through effective brainstorming sessions. A variety of techniques can be adopted. Amongst the most novel are those that use the de Bono concept of the 'different hats', in which participants are asked to adopt different personas and produce appropriate suggestions. They may be asked to adopt the persona of *Mr Silly* or *Mr Sensible*! Evaluation of the options produced relies on ranking, in the form of a numerical sorting scale or decision matrices. Whatever the method used, engagement is the key and the livelier the workshop, the more involved participants become and the better the result.

Post-workshop

Detailed development of the most likely options may need some form of evaluation post-workshop in terms of cost, re-design, implications on time, etc. Where further development work is undertaken, it is important that what is to be done, the time scale for action and who is responsible for the action, are clearly identified.

Without question, there is much to be gained upstream by the application of VM methodologies, but there is little evidence as yet to support the contention that contractors are fully embracing these new techniques (Ellis *et al.*, 2003c).

A case study in value management

The project comprises the demolition of office accommodation in the centre of Leeds and the design and construction of a new seven-storey office block with basement car park, a retail shell unit at ground floor level and new mains services and external works. Davis Langdon, project managers for the development, were responsible for the management of a two-stage tender process using JCT98 with Contractor Design. Following receipt of competitive bids in Stage 1, based upon preliminary costs and an overall overheads and profit margin, HBG Construction were invited to work as part of the project team and 'firm up' the contract sum. Figure 11.3 shows the activities that comprise the Stage 2 contract sum development.

During the second stage process, value management workshops were held with the whole team to identify areas of potential improvement. The workshops were aimed at improving value and various proposals categorised as priorities were designated to members of the team to review and report back. The general process is illustrated in Fig. 11.4.

The benefits arising from the VM process are:

- The alignment of design and construction processes with business needs.
- Improved project definition and an increased certainty of outcome.
- A shared understanding of value and the creation of an environment to optimise value.
- The promotion of innovation and the development of new ideas and team building.

The value management process is acknowledged as having forced parties, who would not ordinarily engage in direct communication, to co-ordinate their efforts. In part this is attributed to the two-stage tendering process, which provides greater opportunities to work with the client, fostering a teamwork approach, and making it possible to draw upon specialist knowledge and focus on 'buildability' issues earlier in the project.

In summary, value management is viewed as a proactive process aimed at maximising the value of a project by managing the development from concept to use. Not only are clients becoming more knowledgeable and conversant with value and risk management but project stakeholders are also recognising the benefits of maintaining client involvement throughout the process.

It would be erroneous to infer that the decision to provide value management

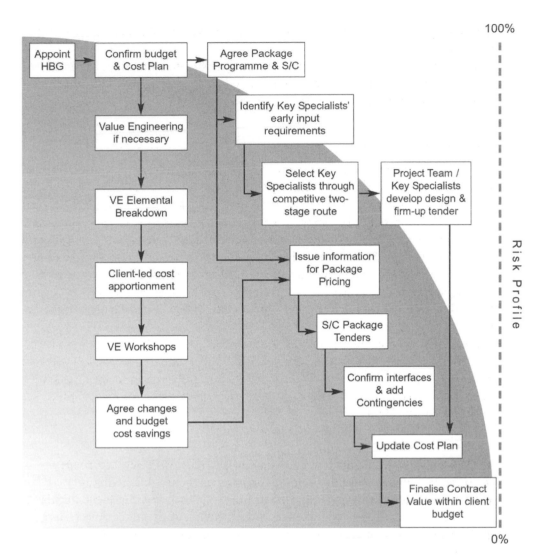

100%

Appoint HBG → Confirm budget & Cost Plan → Agree Package Programme & S/C

Identify Key Specialists' early input requirements

Value Engineering if necessary

Select Key Specialists through competitive two-stage route

Project Team / Key Specialists develop design & firm-up tender

VE Elemental Breakdown

Issue information for Package Pricing

Client-led cost apportionment

S/C Package Tenders

VE Workshops

Confirm interfaces & add Contingencies

Agree changes and budget cost savings

Update Cost Plan

Finalise Contract Value within client budget

Risk Profile

0%

Figure 11.3 Contract sum development.

services resides solely with the contractor. On the contrary, it appears that the client continues to dictate whether or not value management exercises are implemented. Indeed it is frequently the individual within any organisation and their experience of the service that appears to be a determining factor in the uptake of VM.

Risk management

Risk management is normally associated with health and safety. Largely due to CDM regulations, there is an imperative upon all parties to the construction process

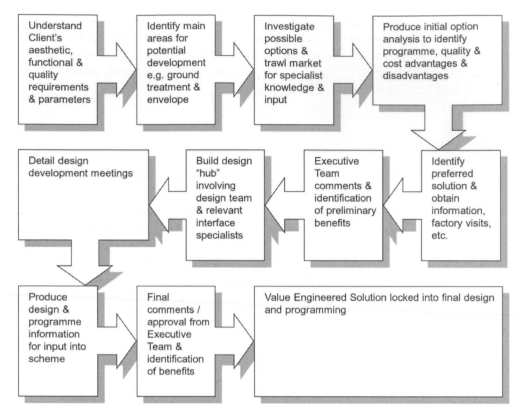

Figure 11.4 Value management process.

to assess risk and develop risk registers. However, there is increasing interest in a more generic interpretation of the risk management process. Since the mid-1980s many authors have suggested that the management of construction projects, large or small, benefits from a greater understanding brought about by the application of risk management techniques. Heightened awareness of risk is evidenced by a broad range of guidance documents for practitioners undertaking risk management procedures, e.g. CIRIA (Godfrey, 1996), HM Treasury (1997), Association for Project Management (1997), Institute of Civil Engineers *et al.* (1998), BSI (2000) and RICS (2000). Perry and Hayes (1985) were among the first authors to recognise the benefits that risk management might bring to the construction industry. They concluded that risk and uncertainty were not the sole preserve of large capital projects, but that factors such as complexity, speed of construction and location also contributed to the inherent risk within a project. They identified a three-stage process that comprised:

● Identification;
● Analysis; and
● Response.

A recent risk management study (Wood and Ellis, 2003) revealed that a broad range of clients, e.g. developers, pharmaceutical and oil companies, banks, insurance brokers, government organisations, transport and utilities, were demanding risk management studies. The latter, in particular water, gas and power, adopted rigorous procedures as a matter of routine.

Risk identification

Commonly, risks are identified in workshops or by interviewing key project stakeholders. Those that surface may, for example, include buildability, health and safety, or logistics. On occasion, at this stage in the process, Probability Impact Analysis charts are used to gauge the expected impact of these risks and the probability of them occurring. However, almost certainly, risk registers are developed that identify the members of the project team who are to be held responsible for mitigating the effects of the risk should they occur. It is widely recognised as an important stage in the risk management process, as it leads to a greater understanding of the project.

Risk analysis

Flanagan and Norman (1993) provide a useful classification of risk analysis techniques, namely, decision trees, sensitivity analysis and probabilistic analysis. The former, based on a series of either/or decisions is seldom used in practice, unlike sensitivity analysis which often finds application at feasibility stage. However, it is the latter which is most widespread. With the aid of risk management software, Monte Carlo simulation considers the likely impact of risks in combination. Programs such as @Risk, Primavera and Crystal Ball are all capable of generating probability distributions, and illustrating the likely effect of risk variables on the economic return of the project, typically involving between 1000 and 5000 iterations. Rarely will these calculations be attempted during the workshop, but the findings are nearly always presented in a report for project team members.

Risk response

Crucially, it is at this stage that action is taken to mitigate risk. Once again, Flanagan and Norman present a model of the process, suggesting that responses may take the form of risk avoidance, transfer, reduction or retention. Yet the distinction between risk reduction and transfer is by no means clear. Rather than classify responses under a series of discrete categories it is helpful to plot the various responses to risk on a continuum (Fig. 11.5) to illustrate the inter-relationships that exist.

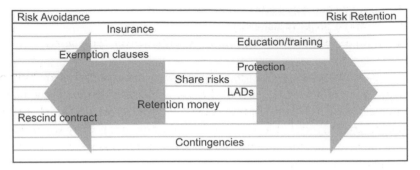

Figure 11.5 Risk response continuum.

Collaborative workshops

Phillips (2002) recognises the considerable potential of a value methodology in promoting a group thinking process. As discussed earlier, both value and risk management approaches rely principally on workshops, which draw together key project stakeholders and encourage participants to reflect on issues that impact upon organisational and project success. These workshops provide what Phillips describes as a very powerful, 'fast tracking' consensus development tool, that aids the transfer of business culture and practice and has the ability to overcome adversarial relationships, resulting in a team approach that leads to joint ownership and a commitment to the end product.

Similar benefits are acknowledged by Bovis Lend Lease during the management of a major capital investment programme for BP (Locke, 2002). Not only were significant value enhancements secured, but they were also able to focus the team on the overall project Vision and Charter through the involvement of all project stakeholders, rather than the contract. The following benefits are cited:

● Long-term relationships work to the advantage of both parties and their associates.
● It provides a catalyst to align objectives, change culture, improve value through innovation.
● Mutual benefit is based on respect and co-operation.
● It builds and maintains the relationship.
● It facilitates bold decision making and change management.

Summary

There appears to be widespread agreement that value management, risk management, supply chain management and partnering all seek to achieve the same overall goals. That is:

- to make projects better value for money;
- to make projects less risky; and
- to make projects that can more reliably deliver client goals.

Yet there is also a view that the vast majority of projects that are built only meet the true functional requirements of the client by pure chance. Value management has greatest impact at the inception stage of a project, but it is widely regarded as being under-utilised – risk management is more dominant. Only when these services are viewed as being central to the project delivery process are construction clients likely to achieve best value.

There is little doubt that value and risk management will increasingly contribute to project success, but much will depend upon achieving real cultural, organisational and attitudinal change and openly and maturely dealing with the realities of buyer/seller relationships within a market economy. Put another way, it is extremely important for all organisations within the supply chain to be able to move towards collaborative approaches, which more effectively balance business relationships and commercial issues.

Student activity

An Employer (E) wishes to construct a state-of-the-art leisure complex. A site has been found, a scheme design has been prepared and funding to a set level based upon contemporaneous 'pre-lets' has been secured.

E needs to move quickly in order to exploit the particular leisure 'niche' before competitors act and capture a larger share of the market.

A pilot project has been constructed elsewhere in the country, upon which the scheme design has been based. It is known by Contractor C, via contacts among sub-contractors involved, that the pilot was completed successfully, although some eight weeks late and significantly over budget. It is also known by C that, although the pilot was completed some eighteen months previously, the final account had not been resolved and there were numerous contractual disputes.

C has tendered in competition with the original contractor who constructed the pilot and there is little between the bids in terms of cost. However, E is faced with the following difficulties:

- Both bids significantly exceed the current budget.
- The spatial requirements of the development appraisal will not fit within the existing scheme design and a substantial redesign is necessary.
- There is no time to redesign from scratch and re-tender, since the development appraisal demands an earliest possible start to trading.

C's was not the most competitive tender, however, and E elected initially to negotiate with the original contractor. Evidently the negotiations did not proceed satisfactorily, however, and E has approached C with a view to trying to negotiate a contract for the construction of the facility.

C's dilemma

C realises that they have an opportunity to negotiate a sizeable contract with an employer who is committed to proceed with a project which will, from day one, be subject to wholesale re-designs leading to a substantial variation account, inevitable delays and disruption to progress, which will lead to loss and expense claims.

On the other hand, C wishes to avoid the disputes, preserving his relationship with E and with his sub-contractors and freeing up his valuable technical resources soon after practical completion. People have said that, by improving relationships between the parties, this can benefit everyone involved: C and his sub-contractors will improve their margins; E will receive his building on time and at a reduced cost and the designers will be left to concentrate upon key issues and increase their productivity accordingly. These people are surely heads-in-the-clouds academics and know nothing of the real world!

Question: How is it done?

Chapter 12
Managing Innovation in Construction

There is consensus among governments, researchers and within the construction industry that a relationship exists between an organisation's efficiency or profitability and its ability to innovate (DTI, 1998; CRISP, 1997; CIC, 1993; Egbu, 1999a, 1999b, 2001a). Innovation is viewed as a major source of competitive advantage and is perceived to be a pre-requisite for organisational success and survival (Egbu, 1999a).

'Innovation is the successful exploitation of an idea, where the idea is new to the unit of adoption' (Egbu, 2001a, p.1). Innovations come from many different sources and exist in many different forms. There is a dichotomy between radical and incremental innovation. Innovation can be radical, in response to crisis or pressure from the external environment, but it can also be incremental where step-by-step changes are more common. Moreover, a common typology distinguishes product and process innovation. Product innovation describes that, where a new product is the outcome, it is seen to focus on cost reduction by obtaining a greater volume of output for a given input. Process innovation denotes innovation where the process by which a product is developed is exposed to new ideas and therefore leads to new, often more sophisticated methods of production. It describes new knowledge, which allows the production of quality superior output from a given resource. From the perspective of sources of organisational innovations, there are emergent (from within), adapted/adopted and imposed innovations.

Rothwell (1992) has developed a historical model tracing the evolution of innovation models since the 1960s. His 'five generations of innovation' model explains the transition from the simple, linear models to the complex and interactive models of innovation. In Rothwell's (1992) fifth generation model, innovation is perceived as a multi-faceted process that requires intra- and inter-firm integration, through extensive networking. Similarly, Wolfe (1994) noted that '... Innovation is often not simple or linear, but is, rather, a complex iterative process having many feedback and feed-forward cycles' (Wolfe, 1994, p.411).

In essence, innovation can be viewed as a process of inter-linking sequences from idea generation to idea exploitation that are not bound by definitional margins and are subject to change. Therefore, it is necessary to understand the complex mechanisms of this process and the context in which the innovation takes place.

Challenges associated with managing innovations in organisations

The dynamics of innovation, which have become increasingly intensive, result in high levels of risk and uncertainty arising, for example, from difficulties associated with accessing, transferring and assimilating knowledge that is external to the organisation. These externalities include the heterogeneity of the knowledge sources which are important to innovation; technological complementarities (including those between product and process innovations); cumulativeness, path dependency and incrementalism; compatibility between innovations; user-producer relationships; inappropriability; and bounded rationality.

The following are some of the main challenges that were observed in a study of four innovative UK construction organisations, Egbu *et al.* (1998).

- The inability to link innovation strategy to the wider organisational business strategy.
- Managing the uncertainty and risks associated with innovation (e.g. risks associated with design and buildability of construction projects, technological risks, financial risks, contractual risks and increased exposure to litigious claims, safety risks, risk of complete failure of the innovation).
- Difficulties associated with being able to scan and search the environments adequately to pick up and process appropriate signals about potential innovation.
- Lack of resources and competencies associated with making strategic selections from potential alternatives and triggers for innovations and implementing the chosen innovation.
- Getting members of the organisation to 'buy-into' and support the innovation idea (commitment of the rank and file).
- Difficulty in getting the market to take up the innovation (opportunism and the readiness of the market for the innovation).
- Difficulty in getting the Building Regulators to accept the innovation (e.g. design and the eventual product).
- Difficulty associated with auditing and measuring the benefits associated with innovation.
- Difficulty in understanding and putting in place an appropriate culture for innovation.
- Difficulty in maintaining innovative advantage over the competition once innovation has been implemented.

The innovation strategies of organisations are, however, strongly constrained by their current position and core competencies as well as the specific opportunities open to them in future. In other words, organisational strategies for innovation are 'path-dependent', as is discussed in the next section.

Knowledge management and innovations: building and maintaining capabilities

The ability to innovate depends largely on the way in which an organisation uses and exploits the resources available to it. A vital organisational resource, at the heart of innovation, is knowledge (Nonaka and Takeuchi, 1995). Knowledge is fast overtaking capital and labour as the key economic resource in advanced economies (Edvinsson, 2000). There is a growing acceptance, in competitive business environments and project-based industries, that knowledge is a vital organisational and project resource that gives market leverage and contributes to organisational innovations and project success (Nonaka and Takeuchi, 1995; Egbu, 1999a, 2000).

Knowledge management (KM) is emerging as a vital activity for organisations to preserve valuable knowledge and exploit the creativity of individuals that generates innovation. Knowledge management is important for a number of reasons. Knowledge management is 'the process by which knowledge is created, captured, stored, shared and transferred, implemented, exploited and measured to meet the needs of an organisation' (Egbu *et al.*, 2001). It is important because the rise of time-based competition as a marketing weapon requires organisations to learn quickly. It is important because of the globalisation of operations and because of the growth in the number of mergers and take-overs where multiple organisations must share knowledge in a collaborative forum. In project-based industries the situation is even more complex. Project-based organisations are characterised by short-term working contracts and diverse working patterns. Knowledge management is important in this context because it brings together diverse knowledge sources from different sections of the demand and supply chains achieving cross-functional integration. Understanding how organisations manage knowledge assets for improved innovations is important. There is, however, a paucity of empirical research on KM and its impact on innovation, especially in project-based industries, such as construction (Egbu *et al.*, 1998; Winch, 1998; Gann, 2000).

Knowledge management is highly associated with innovation because of its ability to convert tacit knowledge of people into explicit knowledge (Nonaka and Takeuchi, 1995; von Krogh *et al.*, 2000). This is grounded in the notion that unique tacit knowledge of individuals is of immense value to the organisation as a whole, and is the 'wellspring of innovation' (Stewart, 1997). Given the close connection between knowledge possessed by personnel of the firm and the products and services obtainable from the firm (Penrose, 1959), it is generally accepted that a firm's ability to produce new products and other aspects of performance are inextricably linked to how it organises its human resources. Grant (1996) and Hall (1993) have argued that it is tacit rather than explicit knowledge that will typically be of more value to innovation processes. Yet, tacit knowledge is knowledge that cannot be easily communicated, understood or used without the 'knowing subject'. The implication of the above discourse is that knowledge management that focuses on creating network structures to transfer only explicit knowledge will be severely limited in terms of its contribution to innovation and organisational and project success.

Construction organisations need to determine their positions in terms of processes, services, products, technologies and markets. Since an organisation's innovation strategies are constrained by their current position, and by specific opportunities open to them in the future based on their competencies, construction organisations will need to determine their technological trajectories or paths. This will involve due cognisance of strategic alternatives available, their attractiveness and opportunities and threats, which lie ahead. The organisational processes that an organisation adopts in integrating the transfer of knowledge and information across functional and divisional boundaries (strategic learning) is essential and needs to be consciously managed. As competitive advantage and financial success are bound up with industry dynamics, it is necessary to place strategic change in a competitive context and identify what kinds of changes lead to strategic innovation, and when these changes result in benefits for the organisation.

Core capabilities and competencies are difficult to imitate and provide competitive advantage for organisations. Leonard-Barton (1995) suggests that core capabilities are built through a knowledge building process that is clustered around four learning styles – present problem solving, future experimenting and prototyping, internal implementing and integrating, and external importing of knowledge. Grant (1995) sees resources and capabilities as keys to strategic advantage and notes that organisations must build and maintain capabilities if they are to innovate. Similarly, for Teece and Pisano (1994), an important capability is the expertise to manage internal and external organisational complementary resources. Through collaboration and by forming long-term relationships, construction organisations are able to learn from projects and transfer knowledge to an organisational base and along supply chains.

Innovation might be thought of as a process of combining existing knowledge in new ways, often termed 'resource combination' (Galunic and Rodan, 1998). Resource combination depends on a cognitive process of 'generativity', which is the ability to form multipart representations from elemental canonical parts (Donald, 1991). This cognitive integration or 'blending' is at the heart of the creation of novelty. Since generativity is in essence a combinatorial process, the more knowledge that we collectively accumulate, the more opportunities there are for the creation of innovative ideas (Moran and Ghosal, 1999; Weitzman, 1996).

Organisational innovations and strategies: critical success factors

A variety of factors have been identified as influencing the rate of innovations in organisations. This can be seen through different schools of thoughts and perspectives. The individualist perspective, which is grounded in social psychology, is predicated on the assumption that the individual is the source of innovation. They are the 'champion[s]' (Madique, 1980) or 'change agents' (Rogers, 1983) in an organisation. In contrast, the structuralist perspective hinges on the idea that the structure and function of an organisation is the fundamental dynamic of innovation. There is, however, a highly charged debate about what components of an organisation have a bearing on innovation and how they are determined. For example, the link between organisational size and innovative capacity is fiercely contested. Some suggest that larger organisations are more innovative, while others stress that a company's size does not matter (Rothwell and Dodgson, 1994). It is also assumed that the organisational characteristics such as structure, strategy and longevity play a central part in organisational innovations. The structural variables of centralisation, formalisation, complexity and stratification have been shown to have contrasting effects at the initiation and implementation stages of the innovation process (the so-called 'innovation dilemma'). Low levels of centralisation and formalisation, and a high level of complexity facilitate the initiation stage of the innovation process. The implementation stage is facilitated by high centralisation and formalisation and low complexity. The consensus view is that a high level of stratification inhibits innovation, because it leads to over preoccupation with status and insufficient freedom for creative thinking. The consensual view points out the deleterious effect on creativity of the 'elevator mentality' of organisations dominated by rigid vertical relationships and 'top down dictate'. Similarly, 'an organic, matrix and decentralised structure could provide the creative individual with freedom sufficient to be creative'.

It is therefore important to take a more multivariate approach to understanding organisational innovation. The integration of both the individual and organisational levels of analysis to achieve a synthesis between action and structure should be encouraged. Attempts to incorporate these diametrically opposed concepts have influenced developments in process theory. The process perspectives on innovation need to recognise the unpredictable and dynamic nature of innovation. It is therefore a complex process with cognitive, social and political dimensions that should be understood in particular organisational contexts.

Van de Ven *et al.* (1989) have argued that there is the need to create an organisational culture by definition of aims, embodying purpose through structure and systems, defending integrity of organisation and ordering internal conflict. For Tatum (1987), a climate favourable to innovation must be achieved by committing resources, allowing autonomy, tolerating failure and providing opportunities for promotion and other incentives. It therefore follows that an organisation must be flexible enough to facilitate the innovation process.

Leadership is an organisational responsibility. The value of institutional leadership is useful in the creation of structures, strategies and systems that facilitate innovation and organisational learning. It should build commitment and excitement, collective energy and empowerment (Van de Ven *et al.*, 1989). Innovation responds to market demands and technological progress. Therefore, the climate of the market has a significant influence on the innovation process. A turbulent environment, where the organisation is in crisis, is likely to induce the adoption and implementation of radical innovations. From the discussion so far, it has been stressed that innovation is a complex, context-sensitive social process. No one best strategy exists or is suitable for managing innovations in every organisation. However, any meaningful innovation strategy should have unequivocal support from the top. Its objectives need to be communicated and be accepted by the rank and file within the organisation. An innovation strategy needs to sit naturally within the overall strategy of the organisation. In addition, it is important that it is monitored and reviewed as appropriate.

An organisation's competitive advantage can come from various sources such as its size or assets. In a study of four innovative UK construction organisations, however, Egbu et al. (1998) observed that construction organisations are able to gain competitive advantage by innovating through their dynamic capabilities – by mobilising knowledge, experience and technological skills. In the main, these have been achieved through one or a combination of the following:

- focusing on a particular market niche;
- novelty – offering something which no other organisation can;
- complexity – there are difficulties associated with learning about their processes and technologies, which keep entry barriers high;
- stretching the basic model of a product/process over an extended life and hence reducing overall cost;
- continuous movement of the cost and performance frontiers;
- integrating the person and the team around the product and service.

Although the building of dynamic capabilities or core competencies is vital for organisational innovations, it is, however, important that core competencies do not turn into 'core rigidities', especially when established competencies become too dominant and important new competencies are neglected or under-estimated.

The issues of culture and climate are vital when considering conducive organisational environments for innovation as indicated earlier. In studying the four innovative construction organisations, Egbu et al. (1998) observed that certain characteristics associated with culture and climate were favourable to innovation. These are:

- The support from top management and the presence of a strong 'innovative champion'.
- 'Flexibility in the lines of communications', allowing top-down, bottom-up and lateral communications within organisations.
- A risk tolerant climate, where it is accepted that lessons could be learned through mistakes.
- A climate where people genuinely feel valued and people feel some form of 'ownership' or involved with the innovation.
- A sharing culture where there is openness and willingness to share information, experience and knowledge across project teams and the organisation.
- A climate where people feel secure in their jobs.

Measuring innovation success

Organisations innovate for many reasons. There are also different drivers that fuel innovation. Organisations might innovate to increase profit share, to enter a new market, to be a leader or first follower in the market, for reasons of status, etc. Organisational strategies for innovation differ from one organisation to another. Similarly, the approaches which organisations put forward for measuring their innovation success, as well as the time frame for judging innovation success, differ greatly. What is perceived to be a highly successful innovation for one organisation may not be seen to be so by another. There are organisations that choose to exnovate after a few years after the release of their innovative products or solutions. There are some that might measure the success of their innovation after a few years. It is therefore important to understand the *modus operandi* of an organisation involved in innovation before the judgement is made as to whether the organisation is successful at innovation or not.

There is still an on-going debate about whether many construction organisations are innovative or not, and whether construction is less innovative than other industrial sectors. Organisations and industrial sectors are impacted on by different constraints and they handle these differently. Understanding the innovation trajectory that an organisation embarks upon gives a better understanding as to whether the organisation has been successful at innovating or not.

The innovation strategies of organisations are strongly constrained by their current position and core competencies, as well as the specific opportunities open to them in the future. In other words, organisational strategies for innovation are 'path dependent'.

The extent to which an organisation is successful in innovation can be measured through different variables. These include:

- The percentage of profit/sales derived from the innovative product/solution.
- The number of new products/solutions introduced over the last 1 to 5 years.
- The number of new/innovative ideas generated within a given period during the course of innovation.
- The average number of man–hour input per new product/solution.
- The average time to market of the innovative product/solution.
- The level of satisfaction of the client/customer of the innovative product/solution.
- The average failure rate of the innovation (during developmental stage, for testability and robustness of the product/solution).
- The extent to which innovation planning is linked to overall organisational strategy.
- The extent to which there are formal mechanisms to capture and share learning associated with the innovation.
- The extent to which the workforce is involved in innovation, supported, recognised and adequately rewarded.

Construction organisations are, in the main, project-based organisations. Networking, Communities of Practice (CoP), story telling, coaching, mentoring and quality circles are important mechanisms for sharing and transferring tacit knowledge in project environments. These should be considered, encouraged and promoted more by construction personnel. Communities of practice are needed to encourage individuals to think of themselves as members of 'professional families' with a strong sense of reciprocity. The human networking processes, which can encourage sharing and the use of knowledge for project innovations, are important. Leaders of construction organisations and projects should also espouse 'the law of increasing returns of knowledge' as a positive way of encouraging knowledge sharing. Shared knowledge stays with the giver while enriching the receiver.

Intuitive knowledge is managed by valuing individuals and not by being heavy-handed through project 'controlled processes'. It is folly to believe that any project organisation can make people have ideas and force them to reveal intuitive messages or share their knowledge in any sustained manner. An individual's intuitive knowledge cannot be manipulated in any meaningful way or controlled without the individual being willing and privy to the process. The process of trying to manipulate or control intuitive knowledge in fact creates their destruction. The issues of trust, respect and reciprocity are vital elements of a conducive environment for managing tacit knowledge. It is through these that individual members of the project can be motivated to share their experiences and exploit their creativity. Leaders of construction projects and organisations would need to recognise, provide incentives and reward knowledge performance and sharing behaviour patterns. Leaders should also take action on poor knowledge performance. The regular communication of the benefits of knowledge management is important in sustaining the co-operation of project team members. A variety of ways exist for doing this, including regular meetings, project summaries, project memos and through project GroupWare/Intranet facilities, where they exist. Every project strategy for KM should consider the training, recruitment and selection of project team members (e.g. sub-contractors and suppliers). It should also pay due cognisance to the team members' competencies, requisite knowledge and their willingness and effectiveness in sharing knowledge for the benefit of the organisations and projects. In addition, the 'absorptive capacity' of the parties involved in the knowledge sharing processes is vital.

Knowledge management and improved innovations: issues of strategy, process and structure

Knowledge management impacts upon organisational and project innovations in many complex ways through a host of inter-related factors. An understanding of these factors and their contribution to innovations is important for the competitive advantage of project-based organisations.

A good internal organisational structure, expressed through the strategies, processes and culture of an organisation, is one that is flexible but supportive of the ideas propounded by employees. The organisational structure should respond just as effectively to external pressures. For example, Drucker (1995) claims that hierarchical structures become deficient in turbulent environments. In contrast, structures determined by core competencies can adapt to chaotic external pressures more easily. Such competencies should be flexible to meet new customer demands or exceed expectations (Prahalad and Hamel, 1990). Quinn (1985) asserts that excessive bureaucracy can stifle innovation because of, for example, the amount of time it takes to approve every idea. While in small organisations this may require minimal bureaucracy, in larger, more complex organisations the process is always cumbersome. Organisational structures need to sustain equilibrium between creativity and formal systems. Bureaucracy can inhibit spontaneity and experimentation and thus threaten the innovation process. However, bureaucratic structures may also assist the 'rapid and continuous transformation of ideas into superior products' (as cited in Bennett and Gabriel, 1999, p.217).

While the centralisation of an organisation's structure for decision making can create a definite medium of control, a more informal and flexible structure is desirable for knowledge generation (Bennett and Gabriel, 1999). Woodman *et al.* (1993) assert that flexible structures encourage better internal communications and a more change-friendly climate where ideas and knowledge are shared freely.

As aforementioned, the tacit knowledge of the individual is an essential component of organisational success. However, such knowledge is often guarded by those who are reluctant to transfer this 'power' from an individual level to the organisational level (Cole-Gomolski, 1997). Therefore, the employee must be sufficiently motivated to share their knowledge, through incentives. Byrne (2001) argues that the organisational structure should play a part in the encouragement of knowledge sharing. He contends that motivation is a key facilitator of loyalty and trust amongst employees and eventually fosters continuous learning.

Every manager has a vision of the organisation that they work for. The importance of expressing this vision to the rest of the organisation is paramount. Sullivan (1999) argues the need for a long-term vision to be incorporated into the corporate strategy of the company. This is only achievable if the context of the organisation is fully understood. Sullivan (1999) identifies three key areas that should be understood. First, what are the real features of the business, i.e. the core competencies? Secondly, what is the external context, such as the socio-political and economic forces of change and their particular impact on the company? Finally, what is the internal context, e.g. the strategy, culture, performance, strengths and weaknesses of the company? In sum, Sullivan asserts the need for effective management of a company's capabilities, e.g. management of portfolio (intellectual property and intellectual assets), competitive assessment and human capital management. This has the potential for improving organisational innovations leading to competitive advantage and market leverage.

Project-based industries, including the construction industry, are undergoing pressure to compete in new ways. Strategic planning and the need for growth are seen to require organisations to develop firm-specific patterns of behaviour, i.e. difficult to imitate combinations of organisational, functional and technological skills (Teece *et al.*, 1997). These unique combinations create competencies and capabilities and occur as the organisation's intangible knowledge is being applied in its business behaviour, especially in the value-adding business process. Competitive advantage stems from the firm-specific configuration of its intangible knowledge. The development of core competencies or capabilities creates an environment of strategic thinking, in which knowledge and ideas are key.

Managing knowledge and organisational learning for innovations

Much of the literature on innovation focuses on the need to establish the right kind of organisational culture. Patel *et al.* (2000) stress that it would be a mistake to underestimate the importance of cultural factors in the adoption of KM and organisational learning. More specifically, Tatum (1987) emphasises that a climate favourable to innovation must be achieved by committing resources, allowing autonomy, tolerating failure and providing opportunities for promotion and other incentives. Thus, an organisation must be flexible enough to facilitate the innovation process (Zaltman *et al.*, 1973).

As innovations, especially radical or 'rule breaking' innovations, are associated with challenging thinking, unlearning as well as learning, entrepreneurial organisations appear to need general learning capacity. The ability to learn from others, from the organisation around oneself and from one's own past, is a critical element in making progress.

In order to establish a knowledge-based organisation there needs to be a supportive organisational culture. The cultivation of a 'learning organisation' is an essential requirement for knowledge managers. If an organisation develops a learning culture there is scope for both formal and informal channels of 'dialectic thinking', where individuals develop their individual capabilities through positive experimentation (Bhatt, 2000). Further theories about organisational culture favour the evolution of a 'community of practice' where social interaction of employees cultivates a knowledge sharing culture based on shared interests, thus encouraging idea generation and innovation (Adams and Freeman, 2000).

In all organisations, the politics of knowledge sharing is an issue. Employees and employers from diverse backgrounds often come into conflict over important decisions. It has been suggested that manipulating these tensions to achieve 'creative abrasion' is a strategy to maximise innovation (Leonard and Strauss, 1997). However, it is a challenging task that involves disciplined management. Leadership is an inherent part of organisational culture, but it also extends into areas of strategy and structure. According to Van de Ven *et al.* (1989) leadership is an organisational responsibility. They emphasise the value of institutional leadership to create the structures, strategies and systems that facilitate innovation and organisational learning. Organisations should build commitment and excitement, collective energy and empowerment. Sullivan (1999) argues the need for a managerial commitment to the long-term strategic vision of an organisation and the motivation to achieve the goals set out. Moreover, empowering employees to generate and share knowledge is the task of management. For example, implementation of rewards and punishment schemes are stimulii for successful KM (Scarborough et al., 1999). Motivating employees to share the knowledge they have involves good people management, where obtaining trust is itself an incentive. The establishment of a psychological contract between employer and employee, for example, is a constructive approach to developing a knowledge-sharing culture (Scarborough et al., 1999).

If knowledge management is to have any real impact on the way construction organisations do business, then it has got to be about making radical changes in the way organisations use knowledge. Knowledge has to be 'made productive'. Managers have critical roles to play in making knowledge productive, in knowledge development and in the exploitation of knowledge for innovative performances. Deepening the understanding and analysis of managers' interest on knowledge is vital to understanding how knowledge management can contribute to improved strategic formulation. For organisations, it is important to make KM an integral part of strategic decisions on profitability and competitiveness of the organisation. In this regard, establish at all levels of the organisation a strategic intent of knowledge acquisition, creation, accumulation, protection and an exploitation of knowledge. The linkages between strategic management and human value need to be carefully examined, and so does the role of a KM orientation in adequately supporting successful strategies. Organisations also need to determine appropriate mechanisms for the effective capture, transfer and leveraging of knowledge. Communication infrastructure needs to be established within and between the different departments and strategic business units (SBU). This should support and enhance the transfer of ideas but at the same time not limit the potential for creativity and the questioning of actual activities that are needed to understand the challenges in the wider environment, and may be a source of new solutions to problems.

Knowledge management is about mobilising the intangible assets of an organisation, which are of greater significance in the context of organisational change than its tangible assets, such as information communication technologies (ICT).

While information technology is an important tool for a successful organisation, it is often too heavily relied upon as a guarantee of successful business. Edvinsson (2000) contends that such tools as the Internet are merely 'enabler(s)' and that the true asset of an organisation is the brainpower of its workforce. He stresses that it is the intellectual capital of an organisation that is the key to success (as cited in Dearlove, 2000, p.6). Thus, KM is not just about databases or information repositories. 'In computer systems the weakest link has always been between the machine and humans because this bridge spans a space that begins with the physical and ends with the cognitive' (McCampbell et al., 1999, p.174). Notwithstanding this, the important role of information communication technologies in knowledge (especially for explicit knowledge) capture and retrieval, and their implication for innovations in the construction industry has been well documented (Egbu and Botterill, 2002; Egbu, 2000).

E-business initiatives and the construction industry

The introduction of the Internet and Internet-related technologies has affected the rules of conducting business (Cannon, 1996; Kalakota, 2001). These technologies now introduce electronic business or e-business opportunities to organisations. Some construction organisations are in the throes of embracing the Internet as a core element of their business strategy.

E-business allows organisations to transfer their business processes on-line to collaborate in real time, internally and externally, with their clients. Our early findings have indicated that the application of e-business initiatives in construction organisations is limited to the use and exchange of information along the construction supply chain (Ribeiro and Henriques, 2001). Moreover the construction industry is slow in taking up e-business initiatives compared with other industries, such as the manufacturing, automotive and aerospace industries.

E-business involves any 'net' business activity involving telecommunication networks that transform internal and external relationships to create value and exploit market opportunities, driven by new rules on the connected economy (DTI, 2000, 2002b; Turban, 2000). The implementation of e-business essentially requires appropriate knowledge assets to be readily available with appropriately trained personnel, hardware and software in place. E-commerce refers to conducting electronic transactions, which is the buying and selling of goods and services on-line (Turban, 2000). This allows organisations to purchase products and services on-line from their supply chain partners. There are three main types of e-commerce – business to consumer (B2C), business to business (B2B) and consumer to consumer (C2C). E-business refers to a broader definition than e-commerce.

Developments in the Internet and e-business are constantly changing and some construction organisations are now realising the benefits associated with e-business as a core element of their business strategy. E-business has the potential to cause significant change in economic activities within the construction sector, by impacting substantial improvements in products and processes. It can also help by reducing transaction costs and extending market reach by allowing round-the-clock trading. It also helps with economy and the speed of construction business and enriches interdisciplinary relationships and organisational culture (Bogdanov, 2001; UKonline, 2002).

With the advent of Internet technology, there is now more room for organisations to compete locally and globally. The widespread electronic linking of individuals and organisations has created a new economic environment, in which space, time and size are less limiting factors (Barnes and Hunt, 2001; Kalakota, 2001). E-business initiatives provide the channel for conducting business as part of an individual's or group of individuals' day-to-day work in decision-making (Fahey, *et al*, 2001; Malhotra, 2002). It could help achieve three important goals, which are: superior external performance, including marketplace and financial returns, superior internal operating performance through operating efficiencies, and an enhanced quality of life for individual members of the organisation.

Some of the processes which KM and e-business can facilitate in the construction industry include the timely exchange of appropriate and accurate information/ knowledge assets to the right person(s). This includes tenders, enquiries, quotes, dispatch notes, invoices, credit notes, valuations and site instructions.

For e-business initiatives to be effectively applied, integration must be implemented throughout the organisation and with external business partners. It enables effective partnering, leading to reduced costs and increased benefits. Some specific tangible benefits that can be achieved by e-business in this way are:

● Tender distribution costs can be cut significantly.
● Tender entry time can be dramatically reduced.
● Invoice registration costs can be reduced.
● Re-keying errors, delays and disputes can be avoided.
● Construction organisations can procure materials, plant and any other resources cheaper and better deals can be sought that may be necessary to perform their business outside of their traditional supply chain. Tender documents can be placed on the Internet and contractors and sub-contractors can price them and return them to the issuing authority. This can be achieved without the use of paper and without delays in delivery.

Challenges facing organisations in using the Internet for business activities

E-business initiatives offer the platform for new forms of market place strategy models for construction organisations. However, the main challenges facing construction organisations in using the Internet to commercialise their business assets are associated with cultural, social, organisational, legal and technical issues. These include the following:

● The change of organisational culture to embrace e-business as a core element of business strategy.
● A constantly/rapidly changing organisational culture of business partners (clients, customers) to cope with changing market conditions.
● The inability to cope with the fast changing tools that are offered on the market. This includes the cost of purchasing and the cost of training to use the tools.
● The management task of aligning business strategies, processes, and applications quickly, correctly, and all at once.
● Agility is the key to survival in the new economy but construction organisations are slow to respond to changes.

- The lack of strong and decisive business leadership with the lack of foresight of future business trends.
- The lack of technical expertise available within organisations, especially small and medium sized enterprises.
- The cost of acquiring Internet-based software and hardware tools to facilitate their business processes.
- The uncertainty and risks of changing from the 'old' business model to the 'new' business model. The uncertainty of doing business on-line, due to the recent failure of dot.com companies. This has made the construction industry, already a high risk industry and very conservative, unwilling to take further risks.
- The legal aspect associated with contractual liabilities of on-line procurement of goods and services.
- The security of on-line procurement.
- The ability to spot trends quickly and create effective business strategies.
- Technology shifting power to the buyers (consumers) as e-business changes the channels through which consumers and businesses have traditionally bought and sold goods.

In order to overcome these challenges, first, there must be a clear vision for change from the top management. Secondly, the organisation must define its business processes and develop a 'business model' that suits the organisational needs for implementing the new method of doing business using the Internet. Thirdly, there must be appropriate awareness and training provided for the employees of the organisation. Should the organisation lack adequately trained personnel to implement the e-business strategy, the organisation can employ skilled personnel as part of the organisation, or engage consultants to provide this service. Alternatively, the organisation may source technological expertise and tools by forming an alliance with members of the supply chain, who may have these systems operating in their organisation. In this way, the organisation can benefit from the successes of the integrated system. This may help improve the rate of organisational learning, thus helping the organisation to make the transition within a relatively short time.

Without an effective strategy in place, it will be difficult for any organisation to implement e-business strategies effectively in a satisfactory and sustainable manner. Effective knowledge management practices also play a vital role in the implementation of e-business initiatives. Today, an organisation's success depends largely on its ability to innovate and integrate new technologies into service offerings. It will be practically impossible to be innovative without proper KM implementation strategies. KM will help executives become proficient 'trend-spotters' when implementing e-business initiatives. A new generation of e-business organisations must be imaginative in order to change the value proposition radically, within and across their industries (Kalakota, 2001, p.65). Furthermore, any initiatives addressing e-business without a proper KM strategy will soon expose the inability to cope with market demand. It is important to understand what e-business has to offer, and how

KM plays an important role in identifying appropriate knowledge assets. This helps to identify suitable e-business tools and strategy for trading over the Internet.

Summary

Organisational innovations play a vital role in organisational success. Innovation depends on strategic priorities of construction organisations. Knowledge management is about the process by which knowledge is created, captured, stored, shared and transferred, implemented, exploited and measured to meet the needs of an organisation. The development of a knowledge management strategy, a supportive organisational structure and culture, and the introduction of appropriate IT, will all contribute in some way to the implementation and exploitation of innovation.

E-business initiatives offer construction organisations the opportunities for new forms of market place strategy models for construction organisations. Organisations that previously conducted business activities during traditional hours can now conduct those procurement activities online 24 hours a day, 7 days a week, 365 days a year. Some of the benefits of e-procurement systems include removal of the intermediaries along the supply chain, thus allowing significant reduction of cost of materials and tighter integration of supply chains. The indirect benefits that drive e-procurement processes include reduced time to market products, growth of the market presence and an increase of quality of products and associated services, an improved forecast of delivery dates and market transparency.

Discussion and questions

Assume that you are a director of a large construction company and a champion of a new radical process innovation emerging from within the company. How might you structure the implementation of this innovation in order to ensure that the minimum possible disruption is made to the structure and culture of the organisation, while fully exploiting the innovation for your organisation's competitive advantage?

Innovation is about managing chaos. It is also about serendipity. There is no one single theory that best describes the management of all innovations in organisations. Do you agree or disagree with this statement? Discuss, giving examples from a construction industry context to support your arguments.

Chapter 13
Managing Supply Chains and Construction Networks

Supply chain management (SCM) is a concept that has flourished in manufacturing, originating from Just-In-Time (JIT) production and logistics. The benefits of collaborative rather than adversarial working relationships within and beyond the organisation, increased market competition, the declining incidence of vertical integration and increased market competition have been offered as reasons for the growth in supply chain management in the last two decades.

The supply chain has been defined as the 'network of organisations that are involved, through upstream and downstream linkages, in the different processes and activities that produce value in the form of products and services in the hands of the ultimate customer' (Christopher, 1992).

However, the construction industry has been slow to employ the concept. Issues of interfaces and interdependence in construction are exacerbated by the traditional rigid separation between parties codified in the extant competitive procurement and commercial practices of the industry. In the construction industry, established approaches to procurement, management of materials supply contracts and subcontracting are generally based on aggressive bargaining between buyer and supplier over issues of price, delivery date and payment date, conducted in an atmosphere of legalistic mistrust (Latham, 1994; Egan, 1998). Again, contracts introduce rigid divisions of labour, which cut across critical processes, units of work, and information flow. Moreover, since contracts are released according to a strictly chronological logic, the supply chain is excluded from the earlier phases of the work and cannot provide any contribution either by proposing innovative solutions or by identifying critical interdependencies and issues.

The nature, types and importance of supply chains in construction

There are different types of supply chain management. First, there are development issues including order information transparency, reduction in variability within supply chains, synchronising of material flows, management of critical resources and configuration of the supply chain (Lin and Shaw, 1988). There are strategies for SCM, including the establishment of stable partnerships, modular outsourcing of components, design for suitability for manufacture, flexible manufacturing technologies, evolution of the supply chain with the product life cycle, and information acquisition and sharing. There are also levels of SCM that can be distinguished, including initial partnership (e.g. building good relations with suppliers and distributors), logistics management, and a real, genuine continuous improvement of all aspects of the entire chain (Giunipero and Brand, 1996).

The supply chain encompasses all those activities associated with moving goods from the raw materials stage through to the acceptance of the product or service by the end-customers. This includes sourcing and procurement, production scheduling, order processing, inventory management, transport, storage and customer service; and all the information systems necessary to support and monitor these activities.

Successful supply chain management co-ordinates and integrates all these activities into a seamless process.

A model of SCM in construction

In the UK construction industry, there exists a 'myriad of construction supply chains' (Cox and Ireland, 2002). The construction industry is characterised by the following major supply chains: construction 'integration', professional services, materials, equipment and labour. These supply chains display significant overlap. Figure 13.1 illustrates the key generic supply chains that are required in the integration and delivery of a typical solution.

During the construction process, the end customer will appoint the construction firm and professional services where needed. Within the generic supply chain, the construction firm plays the major 'integrating' role for all upstream supply chains. Subcontractors also play a vital role in supply chains. This is increasingly the case, given the rise in subcontracting in construction in the last two decades. For each individual element of a construction project, there will be a requirement to source from the respective labour, materials and equipment supply chains. Procurement professionals sourcing from these chains also play a significant role.

It is difficult to quantify the exact number of constituent supply chains that have to be integrated into a typical project; such a project does not exist because of its unique project-specific properties.

Mechanisms used to set up successful collaborative relationships – implementing SCM in construction

Supply chain management is a long, complex and dynamic process. Its implementation requires a thorough understanding of the concept. It is also seen as closely dependent upon the ability to create, manage and reshape relationships between individuals, organisations and networks within the supply chain. Supply chain management requires considerable commitment and resources, and takes time to develop. It also provides difficulties for and challenges to parties within the chain.

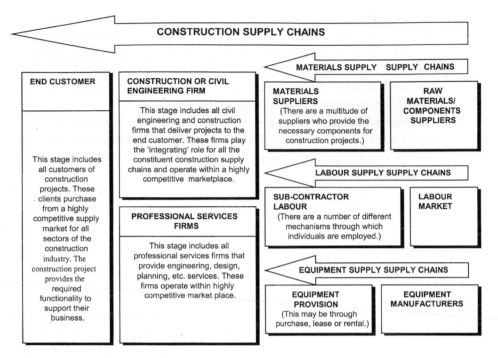

Figure 13.1 The myriad of construction supply chains (source: Cox and Ireland, 2002).

These can include a lack of common purpose, multiple or often hidden goals, power imbalances, different cultures and procedures, incompatible collaborative capability, over dependence, and a continuing lack of openness and opportunistic behaviour (Saad *et al.*, 2002).

Supply chain management has to be introduced progressively and under control, paying due cognisance to the following issues:

- identifying key strategic supply chain partners;
- sharing plans and visions;
- learning from each other;
- being less confrontational;
- becoming proactive rather than reactive;
- exploring joint initiatives and being receptive to change;
- setting SMART objectives for marketing, operations and financial performance.

It is also advantageous, in the chain, to appoint a champion who will have a very positive commitment to the process and its success.

Partnering

For over a decade, partnering as a concept in construction has been promoted within the design and construction community as an effective means of improving the industry's cost effectiveness, profitability, and responsiveness to client needs and expectations. It offers a new paradigm for building owner and construction industry relations. Adopting a partnering approach, all parties agree from the beginning in a formal structure to focus on creative co-operation and teamwork in order to avoid adversarial confrontation. Working relationships are carefully and deliberately built, based on mutual respect, trust and integrity.

Partnering involves two or more organisations working together to improve performance through agreeing mutual objectives. It also provides a way of resolving any disputes and commits parties to continuous improvement, measuring progress and sharing the gains.

In the construction industry, the partnering approach can be project specific (for an individual project, project partnering) or strategic partnering (for more than one project). The early ideas on partnering centred around three main principles: agreeing mutual objectives; making decisions openly and resolving problems in a way that was jointly agreed at the beginning of the project; and aiming to achieve measurable improvements in performance through incentives. The progression towards the adoption of the principles associated with supply chain management is more evident with the emergence in the late 1990s of the 'second generation' style partnering, which includes a strategic decision to co-operate by the key project partners (Bennett and Jayes, 1998). The second generation partnering often places greater emphasis on a more holistic approach based on a wider range of performance criteria in addition to time, quality and cost and acknowledges the strategic importance of such longer-term business relationships.

Partnering is increasingly being seen as a way of developing a culture based on greater co-operation in longer-term relationships (Bresnen and Marshall, 2000), and a way of addressing the industry's fragmentation and lack of integration. Cox and Thompson (1997) have however argued that partnering is still largely misunderstood throughout much of the industry and is therefore not as unified a concept as many of the other forms of procuring facilities in construction. There is also a number of different perceptions of partnering. It is often used to describe the negotiation that takes place in other forms of procurement, such as two-stage competitive tendering, which again causes misunderstandings, further devalues the concept and has engendered a cynical view of it in the industry. It is suggested that whilst partnering has been used to considerable effect by regular, frequent and more informed clients in more routine and repetitive projects, its impact in the case of infrequent clients and unique, one-off projects is clearly more limited.

To date, most partnering has focused on developing collaboration in upstream relationships between regular and frequent clients, consultants and main contractors (Bresnen and Marshall, 2000), with less involvement of organisations such as specialist and trade subcontractors downstream in the process.

Three core principles underpin the partnering relationship: commitment, communication and conflict resolution. The benefits attributable to partnering include the following:

- Improved communications.
- Better working environments.
- Reduced adversarial relationships.
- Less litigation.
- Fewer claims.
- More repeat business/long-term relationships.
- Improved allocation of responsibility, improved value engineering, and decreased schedules.
- Better control over safety and health issues.
- Reduced exposure to litigation through communication and issue resolution strategies.
- Increased productivity because of elimination of defensive case building.
- Lower risk of cost overruns and delays because of better time and cost control over the project.
- Increased opportunity for innovation and implementation of value engineering in the work.
- Potential to improve cash flow due to fewer disputes and withheld payments.
- Improved decision-making that helps avoid costly claims and saves time and money.

Partnering is a proactive process aimed at prevention prior to dispute. For partnering to have its most effective results, all construction team members should participate in the development of the partnering strategy for the project. A partnering workshop should be conducted during the early stages of the contract process.

To ensure partnering success, the following concepts should be considered and incorporated into the process:

- *Commitment.* Commitment to partnering must come from the top management of all construction team members who have a stake in the project and are called stakeholders.
- *Early involvement of stakeholders.* Owner representatives, design professionals, general contractors, subcontractors and local officials need to be part of the evolution of the framework of the partnership charter.
- *Equity.* Every stakeholder's interests must be considered in creating mutual goals. There must be a commitment to satisfying each stockholder's requirements for a successful project by using win–win thinking.
- *Development of mutual goals/objectives.* At a partnering workshop, the stakeholders should identify all prospective goals for the project in which their interests overlap. These jointly developed and mutually agreed goals may include achieving zero injuries during construction, value engineering savings, meeting the financial goals of each party, limiting cost growth, limiting review periods for contract submittals, early completion, minimising paperwork for the purpose of case building or posturing, no litigation, and other goals specific to the project.

- *Implementation.* Stakeholders should develop strategies together for implementing their mutual goals and the mechanisms for solving problems.
- *Continuous evaluation.* In order to ensure implementation, the stakeholders should agree to plan for periodic joint evaluation based on the mutually agreed goals. This will ensure that the plan is proceeding as intended and that all stakeholders are carrying their share of the load.
- *Timely responsiveness.* Timely communication and decision making will not only save money but can also keep a problem from growing into a dispute. Since every project is unique and the particular stakeholders for each project will vary, the process should be tailored by and for the stakeholders for each project. A partnering process can be developed for any type of project and any size of project.

The following points may guide the partnering process.

- Educate your organisation.
- Make partnering intentions clear.
- Commitment from top management at the start.
- Conduct the partnering workshop.

Prime contracting

Prime contracting was selected as the model for procurement of construction and maintenance services for the defence estate in the 1997 Strategic Defence Review when it was decided that a more effective and efficient process was required for the billion pounds a year that the MoD spends on its estate.

Prime contracting is a fundamentally different approach to the traditional MoD works procurement system. Put simply, a single high value 'prime' contract for delivery of property management services is placed with one company/consortium with payment linked to performance and innovation. The prime contractor will co-ordinate and actively manage his or her supply chain (i.e. the sub-contractors) and ensure that lower-tier sub-contractors are all working to the same objectives of meeting standardised targets and improving efficiency.

This may be compared with the traditional method of estate maintenance procurement, which involves many contracts and many contractors (some more capable than others) which are expensive to operate (due principally to duplicated overheads) and difficult to manage. Under this system there is insufficient incentive for the contractors or their subcontractors to improve their performance. Consequently there is, generally speaking, a lack of long-term vision. Prime contracting will address that deficiency in long-term planning and vision as the prime contractor will be able to view the estate as a whole over the long term and plan the resources accordingly. Furthermore DHE (MoD, Defence Housing Executive) expects the prime contractor, during the contract, to generate savings through, for example, reduction in their and their supply chains' costs, standardisation of procedures and the use of their bulk purchasing power.

Prime contracting has been used effectively for high value complex equipment procurement projects for many years and now it is being tailored, commensurate with government policy, to the defence housing requirement.

Prime contracting offers significant advantages in terms of providing an overall single point of responsibility, the co-ordination of a pre-appointed supply chain, the principle of whole service procurement, economies of scale, collaborative working and the adoption of an output specification of requirements. The initiative also emphasises the importance of 'soft issues' such as the ability to manage costs, market awareness, innovation, trust and flexibility.

In the UK, prime contracting contracts have been phased across regions. Regional Prime Contracting, Phase One covers Scotland. Regional Prime Contract South West (RPC SW) is the second of five phases in the implementation plan. Phase Three is planned to cover the South East, Phase Four will cover Central England and, finally, Phase Five will cover the East. The programme will see the whole estate let under the new arrangements by the end of 2005.

Benefits are expected to arise through prime contracting as a direct result of pro-active management by the prime contractor. These benefits include:

- Easier fault reporting (via a free phone number to a call-centre).
- Appointment agreed with the occupant.
- A quicker, flexible service for routine repairs.
- Consistent approach across all estates in England and Wales.
- Continuous improvement and innovation.
- Greater emphasis on quality control and checking as a direct result of a reduction in bureaucracy.

A government policy initiative to improve works contracting mandates prime contracting as the preferred solution where PFI (Private Finance Initiative) is not an option.

Public private partnerships (PPP) and Private Finance Initiatives (PFI)

Public private partnerships (PPPs) are a generic term for the relationships formed between the private sector and public bodies often with the aim of introducing private sector resources and/or expertise in order to help provide and deliver public sector assets and services. The term PPP is used to describe a wide variety of working arrangements from loose, informal and strategic partnerships to design build finance and operate (DBFO) type service contracts and formal joint venture companies. The present UK Government's commitment to PPPs recognises that neither central nor local government in isolation will be able to finance all the investment needed in the country's public sector infrastructure.

PPPs move away from the doctrinal approaches of the past that saw successive governments believing that either the public or the private sector was automatically best. Now, for both central and local government, who does what will be judged, in future, solely on how services are delivered and whether such services are high quality and good value for money for the local community.

The Private Finance Initiative (PFI) is a form of PPP. PFI refers to a strictly defined legal contract for involving private companies in the provision of public services, particularly public buildings. The Private Finance Initiative (PFI) was a creation of the UK Conservative government in early 1992 – but it has been enthusiastically embraced by the present UK Labour government. Governments and local authorities have always paid private contractors to build roads, schools, prisons and hospitals out of tax money.

PFI is also, principally, a form of contracting or procurement. Under a PFI scheme, a capital project such as a school, hospital or housing estate, has to be designed, built, financed and managed by a private sector consortium, under a contract that typically lasts for 30 years.

The private consortium will be regularly paid from public money depending on its performance throughout that period. If the consortium misses performance targets, it will be paid less. The public sector is looking to the private sector for expertise, innovation and management of appropriate risks. The private sector is looking for business opportunities, a steady funding stream and a good return on its investment.

The characteristics of PFI schemes are:

- A long term service contract between a public sector body and a private sector 'operator'.
- The provision of capital assets and associated services by the operator.
- A single 'unitary' payment from the local authority, which covers investment and services.
- The integration of design, building, financing and operation in the operator's proposals.
- The allocation of risk to the party best able to manage and price it.
- A service delivery against performance standards set out in an 'output specification'.
- A performance-related 'payment mechanism'.
- An 'off balance sheet treatment' for the local authority so that any investment delivered through the project does not count against borrowing consents.
- Support from central government delivered through what are known as 'PFI credits'.

To make a success of a PPP requires new attitudes and skills in order to identify when a partnership route might be best for the public and then to make it happen in practice. Similarly, for the partnership to work each party must recognise the objectives of the other and be prepared to build a good, long-term relationship.

A PPP approach may provide better value because it brings economies of scale. An organisation whose core function is to manage and maintain properties nationwide might be able to bring greater expertise to the task than would a single local authority acting in isolation. A PPP may lead to a more vigorous use of assets to deliver benefits to both parties to the agreement. A PPP forged through negotiation may deliver value for money because the competition process provides a spur to the private sector to perform and step forward with their optimum solution. It may also bring in additional expertise or finance that the private sector has developed on a global scale. But benefits are not automatic from PPPs; they only result from well-planned and rigorously appraised schemes.

Public private partnerships (including the Private Finance Initiative), can provide the public sector with better value for money in procuring modern, high quality services from the private sector. The key to a successful PPP deal is the partnership negotiated between the procurer and the supplier.

The attraction of PFI for the government is that it avoids making expensive one-off payments to build large-scale projects that would involve unpopular tax rises. Also, as the risk of PFI projects is technically transferred to the private consortium, it does not show up as increased public borrowing in the Government's accounts.

According to the NHS plan, more than 100 new hospitals will be provided using the PFI by 2010. In 2001–02 the PFI accounted for 9% of public investment.

However, critics claim that, as with any form of hire purchase, buying a product over a long period of time is more expensive than buying it with cash up front. They point out that governments can borrow cash at a cheaper rate than the private sector.

There is also a question mark over how much risk is genuinely transferred to the private sector, given the Government's record of bailing out private companies managing troubled public services. Growing concern has recently been expressed amongst experts about the cost of PFI. Public sector accountants claim that hospitals and schools would be cheaper to build using traditional funding methods. The national audit office described the value-for-money test used to justify PFI projects as 'pseudo-scientific mumbo jumbo'.

Unions point out that the main way private firms involved in PPPs and the PFI achieve 'efficiency savings' is through cutting staff wages. They point to a two-tier workforce created in public services run by private companies, when former public sector workers on terms and conditions protected by law work alongside poorly paid new joiners with no such protection. Even the Labour Party's favourite think-tank, the Institute of Public Policy Research, which had argued that there should be no restriction on the private provision of public service, has since expressed doubts about PFI. It said that there was little evidence so far that the PFI offered increased value for money, especially in providing new schools and hospitals. However it will take at least a further 20 years, when the first PFI contracts have been completed, before the real cost of PFI can be judged.

Discussion and questions

Effective supply chain management provides benefits to construction organisations. However its implementation raises a host of challenges. Discuss this statement.

Discuss the risks and gains associated with PFI and PPP projects, especially in the context of such projects lasting for up to 30 years?

Chapter 14
Personnel Management and HRM

Personnel management or human resources management?

The department which supports line management in dealing with employees and employment relationships has traditionally been called *personnel management*. In the mid-1980s, the term *human resources management* (HRM) started to take over. This coincided with an important change in thinking, in which the role of personnel management shifted from one of mediating between employees and senior management, to one of supporting corporate strategy by integrating business goals and people management (Pemberton and Herriot, 1994).

As these authors point out, the emergence of HRM seemed to offer a lifeline to personnel staff 'who had long felt undervalued by line managers'. The low status of personnel work had arisen from the view of some people, especially line managers, that personnel management was little more than a clerical job – keeping the people records straight and adding little value.

But *human resources* is the language of slave owners of the eighteenth century who treated labour as a disposable resource. Pemberton and Herriot argue that HRM needs a radical re-think that takes it back to its roots. Its strategic role should be based in its established position as broker between senior management and employees, bridging the gap between business concerns and employee needs, in a way which shows how people's potential can best be unleashed for the benefit of both business and employees. This view recognises that people cannot be treated as simply a factor of production. They are at the heart of the organisation and the management of people is more central to business success than the management of materials, money or plant. People can act on resources in a way that resources cannot act on people. For these reasons, the term personnel management is preferred, despite the widespread use of HRM.

Construction personnel

Because much of its work is one-off and it lacks a factory base, construction is labour intensive compared with most industries. Its personnel costs are high in relation to total costs. This is another reason why labour remains an important asset, especially in the building sector of the industry, and effective management of people is a key part of every manager's job. Personnel specialists are increasingly employed to support and advise managers.

Even when construction is capital intensive, as in many civil engineering projects, the management of people is still a critical factor. Studies have repeatedly shown that differences in productivity between companies, and even between departments within a company, cannot be solely explained by variations in manufacturing methods. Rather, they result from differences in the way people are managed.

Technology has made the human factor more important, not less so. Disasters show the negative aspect of this – how human error is magnified as technical scale increases. Technology is no protection against people's mistakes or poor judgement. Understanding human behaviour and how to deal with people is therefore a crucial aspect of management.

Personnel management is often misunderstood and undervalued in the construction industry. This is because personnel work is not very 'visible' and its contribution to the business is difficult to measure. Also, it evolved in a piecemeal and somewhat haphazard way, so that it can lack a clear identity.

Moreover, many aspects of personnel management are not easily separated from general and production management, and rightly so! One of the prime tasks of management is to use people's skills effectively. In this sense, all managers are personnel managers and should work within a well thought-out personnel policy.

The personnel function

Personnel departments hardly exist in many construction firms, but the personnel function is present in every firm. It is the process of channelling human energy and skills into achieving business results. Almost every manager is involved in this.

As organisations have become larger and more complicated, work has been broken down into more manageable, specialised jobs. The jobs which are labelled personnel management are those which specialise in designing and operating systems and procedures for recruiting, employing and developing people.

Because of fluctuations in workload, high labour turnover and casual employment, personnel practices and policies have tended to lag behind those of most industries. Also, the industry includes many small firms that cannot afford to employ full-time personnel staff. However, someone still has to do the personnel work. Normally it will be other managers – line managers concerned with production or general management. Some do the personnel work well, but others admit that they neglect this part of their role because other tasks fill their time. These other tasks seem more urgent or important, or appear to have a more direct impact on productivity. The growing body of legislation on employment and other personnel issues has gradually forced organisations to take the personnel function more seriously.

Even among larger firms which do employ personnel staff, there is no typical personnel department. The form it takes usually reflects the firm's special personnel problems. For instance, most large civil engineering companies have operated safety policies and employed safety officers for many years, because they recognised that they had a safety problem. Similarly, some firms were running training schemes long before the industrial training legislation. Like the football clubs, they had recognised the value of intensive training for getting the most out of their human assets.

Conversely, labour relations have been comparatively good in building, so most firms have not felt the need to employ industrial relations specialists. They have been slow to formulate written labour policies and procedures for consulting with workers and unions.

In nearly every case, firms have concentrated on those aspects of personnel management which have helped them solve their particular problems. Some construction companies, mainly the larger ones, have eventually rationalised their personnel work and brought it under the control of a single manager or director. When this has happened, personnel management has been able to offer a more integrated and long-range contribution to the running of the business. Figure 14.1 shows a possible structure for a well-developed personnel department.

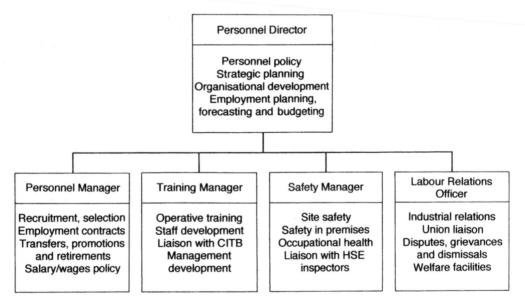

Figure 14.1 Example of a personnel department structure.

The tasks of personnel management

The main areas of personnel management are:

- Employment planning
- Staff development
- Health and safety
- Industrial relations and employment.

These are dealt with in the remaining chapters. However, a well-established personnel department will become involved in other issues, such as:

- Strategic planning
- Organisational development
- Employee remuneration
- Counselling.

These issues are discussed later in this chapter.

Some of the tasks are more strategic than others. For instance, personnel managers increasingly take part in budgetary control and produce staffing budgets. This is important in an industry which relies heavily on labour.

But personnel staff contribute to many day-to-day tasks as well, such as induction, dismissals, grievance handling and advice on pay. The personnel manager has to balance the immediate and tangible operational problems – which can be very time-consuming – with the long-range, more nebulous concerns of senior management.

If the strategic tasks of personnel management are neglected, its potential will not be realised and it will indeed become little more than a clerical function. At its best, personnel management contributes to the overall running of the business, helping managers to use their most important asset to the full.

Personnel policy

A firm which has a system of personnel procedures is likely to have a sound personnel policy. The more people-orientated firms will try to ensure that the policy reflects the needs and ambitions of employees. This means that social as well as economic goals have to be considered when formulating and reviewing company policy. Each company's policy will reflect its particular priorities and problems.

An example of a personnel policy statement for a large construction firm is given below.

Personnel policy statement

This policy recognises that the successful achievement of the company's objectives of profitability and development depends on its ability to provide employees with satisfying and rewarding employment.

The policy will be implemented in accordance with generally accepted employment practices and current employment legislation and the need to avoid unfair discrimination of any kind.

General policy

(1) All employees will be kept informed of the company's practices and policies and of the terms and conditions of their employment.

(2)   The company will establish and maintain suitable procedures through which employees can express their views on all matters affecting their employment.

(3)   The company will create and encourage an atmosphere of mutual understanding and co-operation, in which all personnel feel a sense of involvement and freedom to express constructive views to management.

(4)   Procedures will be established, and made known to employees, governing disciplinary action and the rights of employees to raise grievances and disputes with management.

Employment policy

(1)   The company's recruitment and selection procedures will take account of the need to match individual abilities and preferences to the post concerned.

(2)   The performance standards expected of employees and their progress towards achieving them will be made known to them by their managers.

(3)   Wherever possible, posts will be filled by internal promotion, unless existing personnel are unable to provide the necessary expertise.

(4)   The company will offer alternative employment to, or terminate the employment of, employees who, after adequate warning and the opportunity to improve their performance, fail to reach the company's standards.

(5)   The company will comply with all statutory requirements regarding the employment and termination of contract of all employees.

Staff development policy

(1)   All managers will keep themselves informed of the career expectations and training needs of their employees and will counsel them as necessary.

(2)   Regular appraisal of all employees will be undertaken in order to identify individual development needs and career potential, and to help employees make their best contribution to the company, whilst obtaining maximum job satisfaction.

(3)   The company will provide suitable opportunities for staff development, having regard to individual needs for promotion and increased responsibility, subject to the availability of suitable training opportunities.

(4)   The company will assist and encourage employees who wish to obtain relevant technical, professional and management qualifications.

Industrial relations policy

(1)   The company will recognise the right of a union to represent and negotiate on behalf of a specific group of employees, providing a majority of those employees wishes to be so represented.

(2)   The company will encourage employees to become members of recognised trade unions. Where practicable, such unions will be those taking part in collective bargaining in the industry.

(3)   The company will comply with agreements and procedures established by collective bargaining and contained in relevant working rule agreements.

(4)   The company will operate conditions of employment no less favourable than those provided by competitors.

(5)   Every attempt will be made to maintain good relations and provide proper facilities for consultation and co-operation with union representatives.

Remuneration and employee services policy

(1)   The company will remunerate employees and provide benefits at a level commensurate with performance and responsibility, having regard to current legal requirements, government policies and market forces.
(2)   The company will adequately insure all employees during the period of their service.
(3)   All personnel will be given assistance in periods of sickness or hardship. Pension arrangements on retirement will be on terms no less favourable than those offered by competitors.

Health and safety policy

(1)   The company will maintain a high standard of safety and health and take every practicable step to safeguard the health and safety of its employees.
(2)   The company will comply with all statutory health, safety and welfare requirements.
(3)   The company will provide a high standard of welfare facilities for employees.

Strategic planning and organisational development

One of the strategic jobs of personnel management is to take part, with other managers, in a continual analysis and review of the organisation's structure, culture and operations. The personnel manager can help to develop personnel forecasting and budgeting techniques and to improve administrative functions, as well as supplying forecasts of staffing needs, labour availability, wages budgets, and so on.

Personnel managers can play an important part in identifying the strengths and weaknesses of the organisation and assessing the effects of social, legal, economic and other changes. Personnel staff can help to develop and implement strategies and timetables for organisational change to ensure that the organisation survives and becomes better at doing the things it is designed to do. To achieve this, the personnel manager may recommend improvements to the structure of the organisation, its departments and work groups. He or she will advise on management style, job designs and organisational 'climate'.

The climate of the firm is difficult to analyse, but some measure of it can be obtained by seeing how conflicts are resolved, how people are treated and what levels of trust, co-operation and participation exist. These factors are important, for they can affect efficiency and hence the profitability of the business.

Personnel managers may have more skill – or simply more time – than line managers, for monitoring the organisation and the match between its tasks and people. Because they are not directly involved in operations, they can be more objective.

Employee remuneration

Personnel staff can help the firm to develop effective payment systems and to review them to cope with outside influences, such as government policy, the labour market and wage agreements with the unions. The personnel manager must know how national and local agreements affect the company's employees and see that they are applied.

A salary structure must be established for the many employees not covered by national wage agreements. Guidelines must be laid down for salary increases, benefits and incentives, and how to link these to staff performance.

A system may have to be developed for assessing the relative worth to the company of different people and jobs. This could involve *job evaluation* and *merit rating* techniques, where aspects of a job or an employee's performance are ranked, classified or given a points rating, on which remuneration can be based. In *productivity bargaining* employees agree to make changes in work practices which will lead to greater efficiency in return for improved pay, benefits and working conditions.

Personnel staff may also be asked to produce inter-firm comparisons of salaries and benefits, so that the company continues to offer conditions that will attract the right calibre of employee.

Counselling

Counselling has often been associated with being ill – 'If I need counselling I must have something wrong with me' (Wright, 1998). However, as Wright points out, if words such as 'coaching', 'mentoring' and 'dealing with difficult staff' are used, people become far more comfortable. Today the term counselling is more popular, as in 'investment counselling' and 'career counselling'. Managers have increasingly recognised that counselling their employees is an important part of the personnel function, although their powers over subordinates sometimes makes counselling difficult or impossible.

Personnel staff may be in a better position to counsel employees because there is no 'authority barrier' between them.

Counselling methods are rooted in psychotherapy and owe much to Carl Rogers, who pioneered client-centred therapy in the United States. Rogers (1951) stressed the importance of certain qualities in the counsellor, especially:

- *Empathy* – the counsellor tries to see the problem through the eyes of the 'client', the person being helped.
- *Genuineness* – the counsellor is honest, sincere and puts up no facade.
- *Congruence* – the counsellor uses his or her feelings and is open with the client.
- *Acceptance* – the counsellor regards clients as important and worthwhile, whoever they are and whatever they have done.

These qualities must be conveyed. The person being helped must experience them to benefit from the relationship with the counsellor.

Counselling skills are not easy to separate from general social skills, but experience of counselling has helped clarify our understanding of how warm, trusting relationships develop between people (Hopson, 1984). Counselling embodies the belief that individuals benefit and grow from this kind of relationship and that, properly managed, it helps the individual to become more independent.

Counsellors are unlikely to be successful if they cannot see other people's viewpoints, have radically different values, are poor listeners, make harsh judgements too easily, are unable to be 'open', get emotionally involved, or feel they have to put on an 'act'.

Hopson argues that once counselling relationships have been established, clients will be willing to talk through and explore their thoughts and feelings. This process helps them clarify their difficulties and uncertainties, and to explore options for changing their situations. Given support, people are likely to become more prepared for, and capable of, dealing with their problems.

Unless counselling is properly managed, there is a risk that it may encourage clients to become dependent on their counsellors. They turn to their counsellors every time they have a problem. So, counselling must try to build self-reliance.

Administration and records

Construction firms whose personnel procedures are well developed will have reliable records, providing information for planning purposes and for employee administration. These records must comply with the data protection legislation.

Most firms need records of:

- personal information about employees (such as experience, qualifications, health and the name of a person to contact in the event of illness or accident);
- staffing levels and productivity;
- wages and overtime;
- absence, sickness and accidents;
- statutory requirements and returns.

Personnel staff will be responsible for developing and using suitable methods of data collection, storage and retrieval, including the use of computer-based information systems. They will have to interpret and present information in the way that best facilitates decision-making and control.

Sometimes these tasks lead to a proliferation of records without achieving the intended results. Care is needed to ensure that personnel administration remains a means to an end and does not become a tiresome ritual.

Summary

Human resource management is the term many businesses now use to describe personnel management. HRM embodies the unsatisfactory notion that employees are a resource – a factor of production. In construction, the terms personnel management and personnel manager are still in widespread use.

The personnel function is present in every firm, but a personnel department will normally be found only in the larger firms which can afford to employ specialists. Personnel management has been prone to conflicting assessments, because its contribution to production is not always visible. But in companies which recognise that people are their primary asset, personnel management plays a key role.

The main tasks of personnel management are to obtain and retain employees of suitable calibre, to develop their potential and to help the organisation to manage people effectively. Some personnel tasks are more long term than others and the manager must try to balance immediate demands with more strategic issues. An experienced personnel specialist can help senior management to keep the organisation in tune with changing demands and conditions.

Exercise

Johari Window

Jo Luft and Harry Ingrams (who devised the window in this exercise) believed that we can all enhance our sense of self and get to know ourselves and others better if we take the risk of disclosing ourselves and if we are prepared to hear other people's assessment of us.

Working in groups, each member should complete the Johari Window in Figure 14.2, giving examples in each separate 'pane'. Which window(s) were the most difficult to complete and why?

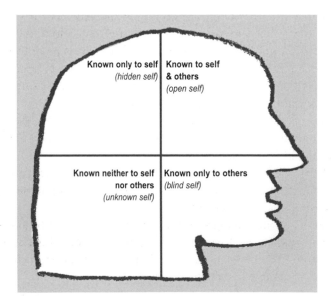

Figure 14.2 Johari Window (adapted from Evenden and Anderson, 1992).

Chapter 15
Recruitment and Staff Development

Employment planning

The purpose of employment planning is to maintain an adequate supply of suitably experienced labour. This can be a major problem for an organisation, especially in the construction industry. The scope for such planning in construction firms varies, but broadly involves the following.

- Analysing and describing jobs and preparing personnel specifications.
- Assessing present and future staffing needs.
- Forecasting labour supply and demand, and preparing budgets.
- Developing and applying procedures for recruiting, selecting, promoting, transferring and terminating the employment of staff.
- Complying with the requirements of employment legislation.
- Assessing the cost effectiveness of employment planning.

Contractors need forecasts of future staffing requirements and the likelihood of meeting them, but the task is extremely difficult. Both future workload and the labour market are highly unpredictable. Most firms have to be satisfied with cautious, short-term predictions and hope that trends don't change too much. However, failure to attempt any forecast of future workload and labour needs leads to staffing problems and organisational inefficiency.

Organisations need people of the right calibre doing the right jobs. This demands reliable recruitment and selection procedures, followed by mentoring, training and monitoring of individuals' career progress.

Staff selection has always relied heavily on judgement and hunch, sometimes based on little more than a short, badly-planned interview. Yet there are other selection methods which can help.

Group problem-solving sessions are sometimes used to assess candidates' skills. They enable selectors to judge how applicants contribute to teamwork and cope with pressure. To build up a realistic picture of the applicant, as many selection methods as possible should be used.

Forecasting and budgeting

Forecasts of future labour needs and the likelihood of achieving them should take account of:

- Natural wastage due to retirement and labour turnover.
- Promotions, creating vacancies at lower levels.
- The company's plans for growth, diversification, etc.
- The availability of labour having the necessary skills in the right location.

One of the problems in forecasting is obtaining reliable information. Managers are often reluctant to make predictions and may be sceptical of forecasting, believing it to be a waste of time. Careful data collection and analysis, including a review of existing personnel, are essential for forecasting both the demand for labour and its supply.

Some of the information must come from outside the firm. The state of the labour market can be assessed from published statistics and help can be obtained from job centres and recruitment agencies.

Employment plans must remain flexible. Events rarely turn out as planned! The demands for the firm's work may fluctuate unpredictably. Economic trends may go into reverse, or technical innovations may force the firm to review its methods. Such changes don't invalidate planning. On the contrary, uncertainty makes planning all the more important if the firm is to survive. Every construction firm needs an accurate picture of its labour force and the labour market. Unfortunately, only the larger companies will have the resources to produce it.

Present labour force

Many operatives are employed on a temporary basis, from project to project, but technical, clerical and managerial staff will be more stable. The firm needs to know quite a lot about its present employees. It helps to have:

- A *skills analysis*, showing where the firm's strengths and weaknesses lie. One person leaving or absent through sickness can create serious problems if no one else has the same skills.
- A *succession plan*, showing who can take over if someone leaves the company. This particularly applies to more senior posts.
- *Training plans*, specifying what training is needed by employees. This will usually be carried out in conjunction with some kind of appraisal scheme.
- A *labour turnover analysis* for each occupation, indicating problems like excessive losses in one department or specialism.

In construction, where there is a rapid movement of labour, the stability of the workforce can be monitored using the ratio:

$$\frac{\text{Number of employees with one year's service or longer}}{\text{Number of employees one year ago}} \times 100$$

The level of detail in such planning will vary from firm to firm.

Employment planning will help an organisation to know if it is overstaffed in some sections and understaffed in others. This makes it possible, with retraining, to transfer employees from one part of the business to another, rather than dismissing and recruiting staff. Whatever the picture, the firm will inevitably have to look outside for some of its labour needs.

External labour supply

There is a lot to consider when assessing the external labour supply, including:

- *Local population profile*. Its density, distribution and occupational composition.
- *Pattern of population movement*. This is important if the people coming into or leaving the area are in the occupations the firm needs to tap.
- *Career intentions of local school and college leavers*. Whether there are suitably trained young people wanting careers in construction.
- *Local employment levels*. How particular occupational groups are affected by the demand for labour.
- *Level of competition for recruits*. Whether the firm is able to attract people of the right calibre.
- *Patterns of travel* and local transport arrangements.

These factors are especially important when a contractor is starting up in a new area.

Producing the plan and an action programme

Mullins (2002) points out that a reconciliation of the supply and demand data not only forms the basis of the plan but of a *personnel action programme*. The latter is the starting point for increasing, decreasing or changing the composition of the work-force, through recruitment and selection, training and staff development, transfers, redeployment and redundancies. Figure 15.1 shows some key relationships between the processes involved. A vital part of planning is the setting of target dates for achieving these actions. In larger firms, computer programs are used to model the organisation's personnel planning options.

Planning for projects

Project staffing must be planned too. The manager must forecast project workforce requirements, taking into account the availability of various kinds of labour (including sub-contract labour), the need to avoid sharp fluctuations in staffing levels and the overall resource pattern for the project.

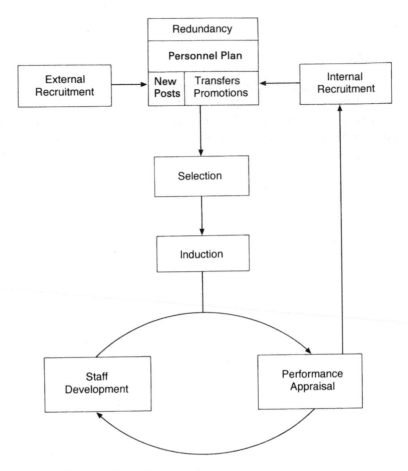

Figure 15.1 Matching people to jobs: some key processes.

Even when unemployment is high, certain types of labour may be in short supply and the organisation may have to train people to meet its needs. A shortage in just one trade or occupation may make nonsense of a contract programme and add considerably to the project duration or cost. Recognising such problems at the outset can lead to better balancing of work and reduce peak labour needs.

Site labour should be built up and run down in a planned way, avoiding sudden changes in the numbers of each trade on site. This means scheduling project activities so that cumulative needs do not exceed the labour available and do not fluctuate too much. Such planning will not guarantee high productivity or good labour relations, but its absence can lead to poor performance and strained relations.

The number of employees that can work simultaneously on a project without productivity falling is limited, but there should not be too much reliance on overtime to make up for labour shortages. Regular overtime is usually expensive and output during overtime working is often lower than that achieved during normal hours.

Recruitment

Job specifications

Before recruiting an employee, the job requirements should be carefully analysed. The purpose of the job should be questioned and whether it might be better to transfer someone from elsewhere in the organisation or reallocate parts of the job to other employees. A vacancy may still emerge from this exercise, but it may bear little resemblance to the job of the previous post-holder, and other jobs may have been rationalised or enriched in the process. For this reason, it is important to draw up an accurate job specification at the outset.

Normally a job specification includes:

- a description of the job;
- a specification of the kind of person likely to do the job well.

Producing a job specification is not a one-off exercise. Jobs change for technical, legal and organisational reasons. In construction, some jobs are more easily defined than others. The work of a plasterer is fairly stable and can be easily measured, but the site manager's role is more difficult to assess. Management work can vary considerably from project to project and it is more difficult to define criteria for assessing it.

However, job specifications are time-consuming to produce and can discourage flexibility. People may be unwilling to do work which is not in the specification. Unions may use these documents to enforce demarcations which are inconvenient and costly to the employer. In construction, the one-off nature of projects means that jobs have to be adaptable and job specifications may therefore be ignored. At best, they are difficult documents to draft and large firms may employ specialist job analysts if they want them prepared properly.

Job descriptions

Typically, a job description will contain some, if not all, of the following details:

- The title of the job
- The title of the job holder's manager
- The job location
- The purpose of the job
- A description of the job content
- A list of responsibilities
- Details of subordinates (if any)
- Standards of performance expected
- Working conditions
- Prospects.

If a firm intends to take job specifications seriously, it must give some thought to standardising the words with which it describes objectives, tasks and responsibilities. If a word is used in different job descriptions to mean different things, much of the value of the description is lost. A job description is only a means to an end. A compromise, therefore, must be reached between a comprehensive but unwieldy description, and a vague summary which makes it hard to distinguish one job from another.

Personnel specifications

These describe the kind of person likely to perform a job well. They are often difficult to produce and the help of a skilled job analyst may be needed when attempting this task for the first time. Various schemes have been developed for analysing personal characteristics and skills, and these usually centre around headings like:

- *Physical characteristics*. Such as strength, health and appearance.
- *Education*. Schooling, further education and qualifications.
- *Job experience*. Previous employment, responsibilities undertaken.
- *Intelligence*. Ability to think analytically, capacity for difficult mental work.
- *Interests*. Inclination towards social, practical, physical or intellectual activities.
- *Personal qualities*. Such as reliability, self-confidence and ability to work with others.
- *Special skills*.

Other breakdowns include factors like impact on others and motivation, but the ability to think creatively is often neglected!

The kind of analysis used will depend very much on the job. Some attributes, such as physical strength, are important for manual construction jobs but not for technical or clerical ones.

Recruitment procedures

The purpose of recruitment is to bring jobs to the attention of job seekers and persuade them to apply. Most firms use several recruiting methods, depending on the type of job. The more common methods are described below.

Personal introductions and contacts

This method has been widely used in the construction industry and with some success. But it cannot be relied on to produce the right applicants at the right time.

Vacancy lists outside premises

This method is used on some construction sites. It is an economical way of advertising vacancies but the information may not reach the right people.

National press advertising

This reaches large numbers of people looking for jobs, but only a small proportion of readers will be suitable or interested. Much of the effort and cost of national advertising are wasted.

Advertisements in the technical press

These reach a specific group and there is less waste. A minimum standard of applicant is more assured. However, some publications are infrequent, causing delay.

Advertisements in the local press

These are mostly read by local people seeking local employment. This may be satisfactory for routine jobs, but may be inappropriate for more specialised posts, for which a wider range of applicants is sought.

Job centres

These can produce applicants quickly and, with computer back-up, from a wide area. They tend to produce applicants who are unemployed, rather than employed people who are looking for a change.

Commercial employment agencies

These have become quite popular in some areas, notably London, and for certain kinds of vacancy. They reduce the administrative burden on the employer, but can be expensive.

Management selection consultants

These are mainly used to obtain applicants for senior posts, often in confidence. The consultant's skills should ensure that a high calibre of applicant reaches the final stage of selection. Again, this service is expensive and not always reliable.

Visits to educational establishments

Some of the larger construction firms regularly visit schools, colleges and universities, to seek out potential employees. This method only produces new entrants to the industry but is a sound, active way of exploring the labour market. It also provides an opportunity to put across a favourable company image.

Recruitment Web sites

On-line employment databases have become a convenient means of reaching a construction industry audience (Figure 15.2). Databases can be searched by job title, location, name of employer or agency, and users are able to enter their own profile and experience to enhance their chances of finding a job.

Internal advertisement

Many construction firms try to fill more senior posts from within the organisation. They prefer senior staff to have had experience of the organisation's methods and culture. This does, however, exclude able outsiders who might bring new ideas and enthusiasm to the business.

Personnel selection

Selecting people for jobs is a very important process. For the applicant, it could be one of the most important events in his or her life. Choosing the right person for a job is not easy. In the past, selection has relied a great deal on experience and judgement, and the results have not always been successful. Firms have tried to take some of the guesswork out of selection by using a wider range of techniques and by giving those involved more training. Personnel selection is only a small part of most managers' jobs and there is a lot they can learn about selection techniques, if given specialist guidance.

The selection process involves the use of:

- Biographical information
- Interviews
- Selection tests and questionnaires
- Group methods
- Work try-outs.

Because the content of jobs differs, it is not always necessary or appropriate to use all these methods but, as a rule, the more techniques used, the more reliable the process is likely to be. This is because each method has different strengths.

Biographical information

Information about the candidate's experience and personal history can be obtained from either a carefully designed application form or curriculum vitae (CV), and from references or testimonials.

The application form provides a basis for comparing applicants and usually gives a reasonably factual summary of what an applicant has done. It will not, however,

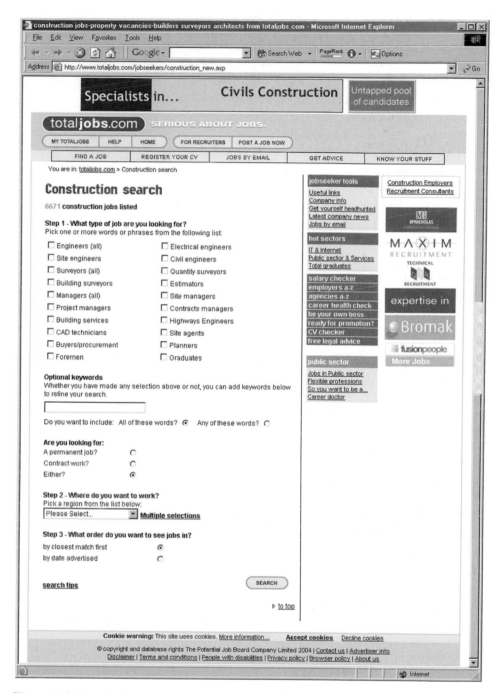

Figure 15.2 Recruitment web site.

indicate how well he or she has done it. Moreover, applicants will emphasise the details they consider most relevant to the job and play down those which are inappropriate or which suggest they may be unsuitable. Occasionally, a questionnaire is used to obtain more depth of information about the candidate's background and experience. Failing this, applicants should be encouraged to give extra information on separate sheets. Application forms are often poorly designed, allowing too much space for some kinds of information and far too little for others.

The organisation can either ask applicants to fill in an application form or invite them to submit a curriculum vitae. The CV has the advantage that it is tailored to the applicant's background and experience. From the organisation's point of view, comparing applicants will be difficult because they will structure their CVs in different ways.

References are a useful source of biographical data if used with care. The referee must be honest and should be familiar with the individual's recent work performance. Many people are reluctant to give unfavourable references and tend to play down the applicant's weaknesses. Selectors should be aware of such biases in references and testimonials. Some companies devise a detailed questionnaire so that the referee has to make specific judgements about the applicant, but this is a time-consuming task and not very popular with referees.

Clearly, biographical details are important in selection, but they should be interpreted with care. Some of the information they contain may not be reliable or relevant to the job.

Interviewing

Interviews are used almost universally in staff selection, although they vary from a casual conversation to a lengthy interrogation. The applicant may be faced with a single interviewer or a panel of interviewers. The strength of a panel interview of, say, three selectors is that a more balanced approach is possible. The one-to-one interview is often more relaxed than the panel kind, but its effectiveness as a selection method relies on one interviewer's ability and judgement. However, panel interviews are often rather formal, making it hard to create rapport with the applicant, who may then have difficulty in talking freely. The success of a panel interview relies a lot on the chairperson's skill in managing the progress of the interview and controlling the others.

There may be more than one interview, especially for senior posts. The purpose of the first interview is to short-list the more promising of the applicants.

The interview can vary in level of formality. One way of describing this formality is the extent to which the interview is structured or open. In a structured interview, the interviewer usually follows a checklist of questions designed to give an overall picture of the applicant in fairly factual terms. The open, unstructured interview does not rely on set questions and the selector tries to get applicants to talk, to find out about their attitudes, motivation and so on.

A combination of these two approaches often gives a good balance between the two kinds of information. Some interviewers prefer to start informally to establish some rapport with the applicant, before switching to a more formal approach to elicit particular information.

Choosing the right questions and giving the candidate the opportunity to answer fully are probably the most important features of an interview. Unfortunately, these features are frequently missing. Questions that merely elicit a yes/no response are unlikely to throw much light on the applicant's suitability. The interviewer should try to ask open-ended questions which force the applicant to give more comprehensive answers. With careful preparation, the outcome of the interview is likely to be more positive.

The interviewer may record ratings of each applicant against the main selection criteria. These may appear in a job specification. Without such ratings, it is difficult to compare candidates from memory after hours of tiring interviews, especially if they have been held over several days. Yet, there is often a fear among interviewers that strict adherence to a scoring system hinders individuals' judgements. Such concerns are understandable but can be largely overcome once again by detailed planning and preparation.

Selection tests and questionnaires

Many of these are psychological tests which give quantifiable answers. The appeal of using them is that some objectivity is possible and applicants' scores can be compared. They are quite widely used in the UK for many types and levels of selection, especially in large organisations. The British Psychological Society (2003) identifies two broad categories of test:

- *Measures of typical performance.* Tests designed to assess dispositions, such as personality, beliefs, values and interests and to measure motivation or drive. They tend to take the form of self-report questionnaires and do not have right and wrong answers.
- *Measures of maximum performance.* Tests designed to measure ability, aptitude or attainment. Such tests consist either of questions with right and wrong answers or tasks that can be performed more or less well.

The construction and refinement of a psychological test is a lengthy and technical process. The value of a test depends on the care with which it has been constructed, administered, scored and interpreted. Used correctly by people trained in administering them and interpreting their results, tests can provide useful information about job applicants, but their use by untrained people has caused concern.

Psychological testing is a form of measurement, but is different from the measurement of physical qualities like length or weight. A test score can only indicate one person's standing relative to others in respect of the attribute being assessed. A score can only be evaluated by comparing it with scores of an adequate sample of the population to which the person belongs. The process of collecting representative scores from different groups is called the *standardisation* of the test. This is an important stage in test construction. Tests vary in their validity and reliability, i.e. the extent to which a test measures what it is supposed to measure and gives consistent results.

Group methods

With this technique, groups of applicants are brought together to discuss a topic or investigate a problem. The theme may be chosen from current affairs or from the firm's business activities. This is a very difficult and stressful task for the applicants who are being asked to co-operate with their rivals!

Selectors observe the group at work to gain insights into applicants' social and problem-handling skills. For instance, it is possible to see whether an individual:

- puts forward ideas effectively;
- adopts a leading or following role;
- persuades others to listen to his or her ideas;
- gets on well with people;
- copes with conflicting views within the group.

Typically, a group session might focus on a broad organisational issue, like productivity, or something specific, such as recycling. Simple discussion exercises will provide the selectors with information about applicants' attitudes. Specific, problem-centred exercises can give valuable information about applicants' skills.

These exercises may take less than an hour, or considerably longer. The selectors analyse each candidate's contribution, taking into account the number and quality of his or her inputs to the group and whether they were well communicated and positive. The extent to which the individual helps the group to make progress with the problem, or prevents it from doing so, may be a useful indicator of the candidate's future performance, if appointed. Aspects of the applicant's personal approach and outlook on life may become apparent, as well as certain qualities, such as self-confidence and initiative.

Selectors should be aware that applicants may not behave normally during the group encounter, which is artificial and sometimes very stressful. Some applicants may perform better under pressure, but many may not achieve their usual performance. This is an important limitation of group methods, for it may discriminate against able candidates who only perform well when they have settled down in the job and are not under pressure. For some jobs, particularly in management, the ability to work under pressure and influence comparative strangers, may be an important quality. Selectors should, however, still be aware that an individual who leads in one group may adopt a quite different role in another.

Work try-outs

Also called work sample testing or proficiency testing, this technique requires the job applicant to perform a task or tasks relevant to the job. In the case of, say, a tiler or mason, typist or engineer, straightforward tasks can be set – fixing some tiles, typing a letter, and so on – and the applicant's performance judged. It is difficult to set a proficiency test of this kind for a managerial job, where performance depends on detailed knowledge of the job and on establishing a working relationship with a team. However, an attempt to introduce this method into managerial selection has taken the form of decision-taking, in-basket exercises, in which applicants tackle simulated managerial or administrative tasks.

Staff development

The terms *staff development* and *human resources development* (HRD) are now used by personnel specialists to describe a range of activities wider than those traditionally linked with education or training. They recognise that learning takes place all the time, as people experience new situations and cope with fresh problems. Learning is not confined to the classroom and, indeed, the most important learning often takes place elsewhere. In staff development, the focus is on *changing* people rather than just teaching them.

Most construction firms acknowledge the need for staff development, but they differ markedly in how they think it should be done. Some firms spend a lot of money and even set up their own development programmes. Other firms simply carry on their business, believing that employees will develop themselves, learning by their experiences and mistakes. There is some validity in this approach too.

Performance appraisal

Performance appraisal is the regular review of the way employees are performing in their jobs. In construction, these reviews are carried out with varying degrees of commitment. In most firms, appraisal techniques are not used. Only among larger firms are formal appraisal schemes likely to be found.

Appraisal objectives aren't always clearly defined and a single scheme may serve several purposes, such as:

- Agreeing performance targets for the next period.
- Identifying employees' strengths and weaknesses, so that training needs can be defined.
- Telling employees how well they are doing.
- Counselling individuals about their job performance, problems and career development.

- Identifying employees with promotion potential.
- Providing employees with the necessary training or development to ensure they can meet the agreed objectives.

There are a number of problems with appraisals. It is difficult to select suitable criteria for the review and to design valid and reliable assessment methods. The manager's ability to make accurate and consistent judgements of subordinates depends on many factors. Many managers are reluctant to carry out appraisals. Specific problems include:

- *Central tendency* in rating employees' performance, where managers are reluctant to give either very favourable or very bad reports, especially the latter.
- *Recent behaviour bias*, where the manager is influenced by the most recent actions of the subordinate.
- The manager's *lack of experience and skill* in forming judgements from observations.
- *Inconsistency in assessment standards*, so that some individuals are appraised more harshly than others.
- *Difficulty in defining the factors* being assessed.
- *Inadequacies in rating scales* and whether managers know how to use them reliably.
- *The purpose of the appraisal* and how the appraiser feels about it. A single appraisal cannot be reliably used for different purposes. Managers tend to vary their assessments depending on the purpose of the appraisal. For instance, the rating given for assessing an employee's salary increase may differ from that given if the purpose is to decide whether or not to make the individual redundant.

External factors also make appraisal difficult, such as the frequency with which a manager can observe subordinates at work. For instance, a chief surveyor, responsible for five surveyors who spend most of their time on different sites, will see comparatively little of their performance.

A case study in staff appraisal

Aker Kvaerner is an international oil and gas engineering and construction group with the capability and resources to undertake some of the world's most challenging projects on behalf of oil companies. Their innovative Personal Performance Programme (P3) is an extension of the traditional appraisal process, which seeks to improve and recognise the performance of all employees.

Crucial to the success of P3 is that everyone understands how their performance contributes to the success of the company, through clear responsibilities and accountabilities for meeting aligned and agreed objectives. The appraisal process is perceived to be a forum for open and honest communication, the setting of development opportunities and recognition of achievements that contribute to both personal and company success.

P3 is a continuous process that is designed to maintain open two-way communication between employees and their team leader. There are three stages in the P3 process:

- The Defining Review takes place in the first two months of the year. Individual accountabilities and objectives are agreed and team deliverables aligned with the company's business plan and strategies.
- Interim Review meetings are held throughout the year. Feedback on employee performance is provided together with discussions on improving future performance.
- The Final Review is a year-end discussion during which the performance for the year is discussed, agreed and recorded.

In practice, the Final Review and Defining Review are amalgamated into one review (see Figure 15.3). It is a natural process to review performance for the previous year and then move on to discuss responsibilities, objectives and individual development needs for the forthcoming year.

There are several performance dimension ratings available for the team leader and options for comments:

E	M	0	Explanation
X			Exceeds expectations in the given dimension
	X		Meets expectations in the given dimension
	X	X	Meets expectations but could pursue further development
		X	Has an opportunity for development

E – Engineers, M – Managers, O – Operatives.

Tangible outcomes of the P3 process are a Development Action Plan and Personal Development Plans. The primary objective of the former is to encourage employees to identify and reflect on work-based development opportunities, which are of maximum benefit to the employee and the company.

The latter recognises employees' long-term ambitions. As such the Development Action Plan is a 12 month snap-shot which contributes to the Personal Development Plan.

Some people claim that formal review meetings are unnecessary, as there should be a continuous dialogue between manager and subordinate. However, other pressures often make it difficult for manager and subordinate to sit down quietly and discuss matters. The formality of the review forces them to tackle the problem thoroughly. Nevertheless, the difficulties inherent in appraisals have led to widely varying opinions about their value.

A special problem with appraisals is how to measure the effect of external factors. The employee's relationship with the manager and with colleagues, and the demands of the job, are just a few of the factors affecting performance.

Appraisals can be one-sided, focusing on the employee's strengths, weaknesses and training needs. A more constructive approach might be to look at a task and ask the work group how they tackled it as a team. What made it easy or difficult

		Personal Performance Programme **(Managers and Engineers)**	
Employee Name		Current Position	
Employee Number		Time in Position	
Date joined company		Review Period	
Project/Department		Defining Date	
Location		Final Date	

1. Roles, Accountabilities and Objectives
A. Duties and responsibilities: *describe scope of work and accountabilities*
B. Objectives: *list of key deliverables, measures and targets established*
C. Supplementary Input:

2. Performance Development
Identify developmental issues and plans relating to current role
Does employee desire a career discussion?
Has a Personal Development Plan been created?

3. Performance Dimensions			
Teamwork	E	M	O
Customer Focus	E	M	O
Communication	E	M	O
Decision making	E	M	O
Continuous improvement	E	M	O
Planning	E	M	O
Ethics	E	M	O
Other	E	M	O

4. Performance Review
A. Final: *summarise on how well roles and objectives were met. Highlight achievements and benefits. Describe areas for improvement and reasons for any performance gaps.*
B. Development: *Comment on employee performance against established plan*

5. Agreement	
Employee signature:	Date:
Supervisor signature:	Date:
Supervisor Line Manager:	Date:

Figure 15.3 Personal performance programme (condensed).

for them? What did they learn from it? This sort of appraisal could become part of *organisational review*. In this climate, people would feel less threatened and be more willing to discuss difficulties. The approach would be especially apt in the context of self-managing teams.

Education and training

The words *training* and *education* are often used rather loosely to describe a variety of ways of helping people become better at their jobs. Education can mean the narrow process of learning a fixed syllabus in order to pass an examination. It can also mean the broad process in which an individual's whole outlook on life is shaped by a succession of varied experiences. The term training usually refers to learning a specific task or job, the skills and behaviours of which can normally be quite precisely defined.

Educational objectives are harder to express in behavioural terms because the learning is often complex and the results difficult to measure. Educational objectives can be expressed in vague, abstract terms, concerned with improving the learner's understanding and self-awareness. The emphasis is on future potential as well as present performance. In a sense, education is person-centred, whereas training is job-centred.

This difference is very apparent when one compares the training of a crane driver with the objectives of a degree course in construction management. Many training objectives are short-term. Indeed, some can be achieved in a few days or even hours. Educational objectives, on the other hand, are long-term and may take months or even years to achieve.

The methods used in training are usually more *mechanistic* than those used in education. Mechanistic learning relies a lot on stimuli and responses, reinforced with plenty of practice, whereas the more *organic* methods of education are less easy to control. They are concerned with developing the individual, and the outcomes are difficult to predict or measure.

Training is essentially practical and job-related. Most of the learning is about work methods, skills and procedures within a firm, trade or profession. Educational activities are broad-based and more conceptual, aimed at developing the individual's critical faculties.

However, these differences should not mask the fact that both education and training are concerned with human development. They are complementary and they overlap. Almost every training activity has some educational impact on the learner, just as many educational programmes help the learner to do a job better. The educational element in the training of an engineer or commercial manager will, however, normally be greater than in the training of a scaffolder or joiner.

Systematic staff development

The Construction Industry Training Board (CITB) report, *Managing profitable construction*, argues the case that construction companies with higher skills make more money, complete more projects on time and have more satisfied clients. Put simply, 'Today's skills will not make tomorrow's profit' (CITB, 2000). It is a view widely accepted by leading construction industry bodies such as the Construction Industry Council, by government sponsored initiatives such as the University for Industry and by Higher Education institutions. However, staff development continues to be given low priority and tackled in a piecemeal fashion, without looking at long-term needs. The result is that time and effort are wasted.

Effective staff development is most likely to be found in firms that recognise its potential for improving company performance.

Systematic staff development involves the following processes.

Identifying development needs

The underlying goal is to make the firm more efficient by making its employees more competent. But it is important to realise that some development objectives are short-term, whilst others are long-term. The organisation needs a supply of reliable, skilled people – operatives, engineers, buyers, and so on – to maintain its present workload. People need to learn the skills for dealing with current projects, because these are the foundation of long-term success.

But the firm also needs people who can steer it into the future; people with the knowledge and skills to recognise new opportunities and to develop and exploit them – people who are adaptable and innovative. Training has always tended to focus on current needs, but the emphasis is steadily switching to preparing people to cope with the future, where different knowledge, attitudes and skills will be needed.

Employees, particularly senior professionals and managers, will increasingly need to understand creativity and creative problem-solving techniques; and they will need a knowledge of future studies methods (see Chapter 10). Training employees in these areas has often been neglected in the past.

The types of development needed and the people needing them must therefore be analysed. Different groups will have different needs.

Planning development programmes

Plans should relate to specific objectives, reflecting the differences between the present and desired knowledge and skills of staff. Thought must be given to how best to meet training needs, keeping in mind any constraints. Development programmes should be flexible enough to meet the specific needs of individuals, provide a timetable for learning activities, and specify where they will take place.

Implementing development activities

Development programmes must be realistic. The award of a new contract may mean that a manager who was to attend a training programme is no longer available. On-site methods of training are usually flexible enough to cope with this kind of problem, but most external courses are not. Efforts must be made to ensure that the methods, content and timing of development activities meet the needs of both the firm and participants. Records should be kept of the progress made by learners so that the programme can be changed if it is too difficult or too easy, or is in some way failing to meet participants' needs.

Evaluating programmes

It is now common practice to evaluate development programmes and incorporate the lessons learned in future activities. Kirkpatrick's (1960) systems-orientated approach remains influential, comprising four levels:

- *Reaction*. Participants are asked what they think about the training they have received. End of course questionnaires or so-called 'happy sheets' are most frequently used by course trainers.
- *Learning*. Changes in participants' knowledge may be quantified using pre- and post-test questionnaires. Such methods, however, are open to criticism for failing to take account of unexpected outcomes and unintended consequences (Smith and Piper, 1990).
- *Application*. This level, which seeks to determine the extent to which learning has been generalised to the work situation, necessarily involves more complex instruments in order to measure behaviour change and improved job performance.
- *Results*. Although this is arguably the most important stage, as the evaluation considers the benefit of training to the organisation, it becomes increasingly difficult to determine the overall impact of the development programme.

A longitudinal evaluation of a novel computer-aided project management training package illustrates the variety of instruments available to the trainer (Ellis *et al.*, 2003a). The problem of collecting valid data during the later stages of evaluation, i.e. Application and Results, is overcome by the use of a self-efficacy evaluation instrument, that measures the participants' self-confidence in being able to complete a task successfully. However, the development of tools such as this is time consuming and relies upon the commitment and support of senior managers to ensure that evaluation is taken seriously.

Approaches to staff development

Staff development has improved over the years, but many development activities have remained rather conventional, with heavy reliance on passive learning, using lectures and other teacher-centred activities. The tutor is expected to know what staff need to learn and how to teach them, putting them through formal courses or programmes of instruction. The content of these has often been rather rigid and general, with the tutor controlling the learning.

The learner is mainly passive; he or she doesn't need to know how to learn, only how to be taught. It is assumed that an engineer, manager or joiner needs to know the same things as other engineers, managers or joiners. Little attention is paid to the *differences* between individual jobs and people.

Regretably, too many courses still assume that the learner:

● is willing to be dependent on tutors;
● respects tutors' authority and trusts their expertise;
● views the learning as a means to an end (such as a diploma);
● accepts a competitive relationship with other learners.

Under these conditions, the learner is expected to do little more than listen (sometimes uncritically) to the tutor, take notes, remember facts and, sometimes, pass examinations.

There is growing concern about this approach, for there is little evidence that it really helps to achieve industry's future supply of competent staff. Many industries have started to move away from the traditional approach, searching for ways of making staff development more active and realistic. Some construction firms followed this trend, but the industry has been rather slow to adapt.

Under various titles, like *work-based learning* and *experiential learning*, attempts have been made to bring about such changes as:

● Shifting the emphasis from the content of learning to the learning process.
● Stressing skills and attitude formation, instead of facts.
● Shifting some of the learning from the classroom to the workplace.
● Stressing skills for coping with new situations, rather than for maintaining current practices.
● Altering the tutor's role from subject specialist to 'facilitator', helping participants to take responsibility of their own learning.

Here, the role of the tutor includes mentoring, providing trainees with resources and acting as a catalyst. Instead of acting as subject expert – telling people what they need to know – the tutor concentrates on creating an effective environment for learning. People learn something because it helps them solve a problem or do their jobs better. They learn at their own speed and have access to resources that complement their preferred learning style.

These approaches do not suit all learners as they demand greater discipline, but many people like it. It brings more learning into the workplace where it can be readily related to real problems and realistic circumstances. It means that tutors increasingly meet the learners on the learners' own ground.

Staff development methods

Coaching and learning from experience

People learn a great deal at work by trial and error and by seeing how others cope with problems. Watching others and taking advice from them are valuable ways of acquiring knowledge and skills. There is no problem of transferring what has been learned from training situation to workplace.

On the other hand, experience can be misleading. Coaching means helping people at work to assess their own performance, think through their difficulties and find suitable solutions. Many people have a contribution to make in coaching others, but the learner's own manager is often in the strongest position, having regular contact with the learner and detailed knowledge of the individual's work.

Construction industry managers value coaching and learning at work, but many managers find it difficult to coach their subordinates effectively, either because of lack of time or because they lack the necessary skills.

Computer-aided learning and beyond

Information technology and telecommunications are playing an important part in education and training. They enable vast amounts of information to be placed at the learner's disposal and can be used interactively to help people develop new skills. But, more importantly, the telecommunications revolution is set to revolutionise all kinds of learning and change the face of training and higher education.

Provision already ranges from the powerful but relatively static standalone CD-Rom to hybrid authored teaching packages and dynamic Internet-based virtual learning environments. However, computer-aided learning (CAL) should not be regarded as a panacea for all training and educational delivery. There is concern among academics that resource issues related to initial development costs, on-going maintenance, administration and institutional support are major obstacles to the uptake of CAL.

What is clear is that so-called e-learning is an exciting development, which offers huge potential both for widening access and promoting an interactive learning environment (Ellis *et al.*, 2003 a,b). Whilst the emphasis in higher education has been towards giving students responsibility for their own learning, care must be taken to ensure that the pedagogic issues associated with these new technologies is carefully thought through. Failure to do so might otherwise lead to disillusionment and foster the belief that CAL is 'second best' education.

Projects and assignments

A wide range of projects and assignments are used in both college courses and in-company training programmes. Companies have used work-based projects for many years in their management development programmes. If carefully designed, projects and assignments are really valuable. They involve the learner – whether student or employee – in actively tackling real-life problems, developing useful skills and learning about processes and procedures in organisations.

One of the strengths of such activities – their realism – is a potential weakness. Learners, immersed in real activities, may not critically analyse their experiences and actions, so that their learning becomes superficial or even misleading. The process therefore needs to be managed by tutors or mentors who really understand what learning from experience is about.

Another problem with assignments and projects, when used in company training programmes, is that they may be undertaken on top of a normal day's work. It is not uncommon in management training programmes for learners to find that they have quite ambitious projects to tackle, but insufficient time to carry out the work. The company must make allowance for this and the tutor or mentor should help the learner to set priorities and exercise time management.

Lectures

The lecture has been widely used in education and training. It has some value for imparting knowledge and changing attitudes, but its contribution to skills training is limited. A major criticism of the lecture is that it requires the learner to be mainly passive.

Many of the functions of a lecture can just as easily be achieved by giving the learner a book or article to read. However, good lectures can give a quick overview of an unfamiliar subject and guide the learner's private study. They give the learner a chance to ask for clarification; books cannot. The lecture may have some 'inspirational' value, although a good book can usually provide this too.

Role-playing and simulation

These techniques are valuable for changing people's attitudes, helping them to see other people's viewpoints, and developing their problem-solving and social skills. The activities usually happen off-the-job but are designed to reproduce the work setting as closely as possible. A useful feature of role-playing is that learners can be asked to 'be themselves' or to adopt another role, thus gaining insights into other people's points of view. Again, feedback is essential, from the tutor, other role-players and, sometimes video recordings.

For role-playing and simulation to be effective, the following would apply.

- The activity used must provide a realistic scenario.
- The activity must be designed to meet different participants' needs, which will not all be the same.
- The trainer must prepare the learning activity thoroughly.
- A clear briefing must be given. Any constraints or rules of the game should be explained.

- The trainer must know how to evaluate participants' performance and should give them feedback.
- The trainer should use video facilities, where appropriate, to provide additional feedback and should help participants interpret that feedback.

A balance needs to be struck between a structured and open learning experience. Some structure may be essential but excellent learning can also occur when there are no objectives and no rules. A lot depends on the group and their perceptions of the activity, the task they are involved in and the whole training set up.

Mentoring

Mentoring describes ways in which one person can help another to learn from their work. It is a process which is quite well established in management development, but it can be valuable in most kinds of staff development or training. The mentor acts as:

- *Role model* (sometimes called a *competence model*) – providing vision and inspiration, setting expectations of performance and demonstrating professional behaviour.
- *Instructor* – passing on knowledge, insights, wisdom or perspective to another or providing challenging tasks or ideas.
- *Coach* – making suggestions for improvement, offering encouragement, building self-awareness and self-confidence, and providing feedback on the learner's performance; helping when things go wrong.
- *Counsellor* – actively listening to the learner's difficulties and offering suggestions.
- *Assessor* – helping to evaluate the learner's performance so that suitable credit can be given either towards a qualification or career advancement.

The use of mentoring increased during the early 1990s, spurred on by a fast-growing interest in learning in the workplace, or *work-based learning*. Clearly, the success of mentoring depends on the skills of the mentor and on his or her relationship with the learner. Among the many skills mentors need are the ability to:

- help people change their ideas, attitudes, values and behaviour;
- encourage planning, analysis, experimentation and increasing autonomy;
- show positive regard for people, even when things go wrong;
- empathise with and show understanding of people's feelings;
- deal effectively with negative behaviour or mistakes;
- give learners scope to think and decide for themselves.

Although it seems obvious that mentoring should include giving advice, this should be done sparingly. Learners can become too dependent on the mentor. They can only learn to become truly independent and think for themselves if they make the most of their own decisions. The mentor should therefore avoid taking over problems and solving them for the learner. In particular, the mentor should encourage the learner to set his or her own targets and goals (Fryer, 1994b).

Selecting an appropriate mentor with the skill, time and authority to facilitate learning support is a key consideration. The mentor is often the manager of the person being trained, though this is not always appropriate or desirable.

Disadvantages of the manager mentor include:

- Lack of frankness in the relationship particularly where a formal appraisal mechanism exists.
- Managers having insufficient time owing to pressure of other commitments.
- Problems caused where disciplinary action is required against the mentee.

However, the mentor's inability to facilitate learning opportunities because of insufficient authority or resources vested in them is solved at a stroke if the manager acts as mentor, though it may still not be possible to provide all that the mentee would desire (Keel, 1995).

Management development

The education and training of managers has expanded rapidly since the 1950s, accompanied by a startling growth in the numbers of management colleges, business schools and courses. Impetus for this came partly from industry and government and partly from professional bodies, such as the Chartered Management Institute and the Foundation for Management Education.

During the period of expansion, many management development programmes were rather conventional. They often embodied the concept of the 'ideal manager' centred around the basic management functions. Their approach reflected what Handy (1975) called an *instrumental* philosophy, which sees learning as deductive, proceeding from theory to application. Teaching tends to be expository. Inductive thinking and creativity are the province of the teacher and researcher; the manager is concerned with deduction and application.

Handy contrasted this approach with what he called an *existential* view. This focuses on helping people to formulate their own ideas and develop their unique talents. Existential tutors tend to talk of teaching people rather than subjects, of giving feedback rather than examinations, of progress rather than achievement, and of general aims rather than specific objectives.

New approaches to management development, such as action learning, embody much of this philosophy. Action learning was largely pioneered in management courses by people like Reginald Revans. Techniques like role-playing and simulation have been widely tested on managers, because of their value in developing interpersonal, problem-solving and information-handling skills.

Two reports, *The Making of British Managers* (Constable and McCormick, 1987) and the so-called Handy Report, *The Making of Managers* (Handy, 1987), were important position statements on management development as the UK approached the 1990s.

Enrolments on business and management qualification courses continued to grow rapidly during the 1990s and in response to these two reports the Management Charter Initiative became the lead body for management occupational competence standards (Brown, 1999). Brown believes that employers have increasingly used formal qualifications as a vehicle for management development, linking them to company strategy. However, corporate in-house qualification programmes, whilst sometimes viewed as being inward looking, offer opportunities to:

- contextualise learning;
- tailor content to the specific needs of the company;
- foster team-working within the company;
- gain the support of senior managers;
- improve the administration and monitoring of the programme.

A recent example of innovative management development is described in Box 15.1.

The practice of downsizing, to create leaner, flatter structures in organisations, has in many cases resulted in real shortages of managerial expertise. McClelland (1994) suggests that this has created a need for a more strategic approach to management development. Brown (2003) defines strategic management development (SMD) as:

> Management development interventions which are intended to enhance the strategic capability and corporate performance of an organisation.

The purpose of SMD is to ensure that, despite management shortages, competent managers are identified and mobilised in pursuit of business goals.

Continuing professional development (CPD)

All employees need to regularly update their knowledge and skills and learn new things. But the need is perhaps greatest in the professions, whose members are affected by numerous and substantial developments in technology, legislation, contracts and many other areas. These developments can be highly complex and take time to assimilate. This has led the major professional institutions in the construction industry to collaborate in finding more reliable ways to provide continuing professional development for their members.

The focus of CPD must include both the short-term and long-range development needs of employees. Development activities can take many forms, ranging from one-day courses to job rotation at work, from distance learning to working party

---

Innovation in management development

Since 1993, the Shepherd Group and Leeds Metropolitan University have collaborated to develop postgraduate courses specifically for Shepherd employees. The courses develop competence, skills and knowledge related to the specialist needs of the group and individual needs of the student. Different courses have been delivered for construction managers, quantity surveyors and mechanical and electrical engineers.

The company selects the participants for the courses and decides the core content of the modules according to its business needs. External consultants and Shepherd employees, who are experts within their own field, deliver the lectures and workshops. The University provides specialist tutoring, assesses the participants' work and has a role in ensuring that academic quality standards are maintained.

Using work-based projects and reflective practice assignments each participant is able to tailor the learning experience to their current knowledge and future work needs. The flexibility of the research and assessment allows students to focus in on an aspect of their work, gathering information and investigating issues in detail. Through the study delegates reflect on their practice, become more aware of the strengths and limitations of alternative approaches and develop a high level of understanding within a specific field. Presentations, group-work and discussions during the course help to explore topics, exchange information and generate ideas. Group discussions also have an impact on the individual, increasing their awareness of different perspectives and encouraging greater reflection when engaging in their own personal study. The course has fostered the development of informal networks allowing employees to seek advice and information from colleagues in different parts of the country. Such interaction encourages the knowledge exchange required to develop a true learning organisation.

The programmes are highly successful with many students achieving postgraduate Certificates, Diplomas and Masters awards from Leeds Metropolitan University. Increasing national interest in work-based learning means that innovative programmes like this are likely to become more popular. The Shepherd Group has received national training awards for its in-house management development programme.

Box 15.1 A unique management development programme.

---

membership. The main concerns are about the effectiveness of these CPD activities and their costs. There has still been rather heavy reliance on short, updating courses, delivered in conventional ways; and employers have to ask themselves whether a one-day conference costing £250 per delegate and comprising a dozen specialists giving a series of short lectures, has really led to any significant behaviour change or skill learning among their staff. CPD needs to shift more of its focus towards life-long learning and encourage learning interventions which actively involve people and don't have them passively listening to others.

Construction Industry Training Board (CITB)

The CITB is a non-departmental government organisation and is the National Training Organisation (NTO) for construction. It is largely funded by an employment levy that is set by parliament every year. The levy is calculated by applying percentage rates to payments to directly employed personnel (currently 0.5%) and payments to labour-only subcontractors (currently 1.5%). Research suggests that firms that directly employ staff are more likely to invest in training for these staff than firms that employ subcontractors. Hence the levy is a fair way for the costs of construction training to be shared between the entire sector (CITB, 2003a).

The CITB (2003b) has a three pronged approach to combating the industry's skills shortage:

- to provide training grants and support to encourage employers to take on apprentices and train existing staff on-site;
- to match young people interested in a career in construction with the best training route for them;
- to actively promote construction careers in schools.

In October 2001 the government announced that NTOs would be phased out and organisations would be asked to form Sector Skills Councils (SSC). As well as delivering qualifications and meeting the basic skills needs of the industry, SSCs will be more strategic bodies. The CITB has become a Sector Skills Council and is known as CITB-Construction Skills.

Summary

Helping employees to achieve their maximum potential is a central activity of the business. Staff development is necessary to maintain existing levels of skills in the business, but it becomes vital when the organisation has to keep pace with change.

Implementing staff development is difficult, costly and hard to evaluate. Benefits are seldom 'visible' in the short-term and are obscured by other changes in the longer term, so long-range planning of development coupled with an act of faith, are essential. Firms differ in the degree of formality with which they tackle staff development. Some, like the farmer, simply encourage their employees to grow; others put them through tightly-structured programmes of instruction.

The development process begins with performance appraisal, in which employees' achievements are reviewed and training needs identified. Development plans are put into action using a wide range of approaches and methods. Increasingly, the methods favoured are those which involve the employee in active, problem-based learning, undertaken at least partly within the realistic setting of the workplace.

New approaches to development concentrate on helping employees to learn appropriate skills and attitudes; they encourage them to take charge of their own development, learn how to identify their own goals, and locate the means for achieving them; they emphasise the need for flexible learning provisions, which can be tailored to the individual's needs and preferences.

CAL in all its many guises offers huge potential, but care must be taken to ensure that exciting new multimedia educational and training packages do not ignore important underpinning pedagogic design issues.

If the firm has a personnel department, one of its responsibilities will be to advise on training priorities, help produce development plans and secure the resources for meeting them. Increasingly, staff development activities will be integrated with the broader processes of strategic management and organisational development. They will focus on helping employees to manage change and improve organisational performance in a turbulent environment.

Mind map tasks

Developed in the 1960s, mind maps provide a quick way of getting your thoughts and ideas down on paper. Using words, symbols or images, mind maps can be used to take notes, plan essays, aid revision or assist in problem solving. Write your main idea in the centre of a sheet of paper, draw lines between key concepts and add further 'branches' as you develop each train of thought (LMU, 2003). Simple but powerful!

Construct a mind-map to represent the key issues associated with employment planning. Consider how you might use the map to highlight inter-relationships between complementary concepts.

Chapter 16
Health and Safety

Safety

Construction sites are dangerous places. The industry remains one of the most dangerous in the UK. The fatal injury rate of 4.2 compares unfavourably with the all industry average of 0.88. Roughly, 80 people die (Table 16.1) on construction sites every year (more than in all the manufacturing industries put together). In the last 25 years, 2800 were killed on construction sites or as a result of construction activities. Many more have been injured or made ill. In the Health and Safety Executive's reporting period 2001/2002, there were about 4600 major injuries and a further 9700 'over three-day' reported injuries in construction (HSE, 2003). These figures include employees and self-employed persons.

The main causes of deaths and major injuries (Table 16.2) are:

- Falling through fragile roofs and roof lights.
- Falling from ladders, scaffolds and other work places.
- Being struck by excavators, lift trucks or dumpers.
- Overturning vehicles.
- Being crushed by collapsing structures.

In June 2002, The UK Government launched an initiative entitled *Revitalising Health and Safety in Construction* (HSE, 2002). This is designed to inject new impetus into the health and safety agenda, 25 years after the Health and Safety at Work Act 1974. There have been other recent initiatives to address the poor safety record of the construction industry. The area of 'competence' is probably where the most visible cultural change has been seen. The industry continues to work towards a fully qualified workforce, and the Construction Skills Certification Scheme (CSCS) and other equivalent training schemes are already being recognised as a minimum industry standard. In *Accelerating Change*, the Strategic Forum states that the move to a fully qualified workforce 'will have a major impact on the number of avoidable accidents caused by a basic lack of site awareness'. From the end of 2003, contractors are not allowed on to sites of members of the Major Contractors' Group (MCG), part of the Construction Confederation (CC) without a relevant training certificate. Of course, the delivery of this objective would depend upon

Table 16.1 Number of fatal injuries to workers and members of the public 1991/92–2002/03(p) (a)

	91/92	92/93	93/94	94/95	95/96	96/97	97/98	98/99	99/00	00/01	01/02	02/03(p)
Employees	83	70	75	58	62	66	58	47	61	73	60	57
Self-employed	16	26	16	25	17	24	22	18	20	32	20	14
Members of the public	6	5	6	5	3	3	6	3	6	8	5	5

Table 16.2 Percentage of fatal injuries to workers by kind of accident 1996/97–2002/03(p) (a)

	96/97	97/98	98/99	99/00	00/01	01/02	02/03(p)
Falls from a height (b)	56%	58%	60%	52%	44%	46%	46%
Struck by moving vehicle	11%	6%	12%	6%	16%	14%	7%
Struck by moving/falling object	12%	15%	12%	21%	10%	16%	15%
Trapped by something collapsing or overturning	7%	5%	5%	2%	17%	5%	7%
Other	14%	16%	11%	19%	12%	19%	24%
Total number of injuries	90	80	65	81	105	80	71

(a) Reported to all enforcing authorities.
(b) Falls from a height include falls from up to and including 2 metres, over 2 metres and height not known.
(c) Non-fatal injury statistics from 1996/97 cannot be compared directly with earlier years due to the introduction of revised injury reporting requirements (RIDDOR 95) in 1996.
(d) The definition of a non-fatal injury to a member of the public is different from the definition of a major injury to a worker.
(p) Provisional
Source: http://www.hse.gov.uk/construction/index.htm [HSE, 2004]

co-operation across industry as many of the site workers are employed by the sub-contractors.

Revitalising Health and Safety in Construction (HSE 2002) identified the drivers for improved performance in Health and Safety, a key aspect of Respect for People, not only as industry-wide leadership, leadership within businesses, client leadership, and corporate responsibility, but, most importantly, the economic levers of socially responsible investors (SRI) and the costs and availability of insurance.

Some, like the Highways Agency, have built improved health and safety into their own overall procurement strategy. *Revitalising Health and Safety* (HSE, 2002) accepts that making changes to improve health and safety also improves 'bottom-line' performance: profitability, productivity and quality. This recognises that step changes in process and attitudinal change impact on everything. OGC Procurement Guidance Nr. 10: *Achieving Excellence through Health and Safety* (HM Treasury, 2002) requires all government departments to be able to relate health and safety performance to overall business performance. Further, the government procurer is encouraged to favour suppliers and procurement routes that have the greatest impact on health and safety.

People, culture and organisational processes are at the heart of health and safety in construction. Any business case must be understood within the context of different organisational characteristics. The need to establish the business case for Respect for People is also called for. While the business case, established in 'best practice' exemplars is important, other issues such as capability and capacity of organisations to change, needs to be addressed. The HSE, in *Changing Business Behaviour: Would Bearing the True Cost of Poor Health and Safety Performance Make a Difference?* (HSE, 2002) discussed a 'social cost' of health and safety failure and this must be considered.

Against this background, managers have to take their safety responsibilities very seriously and it was not surprising that the Construction (Design and Management) Regulations 1994 (CDM Regulations) were introduced, in an attempt to bring a more comprehensive and strategic approach to construction health and safety.

Construction managers must take every reasonable step to protect the health and safety of employees, comply with statutory requirements and improve standards wherever they can. Economic pressures to cut project times and costs militate against this, but managers must make health and safety a high priority. They must encourage responsible attitudes to health and safety and increase people's awareness of the dangers.

This involves the use of many management skills and techniques discussed earlier in this book, including the following.

- Clear communication.
- Good organisation.
- Persuasion and education.
- Rewarding employees for safe practices (operant conditioning).
- Setting realistic targets for employees to achieve, under appropriate working conditions.

Attitudes to safety

The attitudes of construction workers and managers towards safety is undoubtedly a major factor in the poor accident record of the industry. Many managers see construction as a rough job for tough, self-reliant people. Some of them believe that building to tight deadlines at low cost is incompatible with high safety standards.

Construction workers accept that their work is demanding and risky, although they usually underestimate the risk. Group norms may cause individuals to ignore safety measures for fear of appearing cowardly or weak to their workmates.

Some managers and workers try to avoid complying with safety regulations and sometimes make collusive arrangements to avoid them. In one instance, employees were given a bonus to undertake hazardous work without safety equipment, on a

Sunday, when it was known that neither the safety officer nor safety representative would be present (Shimmin *et al.* 1980).

Although examples like this may be uncommon, the use of bonus payments to increase productivity encourages workers to take short cuts and skimp on safety measures. For these workers, the benefits of risk-taking seem to outweigh the potential costs.

It will need more than legislation to make construction safer. Attitudes in the industry have to change. This includes clients' and designers' attitudes towards safety. The CDM Regulations 1994 were intended to bring about attitude change as well as compliance. Certainly there is evidence that clients are taking the health and safety performance of contractors more seriously when awarding construction contracts.

The Health and Safety Executive (HSE) has played an important part in attitude change. For instance, it has argued the case that clients should expect to pay a 'safety premium' on construction; that if a rock bottom price means that people will die or be injured, then the price is too low. The HSE has also helped to change attitudes by arguing that safety measures can be cost-effective. This is because the costs resulting from accidents and fatalities can be higher than the cost of preventive measures.

Team approach to health and safety: the CDM Regulations 1994

To persuade the whole construction industry, including its customers, to take site safety and health more seriously, the EC introduced, in 1992, the Temporary or Mobile Sites Directive, which has been implemented in the UK as the CDM Regulations 1994. The regulations require clients and all the members of project teams to work together to maintain health and safety standards on projects.

The regulations place duties on clients, designers, other professionals, contractors and site workers, so that health and safety must be planned and managed through all stages of a project. Specialist contractors and self-employed sub-contractors must co-operate with the main contractor and provide relevant information about the risks created by their work and how they will be controlled. Under the regulations, all employees on sites should be better informed and have the opportunity to be more involved in health and safety.

Despite efforts by the HSE to promote the regulations, there has been considerable opposition, not least among the professions. Alexander (1995) has argued that the regulations add another costly layer of bureaucracy, leading to confrontation and legal battles, rather than making construction sites safer.

The CDM Regulations apply to all notifiable construction work and to non-notifiable work that involves five or more people on site at any one time. They also apply to all design and demolition work.

The regulations identify the role of *planning supervisor*, the person with overall responsibility for co-ordinating the health and safety aspects of the design and planning stages and the early stages of the health and safety plan and the health and safety file. Clearly, all parties to the project need to understand their obligations and the HSE has published an approved code of practice to aid the implementation of CDM Regulations (HSE, 1995, revised 2002).

Health and safety plans and files

CDM requires that a detailed *health and safety plan* be drawn up before site work begins. There must be a pre-tender safety plan, available to contractors tendering for work. This must identify, among other things, the risks to workers as far as these can be predicted at the tendering stage. The main contractor, once appointed, must draw up a health and safety plan for the construction phase. This must include all the arrangements for managing health and safety throughout the construction stage.

The planning supervisor is responsible for ensuring that a *health and safety file* is compiled during the progress of the work. This is a record for the client/user and identifies the risks that have to be managed during maintenance, repair or renovation. The client must make this file available to anyone who will have to work on any future design, building work, maintenance or demolition of the structure.

Formal enforcement of the CDM Regulations

The Health and Safety Executive said that it would help the industry to cope with the introduction of the CDM Regulations, by giving advice and visiting major clients, professional practices and larger contractors. However, the Executive made it clear that formal enforcement would be used where there was a high risk of accident or ill-health (Nattrass, 1995). Nattrass gives examples of where HSE might issue Improvement Notices – or possibly Prohibition Notices. They include:

- Failure by a client to appoint a planning supervisor in an obviously complex or risky project.
- Failure by a designer to provide adequate information at the design stage.
- Failure by a client to ensure an appropriate construction phase plan has been prepared before construction begins.
- Failure by a principal contractor to cover obviously high risk aspects in a construction phase plan.

These examples clearly demonstrate the HSE's intent to ensure that all parties play their role in maintaining health and safety standards on projects. Interestingly, the growth in project partnering arrangements will complement this process, health and safety being one of the areas of common objectives.

High-risk activities

Some occupations are more risky than others. Not surprisingly, steel erection and demolition account for many fatalities and serious accidents. Collapses of false-work, scaffolds, hoists and cranes also account for many injuries and deaths. But trades like painting and decorating are also hazardous, because of poor workplace access and inadequate working platforms.

Some of the most dangerous activities identified over the years are given below.

Steel erection

Common accidents involve falls from structures and access ladders, collapse of partially-erected structures which are unstable, and materials or tools being dropped from a height. Steel erectors have traditionally resisted wearing safety belts, a problem made worse by lack of anchorage points. Erectors may also be unaware of the extent to which partially-erected structures are stable or unsafe. To prevent accidents from materials falling from structures, barriers and warning notices must be used to limit access beneath steelwork during erection.

Demolition

Accidents include premature collapse of unstable, partially-demolished structures and materials falling from structures (often outside the site boundary, injuring passers-by). Lack of information about the structural character of a building being demolished has been a serious problem, but in 1995, it became compulsory to prepare health and safety files for buildings to comply with the CDM Regulations 1994. This should help to reduce this problem in the future.

Scaffolding

Poor working practices in the erection, maintenance and dismantling of scaffolds have to some extent been overcome by better training. Failure to tie scaffolding and mobile scaffold towers into structures has caused many collapses.

Refurbishment projects can pose special problems, as clients may not wish to have scaffold fixtures penetrating window openings, disrupting user comfort and property security. On multi-contractor sites, the principal contractor must co-ordinate the use of scaffolds and ensure that modifications are carried out competently and inspections performed regularly.

Excavations

Collapse of the sides of unsupported excavations and sudden collapse of structures adjacent to excavations are among the hazards. Uncovering toxic material and striking electric cables are dangers during digging and working in excavations. Unfenced or poorly protected excavations pose risks not only for employees but the public, especially children.

Falsework

Temporary structures aren't always designed as carefully as permanent ones, because they are temporary. This is more of a problem for smaller contractors who lack staff with good engineering expertise. Some of the worst accidents in the industry's history have involved collapses of large, temporary structures to bridges and viaducts.

Maintenance

Because maintenance operations are often short-term, short cuts are taken, like skimping on access equipment. More attention to maintenance at the design stage can reduce such problems, by including proper access ways, cradles and anchorage points for safety equipment in the permanent structure.

Roofwork

Most accidents involve operatives falling from unguarded edges of roofs or falling through roofs that lack loadbearing strength. Roofers often ignore the statutory requirements and work without roof ladders, edge barriers and crawling boards, etc. Injuries commonly involve not only roofers but other trades working on roofs.

Site transport

Many serious accidents are caused by heavy goods vehicles and earth-moving equipment. A lot of incidents happen whilst vehicles are reversing. Other causes are poor site layout, careless unloading, tangling with overhead power lines and people riding on vehicles in insecure positions.

Painting

Painting has a poor safety record. Work proceeds quickly, requiring access equipment to be moved frequently. Safety precautions are neglected to avoid delays. Painters often receive minimal supervision and have little expertise in the use of cradles and access towers. Painting also has its own health risks. Some specialised paints emit toxic and inflammable vapours, causing problems mainly when working in confined, badly ventilated spaces.

Health

The occupational health record of the industry is as bad, if not worse than its safety record, but finding means to measure this and monitor progress is not easy. However, in 2001/2002, it was estimated that 137,000 people whose current or most recent job in the last 8 years was in the construction industry suffered from an illness which they believed was caused by or made worse by the job. In 1995, it was estimated that over 1.2 million working days were lost as a result of work-related ill health (HSE, 2003).

Safety hazards have overshadowed the health risks to construction workers. This is partly because employers and employees have not been fully aware of the health risks and partly because of an attitude among some employers that health is the worker's own responsibility. Moreover, health hazards are difficult to control because site conditions are so variable.

There have been relatively few systematic studies of the health of construction workers, but the main hazards are known. They include dusts, toxic substances, radiation, vibration, noise, changes in atmospheric temperature and pressure, and inadequate welfare and hygiene.

Whilst occupational health hazards are becoming more widely understood, many new materials and processes are exposing employees to fresh health risks. Fortunately, employers are beginning to realise that safeguarding workers' health makes economic as well as human sense.

The industry is aware of the established hazards such as working with lead paint and in compressed air, but other risks have remained hidden. Asbestos, especially the blue type, has received a lot of attention in the industries producing it (mining, milling and processing) but the risks to construction workers handling asbestos were neglected for a long time.

It only became widely known in the 1970s that prolonged exposure to dusts from silica and certain hardwoods can cause serious lung disease. Handling wet concrete and mortar can cause skin complaints like dermatitis, whilst welding and oxyacetylene burning can create toxic fumes.

The structure of the industry, fragmented into many small units, has aggravated the problem because health safeguards often rest with small firms who cannot afford occupational health measures. These firms might, however, be able to agree to share such facilities (Health and Safety Executive, 1983).

Employees have shown little concern for their own health and many new employees are unaware of the risks. Many workers develop rheumatic and arthritic conditions after long exposure to cold and wet weather, because they wear inadequate clothing.

The industry has undoubtedly neglected occupational health issues in the past in comparison with other industries, which have provided much better facilities. This situation is partly a result of the temporary nature of the industry's production base. The position is improving and legislation is helping ensure that at least some of the occupational health problems are tackled. Relevant regulations include the COSHH Regulations 2002, the Workplace (Health, Safety and Welfare) Regulations 1992 and the Personal Protective Equipment at Work Regulations 1992.

Many ailments occur too frequently in construction workers to be explained by non-occupational factors. These include lung cancer, respiratory diseases, stomach cancer, muscle and joint conditions, arthritis and dermatitis. Unfortunately, statistics rarely indicate the causes of these illnesses. Many chronic, slow-developing diseases do not become noticeable for years, by which time they are seldom traced back to their causes.

Many hazards not only affect workers directly involved in risky operations, but others doing harmless jobs nearby. On multi-contractor sites, control is especially difficult.

One of the implications of the Health and Safety at Work Act 1974 is that the industry needs to focus on the general well-being of its workers and not just on preventing accidents. This concern should include the mental and physical welfare of employees and should recognise that many health problems arise from a combination of factors inside and outside the workplace. Changes in work patterns and family life have put more pressure on people. To cope with this, they often eat, drink or smoke too much. This may reduce the effects of stress but exposes them to other health hazards. Many physical and mental illnesses are linked, so a broad approach to health seems essential. In the late 1970s and early 1980s, many medical people were beginning to turn their attention from ways of curing ill-health to strategies for promoting *positive health*.

Control of Substances Hazardous to Health Regulations 2002

The purpose of the COSHH Regulations is to protect employees from the dangers of hazardous substances, such as solvents, glues, cement, plaster, bitumen, fillers, and brick and silica dust. Lead and asbestos are covered by separate regulations.

The regulations require employers to identify the substances employees may be exposed to, assess the hazards they may cause and eliminate or control the risks. They must provide suitable training and monitor the effectiveness of their control measures. Wherever possible, employers should not rely on the use of personal protective equipment as the foremost method of minimising exposure to a hazardous substance.

Employees must take reasonable steps to safeguard themselves from risk, including the use of protective clothing and equipment, reading instructions and COSHH information sheets, adopting safe working practices and familiarising themselves with emergency procedures.

Physical health hazards

Dusts

Dusts from many sources are a prominent hazard in construction. Particles less than 5 microns in diameter pass through the body's defences into the lungs or stomach, affecting the body in various ways. Silica and asbestos dust can permanently damage lung tissue, whilst lead in dust (from rubbing down lead paintwork, for instance) is absorbed into the lungs and enters the blood stream, causing poisoning.

Silicosis is caused by inhaling small particles of quartz-containing stone. Asbestosis, lung cancer and mesothelioma (tumour of the lining of the chest or abdominal cavity) are potential diseases for de-laggers and demolition workers. Even the hazardous blue asbestos or crocidolite, seldom used in new work, is still encountered in demolition and alteration projects.

Cement dust, especially with chemical additives, is a respiratory hazard, although dust levels are usually low, especially if ready-mixed products are used. The risk of cement burns and dermatitis is greater.

Many other dusts irritate the throat and eyes, damage the lungs or poison the body. They include dusts from some untreated hardwoods, as well as timbers treated against rot and insects. Particles of fibre glass, resin-based and plastic fillers, and bricks and blocks which have been cut, can also be harmful.

Toxic fumes

Lead poisoning has become less common, but can still occur during welding, cutting or burning off old lead-painted structures. The lead in fumes is readily absorbed into the body. This has been a hazard for demolition workers. It is less of a problem in new work, but organic chemicals are creating fresh hazards all the time. Building services operatives are at risk from welding, flame-cutting and lead-burning fumes.

The use of internal combustion engines in confined spaces puts people at risk from carbon monoxide poisoning. Paint solvents and cavity-insulation materials can give off toxic fumes. Stringent precautions are needed when using solvents and other chemicals, especially when working in an enclosed area. A forced ventilation system may be needed.

Vibration and noise

The use of vibratory tools can give rise to numbness of the fingers, commonly known as 'white finger' or 'dead hand', and to general systemic effects, such as weakness. The first sign is usually whiteness in the fingertips and numbness when the hands are cold. Some workers are more susceptible than others, but all employees at risk should be given information about the condition. It can slowly spread to other parts of the circulatory system. Severe numbness and weakness, after long exposure to vibration, can cause accidents as well as health problems.

Noise can cause hearing damage, the risk depending on the noise level and length of exposure. Drivers of heavy machines and operators of drilling equipment powered by air compressors are at serious risk, especially if the work is confined, as in tunnelling. There has been considerable progress in reducing noise problems through the use of acoustic enclosures, muffles and ear defenders.

Skin troubles

Individual reactions to substances vary considerably but cement, tar, bitumen, paints and varnishes, some woods (especially hardwoods), epoxy and acrylic resins, solvents, acids and alkalis, are common causes of skin diseases. Tar warts, for instance, are a form of cancer which can become serious if not diagnosed and treated.

Operatives should be made aware of the hazards and encouraged to use protective gloves, barrier creams and skin cleansers. Workers often use solvents and strong detergents to remove chemicals from the skin and unknowingly substitute one source of skin trouble for another.

Dermatitis is quite prevalent among construction workers. Its signs include reddening and blistering of the skin. Early treatment is important.

Radiation

Radiation from lasers and from non-destructive testing and welding is an increasing risk for construction workers. Protection from ionising radiations, such as X-rays, is very important. Although the work should be undertaken by qualified operatives, the manager must ensure that unauthorised personnel stay away from the restricted area.

Local exposure to radiation results in reddening or blotching of the skin, whilst acute general exposure may cause nausea, vomiting, diarrhoea, collapse and even death. Prolonged exposure to small doses can cause anaemia and leukaemia.

Laser beams can damage the skin and particularly the eye. The lens of the eye focuses the intense light on the retina, causing burning. Since lasers are likely to be used increasingly in surveying and other construction operations, their use must be properly supervised.

The effects of many of these hazards can be reduced if the recommendations of the Personal Protective Equipment Regulations 1992 are followed.

Effective communication and managing health and safety

Effective communication of health and safety (H&S) and workforce involvement in health and safety are critical to effective health and safety management. Construction requires a focus on inter-specialisation communication and the need for teamwork, flexibility and co-ordination making strong demands on the communication requirements. Multi-lingual and multi-cultural worksites raise further communication issues.

Effective communication of health and safety could be achieved by well planned, two way communication between management and employees and by involving staff in the development and implementation of health and safety systems, such as training, writing procedures, hazard spotting and workplace risk assessment. Forms of communicating health and safety include H&S notice boards, H&S site inductions, safety briefings (tool box talks), a formal suggestion scheme and safety committee meetings.

Recent HSE sponsored research (Anumba *et al.*, 2002) entitled *Health and Safety in Refurbishment Involving Demolition and Structural Instability* identified three different stages across which the communication of project information and of health and safety requirements has to be organised.

- *First level – client's team.* The communication of design information and of any assumed demolition sequence developed by the structural engineer is vital for the preparation of the pre-tender health and safety plan. This may also result in modifications to parts of the design or in the implementation of specific safety procedures in the pre-construction health and safety plan.
- *Second level – between designer/structural engineer and contractor.* Project information has to be communicated in detail in order to share the engineering knowledge acquired during the design stage. Communication between designers and contractors during design changes is also fundamental. The exchange of information has to be mutual because contractors, through their experience and skills, may add additional considerations to the project. If assumed demolition sequences have been developed they have to be explained to the contractor. The contractor may modify them adding the demolition method selected and the related plant and equipment.
- *Third level – contractor/sub-contractor's team.* Project information has to be communicated to the workforce and the site manager has to ensure that they understand site rules and health and safety procedures related to the activities they are going to undertake. Workers have also to be adequately instructed about not taking any initiative that has not been authorised by site managers.

The case studies conducted during the study showed examples where project information can be better communicated through drawing-based method statements. Instructions can be given to workers through regular briefings; because the instructions to be given to site workers need to be clear and concise (especially to overcome language barriers) and they have to capture the attention of the audience; drawings and pictures are strongly suggested.

The nature of the industry with temporary worksites and sub-contracting presents some challenges in securing workforce participation and warrants more complex communication systems. In an information communication technology era, newer forms of communication are becoming evident, including the Internet and mobile phones.

The common law on health and safety

Since the early 1970s, the important changes in the health and safety at work law have been statutory. However, the common law is important too, as it is the foundation of statute law and gives employees recourse to damages if they are injured or contract disease in the course of their work.

Employers have common law liabilities towards both their workers (personal liability) and third parties (vicarious liability).

Personal liability

An employer who fails to take reasonable care to ensure the safety and health of an employee can be sued for damages in the civil courts, but only if the employee has suffered an injury or loss. This contrasts sharply with the system imposed under the Health and Safety at Work Act 1974, where criminal proceedings can be taken if safety standards are breached, even if no-one is hurt.

Injured workers can only claim damages if the injury happened during the normal course of their work. This would include, for instance, injury caused in a motor accident whilst a worker was being driven to site in the firm's minibus. One employee successfully claimed damages when he got frost-bite driving the company's unheated van!

The employer only has to take 'reasonable care' and is not expected to be over-protective to workers or guarantee their safety. In one case, the court refused damages to a worker injured in a rush to get to the works canteen. The employee has to show that the employer's failure to take reasonable care caused the accident and that it would not have happened otherwise.

The responsibility for accidents cannot always be attributed solely to employer or worker. If both are partly to blame, the damages awarded are reduced by the amount of the worker's contributory negligence. Contributory negligence can apply when a worker is partly to blame for the extent of an injury, even though he or she didn't cause the accident. An example would be an operative working below scaffolding who neglects to wear a hard hat and is struck on the head by a falling object. The courts are, however, generally reluctant to penalise employees for this kind of negligence.

Vicarious liability

An employer is responsible for the negligence of employees if they cause injury or loss to another employee or a third party. The injured person can sue both employer and employee, but has the sometimes difficult problem of proving that the employee caused the injury during the 'normal course of his employment' (Field, 1982). If a mobile crane runs over the architect who is inspecting the site, there will be little difficulty in bringing a claim. But if the crane driver borrows his employer's lorry to fetch his lunch box from home and runs down the architect on a public road, proving the employer's vicarious liability will be more difficult.

In *Conway* v. *George Wimpey* (1951), the employer was not held liable for the injuries received by a hitch-hiker who was given a lift in a company lorry. The giving of the lift was not for the purposes of the employer's business and was in contravention of company rules.

However, it is hard for the employer to guard against such liabilities, even outside working hours. In *Harvey* v. *O'Dell* (1958), an employer was held liable for injuries to an employee who was riding on the back of the storekeeper's motorbike, returning to site from their lunch at a local café. Because the employer had not made proper eating arrangements on this 'outside' job, the trip to the café was held to be 'incidental' to their employment.

Occupiers' liability

Under the Occupiers' Liability Acts 1957, 1984, employers have a liability to ensure that all premises on which their business is being conducted are reasonably safe places for employees, visitors and persons other than visitors. This obligation is important for sub-contractors working on a main contractor's premises and for main contractors working on a client's premises.

Employers must be aware that their duty extends, for example, to trespassers and they must take reasonable care that non-one is injured, for instance, by a guard dog. A special concern is the safety of children, and contractors have to take precautions to prevent children entering construction sites, during or after working hours, where they are at risk. This requirement is underlined in the CDM Regulations 1994.

The Health and Safety at Work etc. Act 1974

Every employer has a common law duty to take reasonable care of employees and anyone else who may be injured as a result of the employer's activities. However, common law does not require the employer to prevent accidents from happening – it simply establishes liability if they do.

Because of this, many statutes have been passed making the employer take positive action to prevent accidents in the workplace and promote the well-being of workers. An employer who fails to comply with the statute law may be criminally liable for dangerous practices even if no accident has occurred. An example of this would be the persistent use of unguarded machinery. Acts like this have included the Mines and Quarries Act 1954, the Factories Act 1961 and the Offices, Shops and Railways Premises Act 1963. Each dealt with one industry or part of an industry.

In 1972, the Robens Committee reported that there should be one piece of legislation, covering all employees in all industries, providing preventive health and safety policies and involving workers in the making of health and safety policy.

The outcome was the Health and Safety at Work etc. Act 1974 (HSW Act). It brought about some fundamental changes. Unlike previous statutes which applied to places like factories, mines and offices, the 1974 Act emphasises the responsibilities of people – employers and employees. It covers virtually all employees.

The Act also protects other people who may be affected by an employer's activities and imposes obligations on the manufacturers and suppliers of equipment and materials.

The powers of the health and safety inspectors were widened by the Act, which also aims to increase awareness of the need for safe and healthy workplaces.

The employer's duties

Under the HSW Act, employers have a duty to protect workers' health and safety and must, so far as is *reasonably practicable*:

- provide and maintain safe systems of work and plant;
- ensure safety and absence of risks to health in using, handling, storing and transporting materials, tools or components;
- give health and safety information, instruction and supervision as needed;
- provide safe, healthy workplaces and access to them;
- provide employees with a safe, healthy working environment, and adequate welfare facilities.

As well as being responsible for premises over which they have direct control, employers have to provide a safe system of work when employees are working in other people's premises. This is important in the refurbishment or alteration of buildings or structures.

The employee's duties

Employees have a duty to take reasonable care for the health and safety of themselves and of other people who may be affected by what they do or fail to do. They must co-operate with their employers over statutory safety provisions. An employee must not interfere with safety measures or misuse health or safety equipment.

Drake and Wright (1983) point out that if employees neglect their duties under the HSW Act, they could be fairly dismissed under the provisions of the employment legislation.

Safety policy

Every firm must prepare a safety policy, keep it up to date, and bring it to the attention of employees. The policy must be in writing, except for firms employing less than five people. Their safety policy can be oral. The HSW Act does not dictate the content of the policy. Its aim is to encourage firms to work out solutions tailored to their own health and safety problems. However, the Health and Safety Executive publishes guidelines. If a serious accident occurs, the employer's safety policy will often be the starting point of the inspector's investigation. Workers' safety representatives may also take a careful look at the policy.

Most safety policies for construction firms begin with a general statement of intent and then detail the responsibilities of the various levels of management. A policy should set standards and specify how they will be achieved. Companies employing more than 20 workers must appoint a safety officer, who will assess risks and organise the firm's safety measures.

The safety policy should define responsibilities for:

- monitoring the firm's safety activities;
- maintaining contact with sources of advice, such as manufacturers, employers' federations and the Health and Safety Executive;
- organising health and safety training;
- responding to the work of safety representatives and committees.

The policy should explain how these responsibilities will be carried out. This will reflect the scope of the firm's work and should include:

- procedures for dealing with risks, including inspections, plant maintenance and guarding of machinery;
- precautions against special risks created by the firm's work;
- accident reporting and investigation procedures;
- provision and use of protective clothing and equipment;
- safe routines for introducing new equipment, materials and methods;
- emergency procedures for dealing with explosion or fire;
- arrangements for communicating with workers about health and safety matters;
- a system for identifying safety training needs and implementing training;
- inspections, audits and other arrangements for checking health and safety measures.

Administration of health and safety legislation and standards

The HSW Act created the Health and Safety Commission (HSC) and the Health and Safety Executive (HSE). Broadly, HSC formulates policy and HSE implements it. The aims of HSC/E include:

- defining a *framework of law and standards*, in particular by proposing reform of existing legislation and participating in standard-setting in the European Community and with other international bodies;
- promoting *compliance with the 1974 Act* and related measures, particularly through inspection, advice and enforcement, thus protecting employees and the public from safety and health risks;
- *investigation of accident and health problems* and related activities, including assessment, research and information services.

A full description of HSC/E aims and activities can be found in the Commission's annual reports (see, for instance, HSC, 1995a).

Role of the health and safety inspectors

There are about 20 HSE area offices in the UK, each of which has a construction team. The inspectors give advice as well as enforcing the legislation affecting construction operations. Consultants are employed to advise on specialist problems.

Although the inspectors are thinly spread in construction, they have considerable powers and can:

- enter premises (including construction sites) at any reasonable time or, if there is a hazard, at any time;
- take with them a police officer, if obstruction is expected;
- direct that the premises or part of the premises be left undisturbed while investigations are carried out;
- make any examinations and investigations, and take whatever measurements, recordings, photographs or samples they need;
- require any involved person to give information, answer questions and sign a declaration;
- inspect or take copies of any relevant document, such as a record or register, or an entry in such a document;
- require any other person to provide assistance.

In addition, inspectors are given any other powers necessary for exercising their duties. They can give information to employees or their representatives to keep them informed about health and safety matters. This may be information about the employer's premises or activities, or about any action the inspectors have taken or intend to take. The same information must be given to the employer.

If an inspector considers that the statutes have been or are likely to be contravened, he or she can serve an *improvement notice* on whoever is responsible. This requires that the problem be remedied within a specified time.

If there is a risk of serious personal injury, the inspector can prevent or stop an activity by issuing a *prohibition notice*. This can take immediate effect or can be deferred if, for example, the hazardous operation is not due to commence straight away. Prohibition notices can stop all work on a site, but more often they apply to specific operations. A notice can be served on an individual if, for instance, he or she is not wearing eye shields whilst using a grinding wheel.

Over 2500 notices were served by construction inspectors in the year 1994/95. About 2200 of these were immediate prohibition notices (HSC, 1995b). Inspectors can prosecute employers and employees and the penalties include fines and imprisonment. In the period above, about 500 convictions were obtained following proceedings by construction inspectors.

The Construction Industry Advisory Committee (CONIAC)

The Health and Safety Commission created advisory committees to look at the particular problems facing the major industries. CONIAC is the committee for construction. Its aim is to give the industry a chance to help identify areas where action is needed and to contribute to practical solutions.

CONIAC working parties have looked at problems like attitudes to safety, the wearing of safety helmets, health and safety in small firms, and the contribution of the design team and client to site safety.

An early concern of the committee was the provision of guidance on safety policies, and it published a guidance leaflet *Safety Policies*, in 1982.

A further CONIAC initiative was *Site Safe 83*, a campaign aimed at bringing about a permanent change in attitudes to safety by creating greater awareness. One of the worries is that, despite campaigns like this, the safety message is not getting through to small firms (HSE, 1983).

In the 1990s, CONIAC was particularly active in the development of the CDM Regulations and their implementation. The committee has had to tackle fears within the industry and its professions that the regulations would create unacceptable bureaucracy and extra work.

Construction regulations

Whilst the HSW Act deals with general duties, there are also detailed regulations which apply to construction. In particular, the Construction (Health, Safety and Welfare) Regulations 1996 (CHSW Regulations) have replaced the bulk of the earlier Construction Regulations which came into force in the 1960s. They don't, however, replace the Construction (Lifting Operations) Regulations 1961. These were dealt with in a separate consolidation of all the UK lifting legislation, the Lifting Operations and Lifting Equipment Regulations 1998.

The CHSW Regulations consolidate, modernise and simplify the earlier requirements, completing the implementation of the EC Temporary or Mobile Sites Directive 1992 and concluding the updating of health and safety in the construction industry. They apply to most construction work and reflect the particular processes, working practices and hazards of the industry, which are wide ranging and complex. Their scope is broad and they give protection to everyone who carries out construction work, and to people other than employees who may be affected by such work.

Whereas the CDM Regulations provide a framework, within which parties to a project can exercise judgement about what is reasonably required, the CHSW Regulations are prescriptive, targeting specific hazards, systems of work, competencies and so on. They can provide a ready-made agenda for drawing up a Health and Safety Plan under CDM (Joyce, 1995).

The CHSW Regulations 1996 cover, amongst other things:

- Safe places of work and safe access to and from work places.
- Prevention of falls; safety of scaffolding, ladders, harnesses, etc.
- Protection of workers and others against falling objects/materials.
- Avoidance of collapse of structures.
- Safety measures for excavations, cofferdams and caissons.
- Safe arrangements for traffic routes, vehicles, gates and doors.

- Prevention and control of emergencies.
- Provision of welfare facilities for washing, changing, resting, etc.
- Training, inspections and reports, and site-wide issues.

The Lifting Operations and Lifting Equipment Regulations 1998 cover the safe use of lifting appliances, including hoists, cranes and excavators; safe workloads, etc.

Other regulations which affect construction include the so-called 'six-pack' regulations, introduced in 1992 to satisfy EC Directives. These are:

- The Management of Health and Safety at Work Regulations 1999
- Workplace (Health, Safety and Welfare) Regulations 1992
- Provision and Use of Work Equipment Regulations 1998
- Personal Protective Equipment at Work Regulations 1992
- Manual Handling Operations Regulations 1992
- The Health and Safety (Display Screen Equipment) Regulations 1992.

Although most of these regulations have wide application, the Workplace (Health, Safety and Welfare) Regulations have only limited relevance to construction (Francis *et al.*, 1995).

There are many other regulations which affect construction and apply to other industries as well. These include regulations dealing with control of pollution, explosives, asbestos, flammable liquids and gases, woodworking machinery, abrasive wheels and noise at work.

Safety representatives and committees

Under the Safety Representatives and Safety Committees Regulations 1977, recognised trade unions can appoint as many safety representatives as they see fit. This flexibility is to allow for the extent of a site, local conditions, the groups of operatives to be covered, the number of unions on the site and any special features, such as shift working. The recognised unions are those which take part in collective bargaining in the industry.

Procedures for appointing safety representatives and safety committees have been agreed through the major national joint councils and are embodied in the national working rule agreements.

Site safety representatives should normally be employed by the main contractor and should have been with the firm, or a similar one, for two years. The employer is required by law to give safety representatives paid time off to carry out their safety duties and undergo safety training. The employer must help safety representatives, by providing equipment like dust and noise meters.

The national joint councils for building and civil engineering have approved a basic training scheme, designed to enable safety representatives to carry out their duties properly.

An employer must set up a safety committee if two or more safety representatives make a written request for one. The employer must consult with the unions party to the national working rules who have members on the site, and make arrangements for the committee, taking account of sub-contract employees on site. Safety committees give representatives a chance to meet management on an equal footing. Committees normally comprise equal numbers of workforce and management representatives.

Functions of safety representatives

The role of the safety representative is to:

- inspect the workplace, or that part of it to which his or her appointment refers, at three-monthly intervals, or more frequently if there have been substantial changes in site conditions;
- investigate potential hazards and look into employees' complaints;
- investigate dangerous occurrences and examine the causes of serious accidents involving deaths, broken limbs, loss of eyesight, and so on, or anything that results in prolonged absence from work or permanent damage to a worker's health;
- receive information from the Health and Safety Executive and its inspectors, and view relevant documents belonging to the employer, except individual medical records which are confidential; any dispute about disclosure of information can be referred to the industry's joint machinery;
- discuss with the employer issues arising from investigations and general matters of site safety, health and welfare;
- represent employees at safety committee meetings, where applicable.

Safety representatives are not legally responsible for health and safety on sites. They cannot be held liable under civil or criminal law for anything they do or fail to do whilst acting as safety representatives. They are, however, liable as operatives when carrying on their normal trade.

Functions of safety committees

The main functions of safety committees are to:

- monitor working arrangements on site with regard to health, safety and welfare;
- help develop site safety rules, safe systems of working and guidelines for especially hazardous operations;
- study accident trends and safety reports;
- investigate the causes of serious accidents;
- examine matters raised by safety representatives as a result of their activities.

The work of safety committees must be properly publicised to all workers and copies of the minutes of meetings made available to everyone. The HSE has recommended that on non-union sites, management should take the initiative in setting up safety committees.

Protective equipment

Employers must provide operatives with special clothing and equipment, such as the following:

- Protective clothing for working in rain or snow, and for working with asbestos products, when adequate ventilation cannot be provided.
- Respirators for working in dangerous fumes and dusts, where proper ventilation cannot be provided.
- Eye shields or protectors when cutting or drilling concrete, bricks, glass and tiles, shot blasting concrete, welding or using hand-held cartridge tools.
- Ear protectors for any noisy operation, especially when using woodworking machinery.

The Personal Protective Equipment at Work Regulations 1992 (known as PPE), part of the series of health and safety regulations which implement EC Directives, were amended in 1994, 1999 and 2002 and replace a number of old and excessively detailed laws. PPE covers safety helmets, gloves, eye and ear protection, footwear, safety harnesses and waterproof, weatherproof and insulated clothing. Self-employed operatives also have a duty to obtain and wear suitable PPE where there is a risk. Training is important; employees must know how to use equipment effectively. Equipment must also be properly maintained and properly accommodated when not in use. These regulations do not apply when PPE is provided under certain other regulations, such as the COSHH and Noise at Work Regulations.

Many of the regulations are difficult to enforce and the manager must be alert. The wearing of safety helmets is covered in the national working rules. Operatives must wear hard hats wherever there is a risk of head injury. Staff not covered by the working rules are usually required by their conditions of employment to wear safety helmets and protective clothing.

Noise should, where possible, be reduced at source using sound-reducing covers, exhaust muffles and screening. Ear defenders should be a last resort. Regular maintenance can help cut down noise. Special care is needed when operatives are exposed to noise for long periods, even a whole day. Workers at risk from adjacent operations may need ear or eye protection too.

Summary

In spite of a modest downward trend in injuries and deaths in the early 1990s, construction work remains very hazardous. The occupational health risks to construction workers also continue to cause concern. An underlying problem is that safety, health and welfare have not been taken seriously enough in the past. This was partly an attitude problem, linked with employees' perceptions of the industry, and partly ignorance of the risks.

Legislation has been necessary to prevent employers and employees skimping on health and safety either for convenience or to cut time and costs. The HSW Act 1974 was a major breakthrough, creating the Health and Safety Executive with its inspectorate and imposing wide-ranging duties on employers and employees. Other legislation has made detailed provision for construction health and safety, reflecting the industry's unique work processes and problems.

The most important development in the 1990s was the implementation of the Construction (Design and Management) Regulations 1994 which aim to tackle the underlying causes of the industry's poor record. They place substantial responsibilities and functions on clients, designers and other professionals involved in construction projects. The role of planning supervisor has been created to ensure that health and safety are considered from inception to completion. The regulations require all those involved in a project to collaborate in planning, co-ordinating and managing health and safety throughout all its stages.

Although many of the industry's health risks are now better understood, there is still a lot of scope for improving occupational health measures. One of the problems is that ill-health resulting from construction work may not show up for years or even decades, making it difficult to link cause and effect.

The HSE and its Construction Industry Advisory Group have tried to tackle the issues of health and safety in smaller firms, where lack of resources, expertise and reduced levels of supervision make it difficult to enforce statutory or even commonsense measures to protect employees, self-employed operatives and the public.

Discussion and questions

Discuss the role of 'culture' and 'people' as important factors to consider in managing health and safety in construction.

Discuss the role of communication in improving health and safety management in construction.

Chapter 17

Industrial Relations

The construction industry has enjoyed a good industrial relations record. Relationships between employers and unions have mainly been quite informal, with few major disputes. Construction firms have operated simple labour policies, with few written rules and procedures. Some people believe that industrial relations have been good because construction work is varied and interesting. Others attribute it to the fragmentation of bargaining power caused by the industry's structure and employment policies.

However, since the 1960s the industry has had to take industrial relations more seriously, because of the employment legislation and, to a lesser extent in the 1990s, union pressure on larger projects. Managers have realised that labour relations means much more than coping with isolated disputes and strikes. It involves a whole range of problems stemming from the relationship between management and the workforce.

Usually there is some conflict of interests between the two, which can become apparent in various ways. Grievances can show up as action by individual employees, such as absenteeism, bad timekeeping, restriction of output and even sabotage. If dissatisfaction is widespread, conflict may become organised. Stoppages, overtime bans and working to rule are forms of organised conflict, although these were less common in the early 1990s.

In UK industries, between the mid-1940s and mid-1960s, power in the unions gradually shifted down the line to the shop stewards. Unofficial strikes became common.

The accent has since shifted and the legal position of individual workers has improved considerably. Employees are now better protected by statute and have less need to turn to their unions to fight for their basic rights. The unions therefore have more time to engage in productive discussions with management.

In 1963, employees were given the right to receive the main terms and conditions of their employment in writing. In 1965, the system of redundancy payments was started and 1968 saw the first legislation covering race relations at work. Job security improved in 1971, when employees were given the right not to be unfairly dismissed and in 1975 laws appeared covering sexual discrimination at work. Following entry into Europe, Britain has acquired laws on equal pay (1970) and union consultation prior to redundancy (1975). In 1974, the

Health and Safety at Work etc. Act brought practically all employees under the same safety code.

There are different opinions about how far the law should intervene in industrial relations, or whether it should intervene at all. Despite the increase in legislation, the backbone of British industrial life is still the system of collective bargaining, which is more flexible and responsive to change than the law. However, the law does provide a framework – a floor of rights – on which bargaining can be based.

Many techniques are used to regulate industrial relations, ranging from joint consultation and site bargaining to job enrichment and human relations training. Some activities are aimed at settling disputes, whilst others are intended to prevent them arising.

The downward trend in industrial action since the early 1970s seems to reflect not only developments in employment law, but a number of other changes, including government measures to restrict union power, rising levels of unemployment and changes in union tactics (Langford *et al.*, 1995).

Employers' associations

Most employers' associations have several functions, of which industrial relations is usually an important one. They may also act as trade associations and provide their member companies with a wide range of legal and commercial advice. They also represent the interests of employers in dealings with government committees.

In the construction industry, these associations are usually known as *employers' federations*. There are a large number of federations representing specialist trades within construction. Over the years, some have merged, thus strengthening their resources and their bargaining power.

Construction Confederation

The Construction Confederation is the leading representative body for contractors, representing some 5000 companies who in turn are responsible for 75% of the industry's turnover. Formerly known as the Building Employers Confederation, the confederation comprises six organisations:

- National Federation of Builders
- Civil Engineering Contractors Association
- Major Contractors Group
- National Contractors Federation
- British Woodworking Federation
- Scottish Building.

As well as providing services to member firms, it deals with major issues beyond the scope of individual companies, such as political and economic pressures on the industry. It also exerts influence on the UK government and the European Union on behalf of its members. In recent years, the Construction Confederation has been involved in debates and contributing to policies in areas such as: agency work directives, aggregates levy, the black economy in construction, cowboy builders, corporate killing proposals, EU procurement rules, insurance premiums for construction firms, late payments and retention.

The Federation of Environmental Trade Associations (FETA)

This is the recognised UK body that represents the interests of manufacturers, suppliers, installers and contractors within the heating, ventilating, refrigeration and air conditioning industry. It is split into four principal Associations: BFCMA, BRA, HEVAC and HPA.

- British Flue and Chimney Manufacturers Association (BFCMA)
  Represents factory-made chimney products.
- British Refrigeration Association (BRA)
  Represents manufacturers, wholesalers, distributors, contractors and end-users of refrigeration plant.
- Heating, Ventilating and Air Conditioning Manufacturers Association (HEVAC)
  Represents manufacturers and distributors in all sections of the Heating, Ventilating and Air Conditioning Industry
- Heat Pump Association (HPA)
  Represents leading companies in the field of heat pump technology, using waste heat (from the ground or water) and turning it into heating and cooling.

Apart from the services it provides for its members, the FETA is involved in international activity via the European Partnership for Energy and Environment. They also compile market statistics for air conditioning, air distribution and refrigeration equipment.

Trade unions

The role of a trade union is to promote the interests of its members, mainly by negotiating better terms and conditions of employment. Some unions are *craft* or *general* unions, representing one or more groups of craft workers or labourers. Others are (or attempt to be) *industrial* unions, representing the interests of all workers in a particular industry. Over the years, there have been many mergers between unions as they have struggled with difficult problems and attempted to strengthen their bargaining power.

Union of Construction, Allied Trades and Technicians (UCATT)

The general aim of the union is to promote the social and economic well-being of its members and of construction workers generally. It is strongly committed to strengthening free collective bargaining and has tried to achieve the position of industrial union for construction workers. Whilst many now recognise UCATT as the principal building union, others are yet to be convinced.

UCATT was formed when several building trades unions, going through a crisis of survival, decided to amalgamate. A merger in 1970 brought together the Amalgamated Society of Woodworkers, the Amalgamated Society of Painters and Decorators, and the Association of Building Technicians. UCATT was completed in 1971 by a merger with the Amalgamated Union of Building Trades Workers.

UCATT had a difficult time at first, but it did prevent the disintegration of trade unionism in the industry caused in large measure by the growth of labour-only sub-contracting (the lump).

UCATT is the UK's principal trade union specialising in construction; with 125 000 members spread throughout England, Wales, Scotland and Northern Ireland. UCATT also organises building workers in the Republic of Ireland.

In 1996, the union initiated discussions with the major contractors, aimed at moving self-employed operatives back to direct employment. This coincided with the Inland Revenue clampdown on the employment status of the holders of 714 and SC60 tax certificates.

UCATT, it seems, has started to make progress on a fully registered workforce following the launch of the Construction Skills Certification Scheme in 1995. A skills register for building workers was one of the key demands of UCATT during the previous two decades as a means of stamping out the cowboy element of the industry which has prospered as a result of casualisation.

As the UK's premier trade union specialising in construction, UCATT is at the forefront of negotiations concerning pay, terms and conditions of employment with employers in all the main agreements covering the construction industry.

Transport and General Workers Union (TGWU)

The TGWU came into being in 1922 with the amalgamation of 11 unions. This general union has grown at an astonishing rate, more than 70 unions having joined it, making it the largest union in Britain. Since the early 1960s, its influence has spread across nearly all occupational interests, including building and civil engineering.

The union has seats on the National Joint Council for the Building Industry but has less influence than UCATT in the building sector. But in civil engineering, the TGWU tends to dominate the bargaining scene on the union side.

The TGWU is primarily a labourers' union and has strong support among unskilled and semi-skilled construction workers. It has made some progress in expanding its representation of craft trades.

General, Municipal, Boilermakers and Allied Trades Union (GMB)

A signatory to both the building and civil engineering working rule agreements, the GMB has about 20 000 civil engineering and building-related members, mostly public sector employees. They represent about 2.5% of the union's total membership.

The union has received media praise for 'its enlightened approach to industrial relations' and its expertise on health, safety, recruitment and their EC regulatory backdrop. In 1994, the union was involved in an innovative 'partnership deal', in which four unions were invited to join a European-style consortium, led by Hochtief, Siemens and Costain, in a bid to build and operate the high speed Channel Tunnel Rail Link.

In 1995, the GMB launched a recruitment campaign aimed at the half a million plus directly employed and self-employed construction workers who are not currently in a union; and it negotiated with major contractors bidding for design-build-finance-operate contracts to discuss drawing up new working agreements covering both construction and operation.

Collective bargaining

The Trade Union and Labour Relations (Consolidation) Act 1992 describes collective bargaining as negotiation between one or more trade unions and employers associations, relating to one or more of the following:

- Terms and conditions of employment.
- Engagement, non-engagement, termination of employment or suspension.
- Allocation of work or duties of employment.
- Matters of discipline.
- Workers' membership or non-membership of a trade union.
- Facilities for trade union officials.
- Machinery for negotiation or consultation.

In the building sector, collective bargaining between the employers and unions takes place through the National Joint Council for the Building Industry (NJCBI). This body evolved from earlier negotiating councils in 1926 and gives equal representation to employers and unions.

Bargaining usually centres around two types of agreement, which can be made at national, regional, local or site levels:

- *Substantive agreements* relate to wages, conditions, allowances and holiday entitlements.
- *Procedure agreements* relate to methods for resolving disputes and differences, and cover other matters like redundancies, dismissals and union representation.

Agreements can sometimes include elements of both. For instance, site stewards may negotiate a bonus scheme (substantive) in which arrangements for agreeing revised bonus targets are laid down (procedural).

Substantive agreements also deal with such matters as overtime rates, special rates, pension schemes, apprenticeships and paid leave of absence. Because employees have to travel to sites or work away from home, travelling, subsistence and lodging allowances are also an important part of the wage.

The two main building employers' associations – the Building Employers Confederation (now Construction Confederation) and the Federation of Master Builders (FMB) – have been competing with each other in the labour relations field. The FMB is the UK's largest body representing quality and standards within the building trade. With over 10 000 members they have helped to facilitate over £200 million-worth of trade in 2003. The Construction Confederation represents employers' interests on NJCBI and is arguably a more influential association in industrial relations. The FMB has tried to represent smaller builders in labour relations but has had difficulty in gaining suitable representation on the NJCBI.

As the UK's only trade union specialising in construction, UCATT is at the forefront of negotiations concerning pay, terms and conditions of employment with employers in all the main agreements covering the construction industry.

Lack of unity on the employers' side is largely caused by the wide range in size among building firms, from the small, local maintenance and jobbing builders to the large, international contractors.

When its attempts failed in the late 1970s to gain representation on the NJCBI, the Federation of Master Builders reached an agreement with the Transport and General Workers Union to set up an alternative joint negotiating body, the Building and Allied Trades Joint Industrial Council (BATJIC). This fragmented the industrial relations system of the industry still further. BATJIC's first national agreement was published in 1980.

The unions have problems too because of the wide spectrum of skills and tasks they represent. The principal craft union, UCATT, is comparatively weak compared with the labourers' main union, the TGWU, and this has made it difficult to agree a sound wages structure in the industry. National negotiations take place against a backcloth of conflicting interests within and between the unions.

National working rule agreements

Some contractors have direct collective agreements covering wages and conditions, but most follow one of the national working rule agreements. These are formulated through collective bargaining at national level. A few contractors operate outside any agreement and offer a 'catch rate' based on local supply and demand for labour.

In construction, the National Working Rule Agreement for the Building Industry is the outcome of many years of negotiation. It establishes minimum rates of pay, conditions and hours of work of the building trades and for unskilled and semi-skilled operatives. There are regional variations for some rules.

The Construction Industry Joint Council (CIJC) agreement sets rates of pay for around 600 000 construction workers on the country's major building and infrastructure sites. This agreement covers a substantial number of operatives in the industry.

The rule also covers a host of areas, including agreement in the costs of the travel of operatives to work. When travelling to a place of work which is more than 6 kilometres (3.73 miles) from home, workers employed under the Working Rule Agreements of the National Joint Council for the Building Industry (NJCBI) and the Civil Engineering Construction Conciliation Board (CECCB), or their successor from 29 June 1998 the Construction Industry Joint Council (CIJC), are entitled to receive the following allowances:

(1)   a daily fares allowance, on a scale revised annually, for each kilometre (0.62 miles) or part kilometre in excess of the first 6 kilometres, and
(2)   a daily travelling allowance for each kilometre (0.62 miles) or part kilometre in excess of the first 6 kilometres.

Where transport is provided free by the employer, only the travelling allowance ((2) above) is payable. Such travelling allowances are equivalent to travelling time payments, and are taxable in all circumstances. Fares allowances ((1) above), however, are not regarded as taxable, provided they are paid at the rates currently in force, and may be paid in full.

Negotiation in bargaining and conciliation

Effective negotiation depends on a mix of toughness and friendliness, formality and informality. Appeals for reasonableness and co-operation are among the approaches used in bargaining. Negotiators who have reasonable goals often do better than those with extravagant goals. Some understanding of the other party's problems and viewpoint is essential if satisfactory agreements are to be reached.

Maddox (1988) identifies six steps in a negotiation:

- The parties take some time to get to know each other.
- Each party states its goals and objectives in general terms.
- Negotiations are started; specific issues are presented and discussed.
- Conflicts emerge; each side tests how far the other will give way.
- Issues are reassessed and there is a move towards compromise.
- Agreement is reached and affirmed.

Experienced negotiators often rely heavily on friendly, behind-the-scenes talks and sometimes the final agreement is carefully stage-managed, so that neither side loses credibility with its members.

Employee participation and industrial democracy

Employee participation and industrial democracy are approaches to the empowerment of people, a concept outlined in earlier chapters. The potential gains include greater employee commitment and satisfaction; and enhanced decision-making – leading to higher productivity and improved competitiveness. The underlying issue is raised *expectations*. The general rise in education standards, coupled with other social changes, have led many employees to expect closer involvement in the decisions which affect their working lives.

There are several approaches to employee participation. Employers generally favour methods which involve employees at an individual or small group level and with issues close to the work face. Most EC member states have some statutory provision or agreed systems for employee participation at the workplace. Thus, employees participate in quality circles, joint committees on work processes, TQM initiatives and so on (Farnham, 1993). Germany has the most institutionalised system of workplace participation, embodied in the process of *co-determination*.

But, as Farnham points out, unions prefer the more global approach of collective bargaining, promoting a concept closer to industrial democracy. UK employers are generally not keen on the idea of collective bargaining which shifts its focus towards corporate decision-making. True industrial democracy goes beyond collective bargaining and involves employees in the corporate management of their organisations, as in the appointment of worker directors.

These concepts can be difficult to apply in construction because of its employment practices – employee mobility, casual employment, use of self-employed labour and labour-only sub-contractors. These practices mean that many employees are not sufficiently committed to one employer to have any sustained interest in participation.

One technique which can work in such situations is joint consultation. A committee is usually set up by the employer and unions as a forum to bring representatives of management and the workforce together, usually in equal numbers. Membership can change as people move on. Mostly, joint consultation focuses on workplace issues like methods of working and work flow, standards of work, targets and incentives, and job restructuring. It rarely deals with strategic issues like marketing decisions, investment plans and capital financing.

Effective consultation can create enough employee involvement to satisfy the needs of most construction workers and also be valuable for exchanging information, feedback and ideas about a wide range of issues. For professional and managerial staff, who may have a greater sense of being stakeholders in the business, other approaches to participation are needed.

Employment and workplace relations

Contracts of employment

A contract of employment can be oral or in writing, but most contracts these days are in writing and the Employment Protection (Consolidation) Act 1978 requires that, within 13 weeks from the commencement of employment, an employer must give an employee a written contract of employment, stating:

- the names of the employer and employee and the date on which employment began; and, if any previous employment counts towards continuous service, the date on which the continuous period of employment began, must be given;
- the title of the job;
- the wage rate or pay scale, or method of calculating pay;
- the intervals at which the employee is to be paid (e.g. monthly, weekly);
- details of hours of work;
- holiday entitlements, including public holidays and holiday pay;
- details of any sick pay and pension rights;
- the length of notice the employee is entitled to receive and obliged to give, to end the contract;
- whether there is a contracting out certificate for the State Pension Scheme.

Instead of supplying a written statement, the employer can refer to a document containing these details, but it must be accessible. Usually, this document will be an agreement arrived at by collective bargaining and this is the case for construction workers covered by national working rules.

The contract of employment should outline the relevant disciplinary rules or refer to a document containing them. It should also specify a person to whom the employee can appeal if he or she has a grievance. Written particulars need not be given to employees who work less than 16 hours a week.

A contract of employment states the *terms* and *conditions* of an individual's employment. The terms are bilateral – part of the agreement between the employer and the worker. The conditions are unilateral. They are rules or instructions imposed by the employer. The firm can change a condition at any time, but a term can only be changed if both parties agree.

Terms are usually stated in agreements (including collective agreements) and in certain statutory provisions. Conditions are usually found in the firm's procedures, rules and job descriptions.

Terms and conditions can easily be confused, because they often relate to the same aspect of employment. For instance, a term of contract is that the worker is entitled to, say, four weeks' annual holiday. It is a condition of contract that he or she has to take one or more of those weeks in the winter. A term will specify that employees work 37 hours a week, but it is a condition that they start work at 8 a.m. and finish at 4.30 p.m.

Express and implied terms

Express terms are those stated in the agreement between employer and employees. They cover pay, bonus payments, working hours and overtime, etc. However, such terms have to be interpreted realistically.

An *implied* term is one which is not expressly stated, often because it is so obvious that the parties did not think it necessary to mention it. Terms may be implied from accepted practices in an industry or by the terms contained in national agreements.

Discrimination and equal opportunities

Legislation has helped remove *some* of the discriminatory behaviour and inequalities of opportunity, reward and treatment which have been prevalent in many industries, including construction. But it has not achieved enough; a fundamental change of *attitudes* within the industry and society is needed. This has begun to happen, but there is a long way to go. Attitudes towards men's and women's roles and about issues like race and disability are deeply rooted in the patriarchal systems of Western society and influence the *culture* of construction. Attitudes and norms of behaviour are passed on from generation to generation through the complex processes of socialisation and education, by parents, teachers, the media and so on (Srivastava and Fryer, 1991).

So, even today, many male construction personnel continue to have very traditional views of men's and women's roles and some still view women predominantly in the roles of homemaking and child rearing. This is very disturbing when one considers that the education system should have embraced discrimination and equal opportunities issues a generation ago when important legislation on discrimination came into force in the 1970s (see below).

Discrimination based on inappropriate attitudes has been responsible for the so-called 'glass ceiling' which has prevented many women, black people and people with disabilities from being promoted to senior positions in most industries, not just construction. Women, black people and disabled people are also under-represented on construction courses. For instance, scarcely an eighth of construction students are female. Architecture fares better; about a third of its students are women.

It is more difficult to comment on the effects of sexual orientation on employment opportunities, since gay, lesbian and bi-sexual people don't necessarily disclose their sexual orientation. But it is clear that some employers and co-workers are prejudiced towards this group and that discrimination may occur if a person's sexual orientation is known to his or her manager.

For a host of reasons, women and other under-represented groups often don't find construction careers appealing, except for certain professions like architecture and landscape architecture. So, fewer members of these groups enter the industry and, when faced with discrimination in the workplace, very few reach the most senior positions.

The cost of this problem is that the construction industry is deprived of a huge number of talented people who could be attracted to it. The problem is now well defined, particularly in the area of women in construction, but is still a long way from being resolved.

Legislation has attempted to eliminate discrimination and promote equality of treatment but with limited success. Statutes include:

- The Equal Pay Act (Amendment) Regulations 2003
- The Rehabilitation of Offenders Act 1974
- The Sex Discrimination Act 1975
- The Race Relations Act 1976 (Amendment) Regulations 2003
- The Disability Discrimination Act 1995.

European law is increasingly subordinating UK legislation and the EC has adopted a number of Directives on equal opportunity matters, including equality of treatment and equal pay. The Equal Treatment Directive, for example, outlaws discrimination on the grounds of sex in recruitment and selection for jobs, working conditions, and training and promotion opportunities.

In 1994, the Employment Department (now the Department for Employment and Learning) published *Equal Opportunities: Ten Point Plan for Employers*, which offered advice on how to provide equality of opportunity to ethnic minorities, women and people with disabilities. The ideas also have relevance to ex-offenders and other groups. It is important to recognise that there is more at stake than equal *opportunities* – there is a real need for equal *treatment* in every aspect of employment relationships.

Women in construction

Andrew Gale has carried out some pioneering work since about 1987 and has made a thorough analysis of the factors influencing the employment of women in construction. Valuable work has also been done by several other researchers, notably, Clara Greed, Angela Srivastava and, more recently, Andrew Dainty.

Gale (1995) points out that discrimination against minority groups can occur in several ways – for instance as earnings differentials or occupational segregation. An example of the latter is the presence of many more women in clerical jobs than in engineering or management. He also reminds us of J. F. Madden's concept of cumulative discrimination, whereby current discriminatory behaviour is caused and sustained by the impact of previous discrimination (whereby, for instance, men occupy nearly all the top management jobs).

Discrimination against women embodies a number of stereotypical views about them – that they cannot do heavy work, that they won't like the rough conditions on site, and so on. There is little foundation to these beliefs, yet it prevents many women from enjoying successful careers in both construction management and the trades. Gale cites evidence that this is an international problem, not unique to the UK – and he refers to a study by a female construction company director who found that even employers who claimed to favour employing women did not employ any women at all in trade or manual jobs.

Srivastava (1996) found that even though individuals and organisations claimed to be encouraging women to enter construction, there remained many obstacles to women's full participation – including being in a minority, the behaviour of co-workers and the attitudes and behaviour of managers. 'Being in a minority' is more important than it sounds. Somehow the absence of a 'critical mass' of women in a work group can make it very hard for the few who are there to achieve equal treatment and have their input to the team taken seriously. Dainty (1998) observes that target recruitment can be effective in attracting women to the industry. However if women are to remain in the sector, he suggests that efforts must be made to ensure an equitable workplace environment (Dainty et al., 2001).

Unless there is a significant change of attitudes towards women and other groups who are under-represented in the industry, no real change in behaviour and organisational culture within construction will take place. The industry will continue to waste valuable human assets and perpetuate, often unwittingly, unlawful discrimination. Change in provision of facilities is important too. For instance, under the Disability Discrimination Act 1995, employers are expected to take any reasonable measures to remove barriers to equal opportunity in their recruitment and employment practices. So, for example, an employer would find it difficult to reject a disabled job applicant on the grounds that there was no suitable access to the workplace, if the problem was the absence of a ramp which could reasonably be constructed adjacent to the steps.

Disciplinary procedures

Guidelines for a sound disciplinary procedure are given in an ACAS code of practice. The code is not legally binding but describes the kind of employment practice which an industrial tribunal would look for if considering a claim for unfair dismissal.

A disciplinary procedure should normally:

- be in writing;
- state the categories of employees it applies to;
- provide for matters to be dealt with quickly;
- describe what actions may be taken;
- state the level of management which has the power to use particular penalties;
- make sure the employee knows a complaint has been made and is able to state his or her case in the presence of a union representative or colleague;
- provide that no employee is dismissed for a first breach of discipline, except for gross misconduct;
- ensure that disciplinary action is not taken until the circumstances have been fully investigated;

- give the employee a full explanation of any penalty imposed and a right of appeal, specifying the procedure.

Disciplinary procedures vary from firm to firm. They usually allow for two spoken warnings before the offender is finally warned in writing. The accent should be on helping the individual to improve, rather than on punishment.

Gross misconduct should be defined in the rule book and contract of employment. Usually it includes theft, drunkenness and insubordination, although it can be difficult to decide how dishonest, drunk or disobedient an employee must be, before his or her misconduct becomes gross.

The manager should always have a witness who can testify that a warning has been given. The employer should keep a full record of disciplinary actions and warnings, because the firm may later have to contest a claim of unfair dismissal.

Grievance and disputes procedures

A system also has to be provided for employees' complaints. Without this, minor irritations can grow into major disputes.

A *grievance procedure* normally deals with individual complaints, whereas a *disputes procedure* applies to group complaints. In both cases, the procedure should be in writing, stating where complaints should be directed and a point of appeal if conciliation fails. It should allow aggrieved employees to be accompanied by union representatives or colleagues when complaining to management and should set time limits for resolving the complaint. Full details should be recorded.

Difficult disputes involving a group of workers may be referred to an outside body if agreement cannot be reached. ACAS will help, and the Construction Confederation has its own conciliation service which its members may try first.

Dismissal

A dismissal can take place in three ways:

- The employer terminates an employee's contract of employment, with or without notice. If the employee resigns, this is not normally a dismissal. In *Elliott* v. *Waldair (Construction) Ltd* (1975), an employee drove a heavy lorry. It was thought that this work was too hard for him, so he was told to drive a smaller van. He resigned because his overtime earnings would have fallen. It was held that the order to drive a different vehicle did not constitute a dismissal.
- The employee terminates the contract, with or without notice, because of the employer's conduct. This is sometimes called a 'constructive dismissal'.
- The employee is employed for a fixed term. Dismissal takes place if the term expires without being renewed, although certain fixed-term contracts are excluded.

Fair and unfair dismissal

When a genuine dismissal has taken place, it is often necessary to establish whether or not it was fair. The Employment Protection (Consolidation) Act 1978 identified five grounds for fair dismissal:

- *Lack of ability or qualifications for the work.* The employer must act reasonably and may be expected to give the employee the opportunity to make good the deficiency in skills or qualifications.
- *Misconduct.* This would include theft, unreasonable lateness or prolonged absence from work. The employee's conduct outside work may also give grounds for fair dismissal if it could harm the employer's business.
- *Redundancy.* Employers must, however, show that they have acted fairly in deciding who to make redundant, have considered providing alternative employment, and have consulted the unions if applicable.
- *The employee would be breaking the law if he or she continued working.* If, for instance, a lorry driver has had his or her driving licence taken away and the employer cannot find other work for the driver, the dismissal would be fair.
- *Any other substantial reason.* This usually involves commercial reasons. In *Farr v. Hoveringham Gravels Ltd* (1972), it was a company rule that employees must live within reasonable distance of the works. They dismissed a manager who had moved to live 44 miles away. It was held that the dismissal was fair because someone in his position might be called out in an emergency.

Termination of contract

Selwyn (1980) identified five ways in which a contract of employment may end, without it amounting to a dismissal:

- *Resignation.* The employee clearly and unambiguously gives notice of his or her intention to resign.
- *Constructive resignation.* The employee acts in a way which shows that he/she no longer intends to be bound by his/her contract.
- *Frustration of contract.* It becomes impossible for the employee to continue working. This would include, for instance, the employee being sent to prison.
- *Consensual termination.* The parties agree that the contract will end if certain events happen. A civil engineer was given a year's unpaid leave of absence to attend a course. It was agreed that if he did not return at the end of the period, his contract would be ended.
- *Project termination.* A person is employed only for the duration of a project.

Summary

In its broader sense, industrial relations covers every aspect of the relationship between employer and employee. This relationship has always been an uneasy one, although construction has had better employee relations than most industries.

Since the 1960s, employers have been under pressure to take labour relations more seriously, partly because of quite a rapid shift in attitudes to work and towards employers, and partly because of a string of new employment laws aimed at improving employees' basic rights. Amongst other things, employees have the right not to be unfairly dismissed and not to be discriminated against on grounds of married status, race, sex and trade union membership. They are entitled to safe, healthy working conditions, time off for public duties, and financial and other help if made redundant.

Since the management of people is central to business success, managers must handle industrial relations skilfully. The techniques used range from collective bargaining at industry level, to local bargaining and grievance handling on site.

In construction, collective bargaining is quite fragmented, because there are many employers' federations and unions, representing numerous occupational skills. This creates many anomalies in wage rates and conditions and there is a need for rationalisation.

Individual employers have to comply with substantial statutory requirements on employment, including provisions relating to employment contracts, discrimination, disputes, disciplinary rules, redundancies and dismissals.

The 1990s saw a wider acceptance of the need for equality of treatment of employees and job applicants, and the stamping out of discriminatory practices. Many organisations have tried to implement an equal opportunities policy, although there are still many obstacles to overcome. One of these is the attitudes of people, in particular of some senior managers whose positions themselves resulted from discriminatory selection procedures in the past.

Discussion and questions

Discuss the challenges that organisations face in dealing with ever-changing legislation – legislation that necessarily impacts upon construction employees and activities.

What are key issues that organisations must address with regard to industrial relations between employers and trade unions?

Chapter 18

Managing Quality and Environmental Impact

The management of quality and environmental impact have become linked in many managers' minds. This is partly because similar approaches to quality and environment standards are identified in the relevant British and international standards and partly because, in the minds of many senior managers, the issues of quality and environment are major threads in the strategic thinking which is expected to guide most organisations into the twenty-first century.

These two subjects share another important feature – their success is rooted in the management of *people*. Ultimately, the improvement of environmental and quality standards depends on the attitudes of managers and employees and their commitment and willingness to change. Technical innovation and organisational change will largely fail, unless people believe in the changes and actively pursue them.

The Construction Industry Research and Information Association (CIRIA) began a research project in 1996 to examine current construction industry practices for integrating the management of quality, environmental impact, and health and safety.

Recent studies on quality management include the works of Barrett (2000) on systems and relationships for construction quality, and Battikha (2003) on highway construction. In addition, Toakley and Marosszeky (2003) have reviewed and established research needs in total project quality.

Quality management

Managers have always been responsible for the quality of goods and services produced by their teams. In this sense, there is nothing new about quality management. But the emphasis given to delivering quality more systematically and in every aspect of the business has certainly grown over the years. This is a reaction to at least three factors:

- Poor quality in components, production processes and service to clients.
- The impact, during the 1980s, of BS 5750 *Quality Systems* and its international successor, the ISO 9000 series.
- A reduction in clients' tolerance of poor quality.

Total quality management

TQM is not the only approach to quality management, but it has been an influential one. Quality guru W. Edwards Deming described TQM as 'the Third Industrial Revolution', despite the fact that quality control ought to be part of every manager's job. Schmidt and Finnigan (1992) have called it 'a new paradigm of management'. Whilst they agree that the elements of quality assurance are well known to managers, Schmidt and Finnigan argue that it is in *combining* the elements that a new way of thinking about managing organisations arises. They also cite a 1989 report by consultants Coopers and Lybrand, comparing TQM with traditional management thinking. The move to the TQM approach has included:

- *Quality definition* – a shift from product specifications to fitness for consumer use.
- *Quality control* – a shift from post-production inspection to building quality into the work process.
- *Errors* – a shift from tolerance of margins of error and wastage to no tolerance (right first time).
- *Improvement* – a shift from technological breakthroughs to gradual, continuous improvement of every function.
- *Problem-solving and decision-making* – a shift from unstructured to participative and disciplined decisions, based on reliable data.

TQM also recognises the concept of the *internal* customer, something missing from conventional management thinking.

TQM is really a business philosophy based on commitment to customer satisfaction; it involves organising the business to deliver consistent customer satisfaction by careful design of products or services; and creating systems that deliver the chosen quality standards reliably. The growth of global markets and tough international competition will ensure that quality remains high on the organisational agenda, but the *overt* expression of quality concerns in concepts like TQM may recede, as thorough quality assurance procedures become routine – internalised in the culture and management systems of the organisation.

Sadgrove (1994) stresses that benchmarking should focus on measurable items. In construction, this could include the number of complaints from clients and building users and the percentage of work (by value) that fails inspection.

Benchmarking

Many firms have introduced benchmarking. It involves studying the best practices and achievements of competitors and others in the field – and adopting them as standards for improving the company's own performance. Benchmarking can be integrated with TQM or used as part of any quality system. It can include looking at the processes in, and product/service features of, other industries. Indeed, this is sometimes where the most creative improvements can be found. So important is this activity in a highly competitive environment that organisations may set up a research department to do their benchmarking activities.

Tackling quality management in construction

The Construction Quality Forum was set up in 1993 to help the UK industry compete with its counterparts in other countries. All sectors of construction are represented in the forum, whose information on defects and failures in design and construction is fed into a computerised database developed by the Building Research Establishment (BRE).

BRE figures published in the mid-1980s attributed 90% of building failures to problems arising during design and construction. Interestingly, these were mainly 'people' related problems. They included:

- Poor communication.
- Inadequate information or failure to check information.
- Inadequate checks and controls.
- Lack of technical expertise and skills.
- Inadequate feedback leading to recurring errors.

Clients' *perceptions of quality* are also very important. Clients quite often assess quality in terms of how they experience the building in use, rather than its components and assembly.

Gaining the industry's acceptance of formalised quality management and processes has not been easy. Certification under the 1987 version of BS 5750 was almost obligatory in many sectors by 1990, but the UK construction industry has been slow to adopt quality assurance. Compared with countries like Germany and Japan, UK construction had a lot of catching up to do in the early 1990s. The situation started to change when some firms began to see the competitive advantage they might gain through BS 5750 certification. The first construction firm to win a British Quality Award was the John Laing Group, in 1991.

However, many firms that experienced quality assurance inspections perceived little, if any, improvement in the services they were offering. Indeed, many firms (including clients) argued that BS 5750 was unsuitable for construction. The industry report *A strategy for quality management systems in the construction industry*, found ten important features of construction work which differed from those embodied in the British Standard. At least one major property developer publicly questioned the relevance of BS 5750 to building, pointing out that it was based on repetitive manufacturing practices, not one-off construction projects. The developer claimed that quality assurance should begin by examining the way a business works, not by imposing a set of predetermined work practices (but most quality managers would start this way anyway). As Baden Hellard (1993) pointed out, the construction industry has tended to misunderstand the procedures embodied in the quality standards. It has viewed BS 5750 (and the International Standard ISO 9000, developed from BS 5750: 1979 and reflecting eight years' subsequent experience of its operation) as being about paperwork systems and certification, whereas the focus is on improving the overall performance of the business.

So, some companies hesitated to become involved, seeing quality standards as an obstacle to business efficiency, forced on them directly by government or indirectly by clients, especially public sector clients. Baden Hellard argued that total quality management can improve *all* aspects of design and construction, if it starts at the top, which is with the building *client*.

Another argument against adopting the British Standard was that it focused on achieving *consistent* standards or *minimum* standards – not necessarily *high* standards. But the proponents of BS 5750/ISO 9000: 1987 argued that the standard did lead to higher quality, because it required organisations to thoroughly examine existing work practices and procedures before any changes were introduced. And the EC standard, developed from ISO 9000 for the service industries, contains a module specifically written for construction consultants. The ISO addresses four areas which are important in achieving long-term high quality in construction – management of human resources, business development, sub-contracting and the importance of feedback.

Nearly all other sectors of UK industry have recognised the importance of quality management and have installed quality management systems. Defining and implementing quality management is more difficult in service industries like construction; and in professional practices, it can be even harder to define and implement quality assurance, because there is no *tangible* product directly attributable to one practice. A practice will try to improve on the skills and competencies it offers its clients, but these are not always easy to measure. The results that the practice achieves are often intertwined with, and will partly reflect, the strengths and weaknesses of other contributors to the project.

A further difficulty is that while price remains a major criterion on which tender decisions are based, firms can continue to argue that the cost of formal quality systems cannot be justified. A powerful counter-argument is that greater efficiency resulting from quality systems *reduces* a firm's unit costs and increases its competitiveness.

Industry action

There is a further benefit. Quality management techniques can help reduce contract conflicts which have been one of the most damaging problems in the industry. CIRIA examined this problem in 1990 and 1991. It argued that current forms of contract do not encourage full use of quality assurance systems. As architects and engineers are under pressure to take on quality management, CIRIA also looked at how it may have a knock-on effect on these professions' conditions of engagement. At the same time, the former Building Employers Confederation took a close look at how it could involve itself and its members more directly in quality management and in developing suitable systems.

Other industry bodies have since contributed to the quality debate. For instance, the joint review of the industry (Latham, 1994) stressed the importance of better quality management. The report called for measures to raise construction standards (such as fairer construction contracts; better procurement practices like partnering; and standardisation and modularistion throughout the construction chain) and improved management and professional training. In an effort to implement these recommendations, the Construction Round Table (representing major public and private sector clients) formed a partnership with the then National Contractors Group to look at a raft of improvements aimed at changing the culture of construction.

Quality benefits

An important approach to obtaining business, from which some professional practices and contractors can benefit, is to sell on quality and not on price, as many successful businesses already do. Many companies in other industries have found it a better policy to go for a higher value-added product or service, than for a low cost, low quality product or service.

Since the industry has long complained about having to cut prices to win contracts on tight margins, perhaps there is a lesson to be learned about the industry attempting to persuade clients that it pays to pay a little more – and get a building that gives more satisfaction and incurs lower running costs. But, importantly, higher quality does not necessarily mean higher costs. There are costs associated with poor quality, examples of which are:

- The management cost of handling clients' complaints.
- Inspecting the work concerned.
- Making good faulty work.
- Replacing sub-standard materials and components.

These costs can be incurred during and after completion of a project. Repair work carried out when a building is in occupation can be difficult and expensive. If these costs are taken into account, improving quality doesn't automatically mean adding to building cost; the reverse may be true. Peters (1989) claimed that in manufacturing industries, putting right poor quality work absorbed as much as 25% of a firm's resources. In service industries, he argued it could account for as much as 40% of total costs, which is quite staggering. It must also be remembered that one of the purposes of quality assurance systems is to improve efficiency. If this succeeds, there should be long-term cost reductions.

Interestingly, housebuilding is a success story in the quality field. The NHBC's third-party quality assurance certification, which has operated successfully for a number of years, gives customers a legal guarantee of the building's performance and quality standards, and insurance cover against deficiencies or building failure (Griffith, 1990).

Attitudes to quality

One of the critical factors in achieving effective quality and implementing good quality control is employees' attitudes towards it. Frequently, employees lack commitment to quality and this shows up in the level of rejects, customer complaints, repairs under warranty, and so on. In construction, it shows up as bad work, long snagging lists and user dissatisfaction. The industry has to realise that quality comes from people – employees who care and are committed. And people will only care about quality if their managers do – and this means managers and professionals paying attention to quality all the time and being proud of what they are doing. As Peters and Austin (1985) put it 'quality is an all-hands-on' proposition. Or in the words of John Laing's quality director, Phillip Ball, 'quality management is all about people – how they work, how well they communicate and how well they develop and implement a process of continual improvement in their every day activities'.

In the end, perhaps, the achievement of total quality management will depend on whether the industry can replace confrontation and conflict with a philosophy of teamwork and co-operation (Baden Hellard, 1993), a thought echoed in the report *Constructing the team* (Latham, 1994).

Installing a quality management system

The essence of a quality management system is that quality is managed in ways which are clearly identified, well documented and efficiently planned, implemented and controlled. So, introducing quality management involves setting up procedures, if these do not already exist, and providing documentary evidence that quality targets are being achieved (see Box 18.1).

It also means that everyone involved must be trained in quality control methods and that there should be incentives to implement the quality control procedures. Peters (1989) suggests that incentives can be extended to suppliers (and therefore sub-contractors), who are paid the premium rate for high quality materials/work, but a lower rate for sub-standard goods and services. Whether or not quality reward systems are being used, contractors are certainly paying much more attention to assessing their sub-contractors and suppliers, monitoring and recording their performance, and listening to their ideas. Some contractors provide training seminars in quality management for their sub-contractors.

It has long been known that incentives can undermine quality because they usually focus on the *quantity* of work achieved. What hasn't been so widely appreciated is that staff performance appraisal and other forms of employee evaluation can affect quality too. Deming (1986) recognised, for instance, that staff

---

Introducing quality assurance

Induction and training of staff in quality assurance matters
Thorough analysis of existing processes and routines
Development and documentation of new processes
Trials of new systems
Modification of systems following trials
Implementation of modified processes

In addition, if accreditation is sought

External audit of systems and procedures
Amendments to meet auditors' requirements

Box 18.1 Introducing quality assurance.

appraisal often lays stress on short-term performance, discourages long-term planning, demolishes teamwork and encourages rivalry. Deming felt that these directly worked against the achievement of quality.

Most companies wanting to install a quality management system appoint a quality manager, often called a quality controller in smaller firms. The design of the quality system usually involves the know-how of a number of people, so a quality group may also be set up. Because an important role of the quality standard is to define responsibilities for quality assurance, this must be built into the documentation of processes and systems.

Three kinds of *auditing* are used in quality management – internal auditing to regularly review achievement in relation to quality targets; auditing of suppliers and sub-contractors; and external auditing by a certification body if the organisation wishes to be certified to ISO 9000. Part 1 certification is for design and production; ISO 9000 Part 2 certification is for production only. A further kind of audit, known as a second party audit, occurs when a client visits the company or its site(s) to assess its quality systems.

Certification under the current quality standard, BS/EN/ISO 9000: 1994 requires organisations to demonstrate that their systems are capable of meeting customer requirements through:

● Effective systems, procedures and working methods.
● Clear communication systems.
● Clear lines of responsibility.
● Thorough documentation of all systems.
● Control of documentation and clear procedures for change.
● Satisfactory training.
● A clear system for auditing quality procedures.

Quality culture

However, as Drummond (1992) points out, one must not forget that the quality standard is just a means to an end; it is the quality that counts, not the systems. The quality standard is simply a basis for a quality *culture* in an organisation. Systems must be built on – and are no substitute for – a quality culture or philosophy. A quality management system must become part of the mind-set of everyone in a firm, practice or project team. This is built on self-respect, pride and dedication in every aspect of the organisation. Indeed, the best quality systems recognise the notion of the internal client, so that departments treat one another as customers and try to observe similar quality criteria to those which apply to their external clients. And importantly, as Drummond points out, a quality culture is not about fanatical workforce commitment, but about abandoning outdated business and management assumptions. Implementing quality systems forces managers to develop a deep understanding of processes within the firm and the difficulties employees face. Employee commitment should follow.

Quality manual

A key document in implementing quality assurance is the quality manual. There is no standard manual; one has to be written to meet each organisation's operating procedures and type of work. The manual normally includes:

- a summary of the firm's policy on quality, suitable for uncontrolled distribution to potential and existing clients;
- an enlarged version of the above, describing the quality management systems and procedures, by department or function as appropriate;
- the firm's detailed operating procedures and standard forms, including purchasing specifications and product or service specifications.

This information may be split into two or three manuals – a systems and a procedures manual and perhaps a work instructions manual (Sadgrove, 1994). The systems manual is a strategic document which may be used by marketing staff in bid presentations to potential clients. Quality system records and forms, together with documents such as codes of practice, may also be bound together in a further manual.

Control of quality depends on the manual being realistic. A manual which is too vague or idealistic is largely useless; and so are operating procedures and work instructions which are over- or under-specified.

Quality control also depends on the existence of objective criteria, such as strength and stability, durability, dimensional accuracy and environmental performance – and on the clear identification of responsibilities of the people involved. If quality cannot be measured in a fairly objective way, improvement will be difficult to achieve. Tom Peters insists that the measurement of quality should be carried out by the people or department doing the work, not by inspectors or auditors, who may cause the process to become bureaucratic and may become the focus for arguments over the interpretation of quality control data.

Perhaps the most important requirement for effective quality control is senior management commitment. Quality must be high on the agenda for such managers and they must have the tenacity to carry on the campaign for high quality, whatever the difficulties.

Quality in service organisations

Construction is a service industry and the quality of a service is less tangible than that of a product. The criteria which clients use to judge a service are often highly subjective. Indeed, Drummond cites evidence that there are elements of service quality that have little to do with either the service itself or the style of its delivery. What one building user views as a comfortable working environment, another will find unbearable. What one client judges as sociable, informal behaviour, another will view as discourteous or impudent.

Building owners and users will consider objective factors such as the specification of the building's components, but many aspects of the building will be judged much more on their subjective responses. Examples include: whether the individual *feels* that the internal environment is comfortable, bright and pleasing; whether the individual *judges* the air conditioning to be satisfactory; whether the internal finishings meet the occupants' *expectations* (with all the subjective overtones of taste, status and self-esteem).

There is also the level of client satisfaction or dissatisfaction which goes beyond the fitness for purpose of the building itself. It relates to the quality of the *delivery* of the service. Effective management of the service delivery of design and construction processes is vital. The interaction which takes place between providers and clients crucially affects clients' perceptions of quality.

The client's evaluation of service quality will be raised if the industry's professionals demonstrate competence, trustworthiness and dependability; show their concern for, and understanding of, the client's needs; and exhibit considerate, friendly and enthusiastic behaviour. The client's perceptions of a contractor or private practice will depend on *consistency* of service and the *confidence* this engenders. But first impressions also count. Research has shown that the initial contacts have a lot of influence on subsequent relationships. The concept of partnering addresses many of these issues.

The industry might do well to examine US achievements in TQM. Schmidt and Finnigan (1992) summarise the success factors in US award-winning TQM companies as follows.

● A very high level of management leadership and commitment.
● Supportive organisational structures and roles.
● Quality-orientated tools and processes.

- Tailored education programmes.
- Innovative reward strategies.
- Full and continuing communication.

In addition, total quality managers:

- give priority to customers' needs;
- empower, rather than control, their team members;
- emphasise improvement, rather than maintenance;
- encourage co-operation rather than competition;
- train and coach, rather than direct and supervise;
- encourage and recognise team effort;
- learn from problems, rather than minimising them;
- choose suppliers on the basis of quality, not price.

Environmental impact

Business and the environment

Quality has always been somewhere on the management agenda, but the environment has not. Years ago, managers could take a largely 'closed system' view of their organisations and ignore environmental factors almost totally. But this has all changed. Today, business survival is largely about understanding the external environment and how it affects the organisation's performance.

This environment is complex. It includes all the interrelated events, changes and decisions taken in the systems of society (some predictable, but many not) which directly and indirectly influence markets, productivity, competitiveness and so on; and it includes the physical environment, and customers' and society's expectations for the future of the natural environment. More importantly, it includes our growing understanding of the long-term damage that organisations are doing to natural systems and the high probability that this damage is irreversible and will, at some point, lead to global ecological changes.

Bennis and his colleagues (1994) have underlined the importance of reassessing conventional business assumptions and beliefs and moving them towards the goal of sustainable development. This requires a major shift in people's attitudes and behaviour. Only when this happens and senior managers commit themselves and their teams to a new business philosophy, will organisations meet the environmental challenge. A few examples of the economic and business assumptions and beliefs needed for sustainable development are as follows.

- The purpose of businesses should be to satisfy all human needs, with minimum consumption of scarce resources.
- The interests and needs of future generations, and of other communities, must not be jeopardised for short-term economic interests.

- Business operations should enhance the environment, rather than damage it, and contribute to ecological balance.
- The well-being of all the other stakeholders in a business is as important as that of its equity shareholders.
- Businesses do not own all the resources they use; they hold them in trust to make the best possible use of them on behalf of society.

In construction, the refurbishment sector is well-placed to meet some of the criteria for sustainable development. The building is, in effect, recycled or re-used and, with good design and management, keeps its consumption of virgin materials and manufacturing energy to a minimum. It also recycles land, extending the useful life of areas already 'de-natured' and reducing demand for green field sites. New build is a different story and many hectares of green space are put under concrete or tarmac every day in the name of progress.

Environmental management and construction

Both the construction and property industries must play a responsible role in managing the environmental impact of development, because the problems stem both from building operations and buildings in use. Infrastructure projects, particularly roadbuilding, also have significant environmental repercussions. The issues affect planners, project owners, designers, project managers, construction managers, material producers and manufacturers, sub-contractors, facilities managers, building users, local authorities, regulatory bodies and others whose decision-making has an impact on natural systems.

New approaches to procurement, such as partnering, reinforce the point that environmental responsibility is a shared one and must be tackled collectively. Solutions to major environmental impact risks can only be achieved through multi-professional, and even pan-industry, collaboration.

By the beginning of the 1990s, most major projects throughout Europe were subject to an environmental assessment and increasing numbers of construction organisations were thinking hard about drawing up environmental policies and plans. They did this in response to new and proposed environmental legislation and because they could see a slow but unstoppable shift in client and public concern about the environmental impact of buildings and the building process (Fryer and Roberts, 1993).

By 1991, a number of construction industry organisations were carrying out research on environmental issues and the actions needed. CIRIA set up the Construction Industry Environmental Forum in collaboration with BRE (the Building Research Establishment) and BSRIA (the Building Services Research and Information Association) to promote awareness and understanding of environmental issues in the industry. At about the same time, BSRIA began a major research study aimed at producing and encouraging the adoption of an environmental code of practice for building services.

More recently, CIRIA started a multi-disciplinary research project to review the industry's practices in the context of ISO 9000 (Quality Systems), ISO 9004 (Environmental Management), the Construction (Design and Management) Regulations and the Health and Safety at Work Act. Important research into specific environmental problems is also being undertaken by many university departments and other organisations with research capabilities.

The effects which buildings and construction processes have on the environment can be stated fairly simply, but the issues are in fact complex and interrelated. CIRIA grouped the issues under these headings:

- Energy use, global warming and climate change
- Resources, waste and recycling
- Pollution and hazardous substances
- Internal environment of buildings
- Planning, land-use and conservation
- Legislation and policy issues.

Recognising the breadth and severity of these environmental imperatives, organisations like the Construction Industry Council are responding to the call for better environmental management. It is at the level of individual firms, especially the smaller ones, that reaction has been slow. The climate in which many of these firms are struggling to survive is an *economic* one.

Construction managers have a special responsibility for the efficient use of energy and resources, waste management and recycling, avoidance of pollution, land contamination and danger from hazardous substances – all within the context of new environmental legislation and their companies' increasingly visible environmental policies.

The construction industry is under increasing pressure to reflect on and assess its impact on the environment and take concerted action. This requires integrity and commitment on the part of all the industry's professions and a thorough understanding of the issues and the burgeoning European legislation, to which whole books can be devoted (see for instance Griffith, 1994).

Environmental management systems

Until the early 1990s and the enactment of the Environmental Protection Act 1990, few construction organisations had taken the environmental impact of their operations seriously. By 1996, when the Environmental Agency was launched, the situation had changed, but was still far from ideal. Client pressure and the publicity given to BS 7750: *Specification for Environmental Management Systems* (1992), contributed to some shift in attitudes. Some firms began to understand the importance of *sustainable development*, a concept which stresses using resources of energy and materials in a responsible way, so that future generations can benefit from them too.

The Environment Act 1995 set up the Environment Agency. The agency amalgamates the National Rivers Authority, HM Inspectorate of Pollution and some 80 Waste Regulation Authorities in the UK. The 1995 Act makes the polluters of land liable for the costs of its remediation, a responsibility which cannot be ignored by the construction industry. Environmental law is a growing and enormously wide-ranging subject, the bulk of UK legislation now emanating from EU proposals and regulations (Francis *et al.*, 1995).

ISO 14000, which is an international standard series for promoting environmental protection and sustainable development, was introduced in September 1996. It specifies the requirements and procedures for establishing an environmental management system. Few construction companies have actively pursued certification to this standard despite having an obligation to implement it, as the services and products they produce directly impact upon the environment. CIRIA's (2000) Report C533 on *Environmental Management in Construction* provides practical assistance to companies in the construction sector that are considering, or already tackling, environmental management. Other studies that provide useful information in the same area are those of Zhang *et al.* (2000) and Walker (2000).

Environmental management policy and strategy

Environmental management involves designing or revising an organisation's practices, processes and structures so that it can achieve its core objectives in an environmentally responsible way. Any company taking its environmental obligations seriously must start with a policy which relates its core business objectives and strategies to its environmental aims. Such a policy must be flexible, because firms differ markedly and their circumstances change (Griffith, 1994). But unless the policy informs the organisation's business strategy, it is unlikely that effective environmental performance will be achieved. In addition to the requirements of environmental legislation, clients increasingly enquire about the environmental policies of the construction firms and practices with whom they enter contracts, so it is realistic to expect that in future the existence of a sound environmental policy and strategy will be a key factor in a firm's competitiveness (Fryer, 1994a).

BS 7750/ISO 9004 provides guidance for a firm wishing to introduce a management system for improving its environmental performance. The standard parallels EC environmental standards and shares many of the management principles embodied in the quality standard BS/EN/ISO 9000. The main building blocks are now in place to allow a full and positive relationship to develop between corporate objectives and environmental needs.

Sustainability has really emerged as an important management issue. This is partly fuelled by ISO 14000. Issues such as the triple bottom line, which focuses on company profit performance (economic issues) being balanced by demonstrating performance on delivering value to society (social issues) while improving the ecological environment (environmental issues), are gaining currency and providing ample challenges for construction organisations. The report by the International

Council for Research and Innovation in Building and Construction (CIB) – Publication 237: Agenda 21 on Sustainable Construction provides a detailed overview of the concepts of sustainable development and sustainable construction from environmental, social and economic perspectives. It also addresses the main issues and challenges of sustainable construction.

Environmental action planning

Practical guidance on formulating an environmental plan are given in the Institute of Management's action checklist No 19, *Taking Action on the Environment*, published in 1996. The checklist advises firms to:

- secure top management commitment;
- identify the environmental laws and regulations;
- designate a senior manager to be responsible for environmental affairs;
- establish and communicate a clear policy;
- work out the environment-business link;
- carry out regular audits;
- develop a procedures manual;
- start an environment training programme;
- publicise environmental objectives internally and externally;
- build in measures and controls;
- communicate environmental benefits internally and externally;
- involve employees to gain their commitment.

As is so often the case in management, effective communication, consultation and training – to encourage appropriate attitude change and to gain the commitment of employees and other stakeholders – are key factors in the successful implementation of management plans.

Environmental impact assessment

Also known simply as environmental assessment (EA), environmental impact assessment is a set of procedures for measuring the probable environmental effects of a project before it is allowed to start. The principles of EA are not new and have been practised in the oil, gas and petro-chemical industries since the early 1970s (Griffith, 1994).

The UK construction industry is affected by an EC Directive aimed at ensuring that all major projects – public and private – are the subject of an environmental assessment before consent is given. The DoE introduced a system in the late 1980s for ensuring that projects conform with the EC Directive and the major output of the EA process is an environmental statement prepared by the developer and submitted to the competent authority – usually the planning authority – ideally *after* prior consultation with that authority, which can provide the developer with valuable advice and information (Roberts, 1994).

Environmental and quality auditing

Many larger organisations have introduced an auditing process for both their quality management systems and environmental management systems. An audit is a systematic, periodic evaluation of a management system in an organisation to assess its effectiveness in meeting key objectives and statutory requirements. The frequency of audits, the procedures used and the methods of reporting need to be carefully thought through. They will probably differ from organisation to organisation.

There is no statutory requirement for auditing, but many firms see it as an essential part of responsible business operation, contributing to the regular review of the organisation's strategy and, where appropriate, being integrated with procedures imposed by statute, such as the COSHH regulations and other health and safety legislation (Roberts, 1994).

Steps to be taken in quality or environmental auditing include the following.

- Setting audit objectives.
- Deciding on the scope of the audit.
- Defining its baseline.
- Selecting an audit team.
- Collecting evidence and information in relation to audit objectives and means of assessment.
- Assessing and evaluating audit results.
- Publishing the results.
- Developing an action plan for change and improvement.
- Monitoring the effectiveness of action taken.

Summary

Since about 1990, quality and environment have become two of the most frequently used words in management. The setting up of both quality and environmental management systems is seen as a high priority in many forward-looking organisations which want to survive and prosper in the global marketplace. Both systems are recognised as having strategic importance and both have necessitated a major shift in attitudes among employees and managers, with accompanying changes in organisation structures and cultures.

Quality is not a new concern for the manager. Quality assurance has always been a recognised management function. What has changed is the emphasis placed on quality in every aspect of an organisation's activities and the formalisation of quality assurance procedures. The first is a result of concepts like TQM, aimed at changing the philosophy of businesses; the second follows from adoption of the new quality standards.

Environmental impact, on the other hand, is a relatively new issue for most managers. Not many people in industry and the professions had given it serious consideration prior to the World Commission on Environment and Development in 1988, the so-called Brundtland Report. Now, the construction industry, like other sectors, is under pressure to respond to demands from clients, governments and other groups to demonstrate commitment to sustainable development; and to meet new standards and statutory requirements emanating from EC directives. The introduction of environmental assessments prior to approval of all major development projects has put environmental protection right at the forefront of planning and development. All the parties involved in the construction and property industries will have to play a part in achieving new environmental standards.

In future, excessive formality in quality management systems, which has beset some firms as they grappled with the new quality standards, may recede. Their quality procedures may become absorbed into the culture, corporate plans and operational routines of their businesses and ways may be sought to remove the bureaucracy which such systems can spawn. The future of environmental management systems is not yet clear. Governments are more likely to legislate on environmental issues than they are on quality, except where the latter affects health and safety. If organisations don't meet new environmental standards, governments may bring about change through further statutes.

Because quality and environmental impact can both impinge on health and safety, a likely development is the integration of management systems used by firms to deal with these three areas. As many of the procedures used are already similar, this is a logical development which will improve efficiency.

Discussion and questions

Effective total quality management provides benefits to construction organisations. However, its implementation raises a host of challenges. Discuss this statement.

The services and products produced by Envirobabes Ltd, a large construction company, directly impact upon the environment. The director of Envirobabes is planning to make an important presentation to the stakeholders and the city, on the implementation of ISO 14000 by the company and the efforts that the company is making in managing the triple bottom line of economic, social and environmental issues of sustainability. Discuss the main issues which the director needs to consider for an effective presentation in terms of breadth and depth of coverage.

References

Adams, E. C. and Freeman, C. (2000) 'Communities of practice: bridging technology and knowledge assessment'. *Journal of Knowledge Management.* **4** (1), 38–44.

Adair, J. (1986) *Effective Teambuilding.* Aldershot: Gower Publishing.

Alexander, G. (1995) 'Holes in the safety net'. *Building* **260**, (17), 34.

Anderson, C. M., Riddle, B. L. and Martin, M. M. (1999) 'Socialization processes in groups'. In: Frey, L. R. *The Handbook of Group Communication Theory and Research.* Sage Publications, London, 139–163.

Andrews, J. and Derbyshire, A. (1993) *Crossing Boundaries: a report on the state of commonality in education and training for the construction professions.* London: Construction Industry Council.

Ansoff, H. I. (1987) *Corporate Strategy.* Harmondsworth: Penguin Books.

Anumba, C; Egbu, C., Gottfried, A. and Marino, B. (2002) Health and Safety in Refurbishment involving Demolition and Structural Instability. *Final Report. November 2002. HSE Research Project.* United Kingdom.

Argyle, M. (1969) *Social Interaction.* London: Methuen.

Argyle, M. (1983) *The Psychology of Interpersonal Behaviour.* Harmondsworth: Penguin.

Argyle, M. (1989) *The Social Psychology of Work.* Harmondsworth: Penguin Books.

Armstrong, P. T. (1980) *Fundamentals of Construction Safety.* London: Hutchinson.

Arsenault, A. and Dolan, S. (1983) 'The role of personality, occupation and organization in understanding the relationship between job stress, performance and absenteeism'. *Journal of Occupational Psychology* **56** (3), 227–40.

Association for Project Management (1997) *Project Risk Analysis and Management.* London: APM.

Averill, J. R. (1993) 'Illusions of anger'. In Felson, R. B. and Tedeschi, J. T. (eds) *Aggression and Violence: Social Interaction Perspectives.* Washington. American Psychological Association, 171–192.

Baden Hellard, R. (1988) *Managing Construction Conflict.* Harlow: Longman Scientific and Technical.

Baden Hellard, R. (1993) *Total Quality in Construction Projects.* London: Thomas Telford.

Bales, R. F. (1950) *Interaction Process Analysis: A Method for the Study of Small Groups.* Cambridge, USA: Addison-Wesley.

Bales, R. F. (1953) 'The equilibrium problem in small groups'. In Parsons, T., Bales, R. F. and Snils, E. A. (eds) *Working Papers in the Theory of Action.* New York: Free Press, 111–163.

Bales, R. F. (1970) *Personality and Interpersonal Behaviour.* New York: Holt, Rinehart and Winston.

Bales, R. F. (1980) *SYMLOG Case Study Kit: With Instructions for a Group Self Study*. New York: The Free Press.

Bales, R. F., Cohen, S. P. and Williamson, A. (1979) *SYMLOG: A System for the Multiple Level Observation of Groups*. New York: The Free Press.

Barnes, S. and Hunt, B. (2001) *E-commerce and V-Business: Business Models for Global Success*. Oxford, UK: Butterworth-Heinemann.

Barrett, P. (2000) 'Systems and relationships for construction quality'. *International Journal of Quality and Reliability Management* **17** (4), 377–392.

Battikha, M. G. (2003) 'Quality management practice in highway construction'. *International Journal of Quality and Reliability Management* **20** (5), 532–550.

Bayley, L. G. (1973) *Building: Teamwork or Conflict?* London: George Godwin.

Belbin, R. M. (1981) *Management Teams: Why They Succeed or Fail*. London: Heinemann.

Belbin, R. M. (1993) *Team Roles at Work*. Oxford: Butterworth-Heinemann.

Belbin, R. M. (2000) *Beyond the Team*. London: Butterworth-Heinemann.

Bell, M. and Lowe, R. J. (2000) 'Building regulation and sustainable housing Part 1: a critique of Part L of the Building Regulations 1995 for England and Wales'. *Structural Survey* **18** (1), 28–37.

Bell, M. and Lowe, R. J. (2001) 'Building regulation and sustainable housing Part 3: Setting and implementing standards', *Structural Survey* **19** (1), 27–37.

Bennett, J. and Jayes, S. (1998) *The Seven Pillars of Partnering*. London: Thomas Telford Publishing.

Bennett, R. and Gabriel, H. (1999) 'Organisational factors and knowledge management within large marketing departments: an empirical study'. *Journal of Knowledge Management* **3** (3), 212–225.

Bennett, S. (2002) *Best Value: Housing and Support*. The Housing Corporation, London.

Bennis, W., Parikh, J. and Lessem, R. (1994) *Beyond Leadership*. Oxford: Blackwell Business.

Betts, M. (1997) *Strategic Management of IT in Construction*. Oxford: Blackwell Science.

Bhatt, G. D. (2000) 'Organising knowledge in the knowledge development cycle'. *Journal of Knowledge Management* **4** (1), 15–26.

Blackburn, P. and Fryer, B. (1995) 'Head of the class'. *New Builder* (3 February), 244, 22–3.

Blackburn, P. and Fryer, B. (1996) 'An innovative partnership in management development'. *Management Development Review* **9** (3), 22–5.

Blake, R. R. and Mouton, J. S. (1964) *The Managerial Grid*. Houston: Gulf Publishing.

Blake, R. R. and Mouton, J. S. (1978) *The New Managerial Grid*. Houston: Gulf Publishing.

Bliss, E. C. (1985) *Getting Things Done*. London: Futura Publications.

Bogdanov, J. (2001) 'E-business: an answer for construction?'. *Construction Monitor*. [http://www.dti.gov.uk/construction/news/conmon/dec01/con4.htm] [Accessed: 30 September 2003].

Boud, D. (ed.) (1988) *Developing Student Autonomy*. London: Kogan Page.

Bowen, P. A. (1993) *A Communication based approach to price modelling and price forecasting in the design phase of the traditional building procurement process in South Africa*, Unpublished PhD Thesis, Department of Quantity Surveying, University of Port Elizabeth.

BRE (2000) *Value for Construction: Getting Started in Value Management*. BRE, Watford.

Bresnen, M. and Marshall, N. (2000) 'Partnering in construction: a critical review of issues, prelims and dilemmas'. *Construction Management and Economics* **18**, 229–237.

British Psychological Society (2003) *Psychological Testing – A Test User's Guide*. Leicester: British Psychological Society [http://www.bps.org.uk].

Brown, P. (1999) 'Client-based management qualifications – a case of win–win?'. *Journal of Management Development* **18** (4), 350–361.

Brown, P. (2003) Seeking success through strategic management development. *Journal of European Industrial Training* **27** (6), 292–303.

Brown, R. (1986) *Social Psychology*. 2nd edn, New York: Free Press.

Brown, R. (1965) *Social Psychology*. New York: Free Press.

Brown, W. and Jaques, E. (1965) *Glacier Project Papers*. London: Heinemann.

Bruner, J. S. (1966) *Towards a Theory of Instruction*. Cambridge, Mass.; Harvard University Press.

BSI (2000) BS 6079-3:2000 *Project Management – Part 3: Guide to the Management of Business Related Project Risk*. London: BSI.

Building (2003) '50 top clients – a building directory'. *Building* (Supplement), February.

Burns, T. and Stalker, G. M. (1966) *The Management of Innovation*. London: Tavistock Publications.

Business Impact Task Force (BITF 2000) *Winning with Integrity*, Business in the Community [http://www.bitc.org.uk].

Byrne, R. (2001) Employees: capital or commodity? *The Learning Organization* **8** (1), 44–50.

Cannon, T. (1996) *Welcome to the Revolution: Managing Paradox in the 21st Century*. London: Pitman Publishing.

Capers, B. and Lipton, C. (1993) 'Hubble space telescope disaster'. *Academy of Management Review* **7** (3), 23–27.

CBPP (1998) *Fact Sheet on Value Management*. October [Internet http://www.cbpp.org.uk].

CBPP (2002) *Directors' Briefing – Research for Your Marketing*. Business Hotline Publications Ltd [http://www.cbpp.org.uk].

Christopher, M. (1992) *Logistics and Supply Chain Management: Strategies for Reducing Costs and Improving Services*. London: Pitman Publishing.

Cicmil, S. J. K. (1997) Critical factors of effective project management. *The TQM Magazine* **9** (6), 390–396.

CITB (2000) *Managing Profitable Construction: The Skills Profile*. CITB, July.

CITB (2003a) *Key Corporate Messages*, CITB, 24 July.

CITB (2003b) *Reducing Skills Shortages in Construction*, CITB Fact Sheet.

Clark, N. (1994) *Team Building*. Maidenhead: McGraw-Hill.

Cline, R. J. W. (1994) 'Groupthink and the Watergate cover-up: The illusion of unanimity.' In Frey, L. R. (ed.) *Group Communication in Context: Studies of Natural Groups*. New Jersey: Lawrence Erlbaum associates, 199–223.

Cole-Gomolski, B. (1997) 'Users loathe to share their know-how'. *Computerworld* **31** (35), 49–50.

Constable, J. and McCormick, R. (1987) *The Making of British Managers*. London: BIM and CBI.

Construction Industry Council (1993) *Profiting from Innovation – A Management Booklet for the Construction Industry*. UK. ISBN 1 898710 X.

Construction Industry Research and Information Association (CIRIA, 2000) *Environmental Management in Construction*, authored by S. Uren and E. Griffiths. *CIRIA Report No. C533*. UK.

Construction Industry Research and Information Association (CIRIA, 1998) *Selecting Contractors by Value*. SP150 – Jackson-Robbins, A.

Construction Research and Innovation Strategic Panel (CRISP, 1997) *Creating a Climate of Innovation in Construction*. A document produced by the CRISP Motivation Group, February, 1997, UK. (CRISP reports to the UK Construction Industry Board (CIB) on research and innovation issues.)

Cooper, C. L. (1978) 'Work stress'. In Warr, P. B. (ed.) *Psychology at Work*. Harmondsworth: Penguin Books, 286–303.

Cooper, C. L. (1984) 'Stress'. In Cooper, C. L. and Makin, P. (eds) *Psychology for Managers*. Leicester and London: British Psychological Society and Macmillan Press, 239–58.

Cooper, D. (1995) 'Motivation: determining influences on behaviour'. In Hannagan, T. (1995) *Management Concepts and Practices*. London: Pitman Publishing.

Cox, A. and Ireland, V. (2002) 'Managing Construction Supply Chains: The Common Sense Approach'. *Engineering, Construction and Architectural Management* **9** (5/6), 409–418.

Cox, A. and Thompson, I. (1997) ' "Fit for purpose" contractual relations; determining a theoretical framework for construction projects'. *European Journal of Purchasing and Supply Management* **3**, 127–135.

Cox, C. J. and Cooper, C. L. (1988) *High Flyers: An Anatomy of Managerial Success*. Oxford: Basil Blackwell.

Dainty, A. R. J. (1998) *A grounded theory of the determinant of women's under-achievement in large construction companies*. PhD Thesis, Loughborough University, Loughborough.

Dainty, A. R. J., Bagihole, B. M. and Neale, R. H. (2001) 'Male and female perspectives on equality measures for the UK construction sector'. *Women In Management Review* **16** (6), 297–304.

Dainty, A. R. J., Bryman, A. and Price, A. D. F. (2002) 'Empowerment within the UK construction sector'. *Leadership and Organisation Development Journal* **23** (6), 333–342.

Davis, H. and Scase, R. (2000) *Managing Creativity: the Dynamics of Work and Organization*. Buckingham: Open University Press.

Day, D. (1994) *Project Management and Control*. Basingstoke: Macmillan.

Dearlove, D. (2000) 'Value your company's brainpower'. *The Times*, UK, October 26th.

De Bono, E. (1977) *Lateral Thinking: A Textbook of Creativity*. Harmondsworth: Penguin Books.

Defence Estates Organisation (1998) 'Value Planning and Management'. *Technical Bulletin 98/26*, DEO, Sutton Coldfield.

Deming, E. (1986) *Out of the Crisis*. Cambridge: CUP.

Department of the Environment, Transport and Regions (DETR, 2000) '*KPI Report for the Minister of Construction*'. A report produced by the KPI Working Group, January 2000. DETR, UK.

Department of Environment, Transport and the Regions (DETR, 2000) '*E-business in construction: status, opportunities and the role of DETR*'. Prepared by Davis Langdon Consultancy (DLC), UK.

Department of Trade and Industry (DTI, 1998) '*Innovating for the future: investing in R&D*'. Budget '98 – A consultation document. HM Treasury, DTI, UK.

Department of Trade and Industry (DTI, 2002a) '*The impact of E-business in UK Construction*'. Prepared by Davis Langdon Consultancy (DLC), UK, January.

Department of Trade and Industry (DTI, 2002b) *A Review of government R&D. Policies and Practices. Construction Research, Innovation and Best Practice – Rethinking Construction Innovation Research*. The Fairclough Report (Sir John Fairclough), UK.

Department of Trade and Industry (DTI 2003) *Our Energy Future – Creating a Low Carbon Economy*. Department of Trade and Industry White Paper, Cm 5761, London: The Stationery Office.

Donald, M. (1991) *Origins of the Modern Mind: Three stages in the Evolution of Culture and Cognition*. Cambridge, MA: Harvard University Press.

Drake, C. D. and Wright, F. B. (1983) *Law of Health and Safety at Work: The New Approach*. London: Sweet & Maxwell.

Drennan, D. (1989) 'Are you getting through?', *Management Today*, August, 70–72.

Drucker, P. (1968) *The Practice of Management*. London: Pan Books.

Drucker, P. E. (1995) 'The information executives truly need'. *Harvard Business Review* January–February, 54–62.

Drummond, H. (1992) *The Quality Movement*. London: Kogan Page.

Edvinsson, L. (2000) 'Some perspectives on intangibles and intellectual capital 2000'. *Journal of Intellectual Capital*, **1** (1), 12–16.

Egan, J. (1998) *Rethinking Construction – Report of the Construction task force on the scope for improving the quality and efficiency of UK construction*. London: Department of the Environment, Transport and the Regions, UK.

Egan, J. (2002) *Accelerating Change*. Rethinking Construction ISBN 1 898671 28 1.

Egbu, C. O. (1999) 'Skills, knowledge and competencies for managing construction refurbishment works'. *Construction, Management and Economics* **17**, 29–43.

Egbu, C. O. (1999a) 'Mechanisms for exploiting construction innovations to gain competitive advantage'. *Proceedings of the Fifteenth Annual Conference of the Association of Researchers in Construction Management (ARCOM)*. September 15–17, John Moores University, UK, Vol. 1, 115–123.

Egbu, C. O. (1999b) 'The role of knowledge management and innovation in improving construction competitiveness'. *Building Technology and Management Journal* **25**, 1–10.

Egbu, C. O. (2000) 'The role of information technology in strategic knowledge management and its potential in construction industry'. *Proceedings of a UK National Conference on Objects and Integration for Architecture, Engineering and Construction*, 13–14 March, BRE, Watford, UK, 106–114.

Egbu, C. (2001a) 'Managing innovation in construction organisations: an examination of critical success factors'. In Anumba, C. J., Egbu, C. and Thorpe, A. *Perspectives on Innovation in Architecture, Engineering and Construction*. Centre for Innovative Construction Engineering, Loughborough University.

Egbu, C. O. (2001b) 'Knowledge management and human resource management (HRM): the role of the project manager'. In *Proceedings of Fourth European Project Management Conference*, 6–7 June, London.

Egbu, C. and Botterill, K. (2002) 'Information technologies for knowledge management: their usage and effectiveness'. *Journal of Information Technology in Construction* **7**. Special Issue ICT for Knowledge Management in Construction, 125–137 [http://www.itcon.org/2002/8].

Egbu, C., Bates, M. and Botterill, K. (2001) *The Impact of Knowledge Management and Intellectual Capital on Innovations in Project-based Organisations*. Leeds, Centre for Built Environment (CeBE), Leeds Metropolitan University, UK. CeBE Report No. 11, November.

Egbu, C. and Gorse, C. A. (2002) 'Teamwork'. In Stevens, M. (ed) *Project Management Pathways*. High Wycombe: The Association for Project Management Section 71, 1–16.

Egbu, C. O., Gorse, C. and Sturges, J. (2000) 'Communication of knowledge for innovation within projects and across organisational boundaries'. *Congress 2000. Fifteenth World Congress on Project Management*, 22–25 May, Royal Lancaster Hotel, London, UK, Session 5: Project management at all levels.

Egbu, C. O., Henry, J., Quintas, P., Schumacher, T. R. and Young, B. A. (1998) 'Managing organisational innovations in construction'. *Proceedings of the Association of Researchers in Construction Management (ARCOM) Conference*, 9–11 September 1998, Reading, UK.

Elliman, T. and Orange, G. (2000) 'Electronic commerce to support construction design and supply-chain management: a research note'. *International Journal of Physical Distribution and Logistics*, **30** (3/4), 345–360.

Ellis, D. G. and Fisher, B. A. (1994) *Small Group Decision Making: Communication and the Group Process*, *4th* edn. New York: McGraw-Hill.

Ellis, R. C. T., Thorpe, A. and Wood, G. D. (2003a) 'eLearning for project management'. *Proceedings of the Institution of Civil Engineers*. **156** (3), 137–141.

Ellis, R. C. T., Thorpe, A. and Wood, G. D. (2003b) 'Distance learning and postgraduate education in the built environment'. In *BEAR 2003 Conference Proceedings*, University of Salford.

Ellis, R. C. T., Wood, G. D. and Keel, D. A. (2003c) 'An investigation into the value management services offered by UK cost consultants'. In *COBRA 2003 Proceedings*, University of Wolverhampton.

Emmitt, S. and Gorse, C. A. (2003) *Construction Communication*. Oxford: Blackwell Science.

European Commission (2003) *Directive on the Energy Performance of Buildings*. Brussels: European Commission [http://www.europa.eu.int/eur-lex/dat/2003/1 001].

Evans, H. (1972) *Newsman's English*. London: Heinemann.

Evans, M. G. (1970) 'The effects of supervisory behaviour on the path-goal relationship'. *Organization Behaviour and Human Performance* **55**, 277–98.

Evenden, R. and Anderson, G. (1992) *Making the Most of People*. Addison-Wesley.

Fahey, L., Srivastava, R., Sharon, J. S. and Smith, D. E. (2001) 'Linking e-business and operating processes: the role of knowledge management'. *IBM Systems Journal* **40** (4).

Farmer, S. M. and Roth, J. (1998) 'Conflict handling behaviour in work groups: effects of group structure, decision process and time'. *Small Group Research* **29** (6), 669–689.

Farnham, D. (1993) *Employee Relations*. London: IPM.

Festinger, L. (1957) *Theory of Cognitive Dissonance*. Evanston, Illinois: Row, Peterson.

Fiedler, F. E. (1967) *A Theory of Leadership Effectiveness*. New York: McGraw-Hill.

Field, D. (1982) *Inside Employment Law: A Guide for Managers*. London: Pan Books.

Fitts, P. M. and Posner, M. I. (1973) *Human Performance*. London: Prentice-Hall International.

Flanagan, R. and Norman, G. (1993) *Risk Management and Construction*. Oxford: Blackwell Science.

Flowers, R. (1996) *Computing for Site Managers*. Oxford: Blackwell Science.

Francis, S., Shemmings, S. and Taylor, P. (1995) *Construction Law and the Environment*. London: Cameron May.

Frey, L. R. (1996) 'Remembering and "re-membering": a history of theory and research on communication and group decision making'. In Hirokawa, R. Y. and Marshall, S. P., *Communication and Group Decision Making*. London: Sage, 19–51.

Frey, L. R. (1999) *The Handbook of Group Communication Theory and Research*. London: Sage.

Frye, R. L. (1966) 'The effect of orientation and feedback of success and effectiveness on the attractiveness and esteem of the group'. *Journal of Social Psychology* **70**, December, 205–211.

Fryer, B. (1979) 'Managing on site'. *Building* **236** (24), 71–2.

Fryer, B. (1994a) 'Business planning and sustainable development: a construction industry perspective'. In Williams, C. and Haughton, G. (eds) *Perspectives Towards Sustainable Environmental Development*. Aldershot: Avebury Studies in Green Research.

Fryer, B. (1994b) *Successful Mentoring: how to be a more skilful mentor*. Leeds Metropolitan University.

Fryer, B. (1995) 'Mentoring experience'. In Little, B. (ed.) *Supporting Learning in the Workplace: Conference proceedings*. London and Leeds: The Open University and Leeds Metropolitan University.

Fryer, B. (2002) *Creativity: A Team Thing*. The Creativity Centre (website).

Fryer, B. and Douglas, I. (1989) *Managing Professional Teamwork in the Construction Industry*. London: CPD in Construction Group.

Fryer, B. and Roberts, P. (1993) 'Seen to be green'. *Chartered Builder* **5** (2), 12.

Fryer, M. (1983) 'Can psychology help the manager?' *Building Technology and Management* **21** (10), 15–16.

Fryer, M. (1994) 'Attitudes to creativity and the implications for management'. In Geschka, H., Rickards, T. and Moger, S. (eds) *Creativity and Innovation: the Power of Synergy*. Darmstadt, Germany: Geschka & Partner Unternehmensberatung, 259–64.

Fryer, M. (2002) 'Using creativity to improve organisational performance', *Public Service Magazine*. November.

Fryer, M. and Fryer, B. (1980) 'People at work in the building industry'. *Building Technology and Management* **18** (9), 7–9.

Gale, A. (1995) 'Women in construction'. In Langford, D. *et al. Human Resources Management in Construction*. Harlow: Longman.

Galunic, D. C. and Rodan, S. A. (1998) 'Resource recombinations in the firm: knowledge structures and the potential for Schumpeterian innovation'. *Strategic Management Journal* **19** (12) 1193–1201.

Gameson, R. N. (1992) '*An investigation into the interaction between potential building clients and construction professionals*'. Unpublished PhD thesis, Department of Construction Management and Engineering, University of Reading.

Gann, D. M. (2000) *Building Innovation: Complex Constructs in a Changing World*. London: Thomas Telford Publishing.

Gardiner, P. D. and Simmons, J. E. L. (1992) 'Analysis of conflict and change in construction projects'. *Construction Management and Economics* **10**, 459–478.

Garnett, N. and Pickrell, S. (2000) Benchmarking for construction: theory and practice. *Construction Management and Economics*, **18**, 55–63.

Gibb, J. R. (1961) 'Defensive communication'. *Journal of Communication* 141–148.

Giunipero, L. C. and Brand, R. R. (1996) 'Purchasing's role in supply chain management'. *International Journal of Logistics Management* **7** (1), 29–37.

Godfrey, K. (1996) *Partnering in Design and Construction*. London: McGraw-Hill.

Goffman, E. (1971) *The Presentation of Self in Everyday Life*. Harmondsworth: Penguin Books.

Gorse, C. A. (2002) '*Effective interpersonal communication and group interaction during construction management and design team meetings*'. Unpublished PhD thesis, School of Management, Faculty of the Social Sciences, University of Leicester.

Gorse, C. A. and Whitehead, P. (2002) The teaching, learning, experiencing and reflecting on the decision-making process. In Khosrowshahi, F. (ed.) *DMinUCE. The Third International Conference on Decision Making in Urban and Civil Engineering*, 6–8 November 2002, London (CDRom).

Gorse, C. A. (2003) 'Conflict and conflict management in construction'. In *Proceedings of the Nineteenth Annual Conference ARCOM*, 3–5 September, University of Brighton, 173–182.

Gould, J. D. (1965) 'Differential visual feedback of component motions'. *Journal of Experimental Psychology* **69**, 263–8.

Grant, R. M. (1995) *Contemporary Strategy Analysis*. Cambridge, Mass., USA: Blackwell.

Grant, R. (1996) 'Towards a knowledge based theory of the firm'. *Strategic Management Journal*. **17**, 109–122.

Green, S. D. (1994) 'Beyond value engineering: SMART value management for building projects'. *International Journal of Project Management* **12** (1), 49–56.

Griffith, A. (1990) *Quality Assurance in Building*. Basingstoke: Macmillan Press.

Griffith, A. (1994) *Environmental Management in Construction*. Basingstoke: Macmillan Press.

Grundy, T. (1994) *Strategic Learning in Action*. London: McGraw-Hill.

Hackman, J. R. (1992) Group influences on individuals in organizations. In Dunnette, M.D. and Hough, L. M. (eds) *Handbook of Industrial and Organizational Psychology*, *2nd* edn. California: Consulting Psychologists Press.

Hall, R. (1993) 'A framework of linking intangible resources and capabilities to sustainable competitive advantage'. *Strategic Management Journal* **14**, 607–618.

Hammer, M. and Champy, J. (1994) *Reengineering the Corporation: a manifesto for business revolution*. London: Nicholas Brealey.

Handy, C. (1975) 'The contrasting philosophies of management education'. *Management Education and Development* **6** (2), 56–62.

Handy, C. (1979) *Gods of Management*. London: Pan Books.

Handy, C. (1985) *Understanding Organizations*. Harmondsworth: Penguin Books.

Handy, C. (1987) *The Making of Managers*. London: MSC, NEDC and BIM.

Handy, C. B. (1993) *Understanding Organizations*, *4th* edn. Harmondsworth: Penguin Business Library.

Handy, C. (1994) *The Empty Raincoat*. London: Hutchinson.

Handy, C. (1995) *The Age of Unreason*. London: Arrow Books.

Hannagan, T. (2002) *Management Concepts and Practices*. London: Pitman Publishing.

Hare, A. P. (1976) *Handbook of Small Group Research*, *2nd* edn. New York: The Free Press.

Harris, F. and McCaffer, R. (2001) *Modern Construction Management*. Oxford: Blackwell Science.

Harrison, F. L. (1992) *Advanced Project Management*. Aldershot: Gower Publishing.

Harry, M. (1995) 'Information management'. In Hannagan, T. (ed.) *Management Concepts and Practices*. London: Pitman.

Hartley, P. (1997) *Group Communication*. London: Routledge.

Harvey-Jones, J. (1993) *Managing to Survive*. London: Heinemann.

Haslett, B. B. and Ruebush, J. (1999) 'What differences do individual differences in groups make? The effects of individuals, culture and group composition'. In Frey, F. R. *The Handbook of Group Communication Theory and Research*. London: Sage, 115–139.

Hastings, C., Bixby, P. and Chaudhry-Lawton, R. (1986) *Superteams: A Blueprint for Organisational Success*. London: Fontana/Collins.

Hawkins, K. and Power, C. B. (1999) 'Gender differences in question asked during small decision-making group discussions'. *Small Group Research* **30** (2), 235–256.

Hawley, R. (2003) *Creativity, Science, Engineering and Technology*. The Foundation for Science and Technology, 14 May.

Hax, A. and Majluf, N. (1994) 'Corporate strategic tasks'. *European Management Journal* **12** (4), 366.

Health and Safety Commission (HSC 1995a) *Annual Report 1994/95*. Sudbury: HSE Books.

Health and Safety Commission (HSC 1995b) *Health and Safety Statistics 1994/95*. Sudbury: HSE Books.

Health and Safety Executive (HSE 1983) *Construction Health and Safety 1982–83*. London: HMSO.

Health and Safety Executive (1995) *Guide to Managing Health and Safety in Construction*. London: HMSO.

Health and Safety Executive (HSE, 2002) *Revitalising Health and Safety in Construction*. HSE, UK.

Health and Safety Executive (HSE, 2002) *Changing Business Behaviour: Would Bearing the True Cost of Poor Health and Safety Performance Make a Difference?*. Contract Research Report (CRR) No. 436/2002, UK.

Health and Safety Executive (HSE, 2003) *Health and Safety performance in the construction industry. Progress since February 2001 Summit*. Second Report by Kevin Myers, Chief Inspector of Construction, HSE, UK.

Hersey, P. and Blanchard, K. (1982) *Management of Organizational Behavior*. Englewood Cliffs, New Jersey: Prentice-Hall.

Hersey, P., Blanchard, K. and Johnson, D. E. (1996) *Management of organizational behaviour. Utilizing human resources, 7th* edn. London: Prentice Hall.

Hertin, J., Berkhout, F., Gann, D. M. and Barlow, J. (2003) Climate change and the UK house building sector: perceptions, impacts and adaptive capacity. *Building Research and Information* **31** (3–4), 278–290.

Hewison, R. (1990) *Future Tense*. London: Methuen.

Higgins, M. J. and Archer, N. S. (1968) 'Interaction effect of extrinsic rewards and socio-economic strata'. *Personnel and Guidance Journal* **47**, 318–23.

Hirokawa, R. Y., Erbert, L. and Hurst, A. (1996) 'Communication and group decision-making effectiveness'. In Hirokawa, R. Y. and Marshall, S. P. (1996) *Communication and group decision making, 2nd* edn. London: Sage, 269–300.

Hirokawa, R. Y., Gouran, D. S. and Martz, A. E. (1988). 'Understanding the sources of faulty group decision making: a lesson from the Challenger disaster'. *Small Group Behavior* **19**, 411–433.

Hirokawa, R. Y. and Poole, M. S. (1996) *Communication and Group Decision Making*, 3rd edn. London: Sage.

HM Treasury (1996) CUP Guidance No. 54 – Value Management, HM Treasury Central Unit on Procurement, London.

HM Treasury (1997) *Procurement Guidance No. 2 – Value for Money in Construction Procurement*. London: HMSO.

HM Treasury (2002) *OGC Procurement Guidance No 10: Achieving Excellence – through Health and Safety*, HM Treasury, UK

Hollander, E. P. (1978) *Leadership Dynamics*. New York: The Free Press.

Hollander, E. P. and Julian, J. W. (1970) 'Studies in leader legitimacy, influence and innovation'. In Berkowitz, L. (ed.) *Advances in Experimental Social Psychology Vol. 5*, New York: Academic Press, 33–69.

Hopson, B. (1984) 'Counselling and helping'. In Cooper, C. L. and Makin, P. (eds) *Psychology for Managers*. Leicester and London: British Psychological Society and Macmillan Press, 259–87.

House, R. J. (1971) 'A path-goal theory of leader effectiveness'. *Administrative Science Quarterly* **16**, 321–38.

Hughes, T. and Williams, T. (1995) *Quality Assurance*. Oxford: Blackwell Science.

Hull, C. L. (1943) *Principles of Behavior*. New York: Appleton-Century-Crofts.

Hunt, J. (1992) *Managing People at Work*. London: McGraw-Hill.

ICE & Faculty and Institute of Actuaries (1998) *Risk Analysis and Management for Projects*. London: Thomas Telford.

Institute of Civil Engineers (1996) *Creating Value in Engineering*. London: Thomas Telford.

Jacobs, A., Jacobs, M., Cavior, N. and Burke, J. (1974) 'Anonymous feedback: Credibility and desirability of structured emotional and behavioral feedback delivered in groups'. *Journal of Counselling Psychology* **21** (2), 106–111.

Jahoda, M. (1959) 'Conformity and independence – a psychological analysis'. *Human Relations* **12**, 99–120.

Jarboe, S. (1996) 'Procedures for enhancing group decision making'. In: Hirokawa, R. Y. and Marshall, S. P. *Communication and Group Decision Making*. London: Sage, 345–383.

Joyce, R. (1995) 'The proposed Construction (Health, Safety and Welfare) Regulations'. *Construction Law* **6** (5), 167–70.

Kahn, R. L. (1981) *Work and Health*. New York: Wiley.

Kalakota, R. M. R. (2001) *E-Business 2.0: Roadmap for Success*. New Jersey: Addison-Wesley.

Kanter, R. M. (1983) *The Change Masters: Innovation and Entrepreneurship in the American Corporation*. New York: Simon and Schuster.

Kanter, R. M. (1990) *When Giants Learn to Dance – Mastering the Challenges of Strategy, Management, and Careers in the 1990s*. Unwin Hyman Ltd.

Katz, R. L. (1971) 'Skills of an effective administrator'. In Bursk, E. C. and Blodgett, T. B. (eds) *Developing Executive Leaders*. Harvard: Harvard University Press, 55–64.

Keel, D. K. (1995) *Mentoring vs. Managing*. Mentoring – The 'Working for a Degree' Project, Leeds Metropolitan University, January, Issue No. 3.

Kelly, J. and Male, S. (1993) *Value Management in Design and Construction: The Economic Management of Projects*. London: E&FN Spon.

Kelly, J. and Male, S. (1998) *A Study of Value Management and Quantity Surveying Practice*, RICS Occasional Paper. London: Surveyors Publications.

Kelly, J. and Poynter-Brown, R. (1990) In Brandon, P. S. (ed.) *Value Management In Quantity Surveying Techniques: New Directions*. Oxford: BSP.

Kelly, J., Male, S. and MacPherson, S. (1993) *Value Management – A Proposed Practice Manual for the Briefing Process*, Paper No. 34. London: RICS.

Kelly, J., Morledge, R. and Wilkinson, S. (2002) *Best Value in Construction*. Blackwell Publishing.

Kennedy, C. (1993) *Guide to the Management Gurus*. London: Century Business.

Keyton, J. (1999) Relational communication in groups. In Frey, L. R. (ed.) *The Handbook of Group Communication Theory and Research*. London: Sage, 192–221.

Keyton, J. (2000) 'Introduction: the relational side of groups', *Small Group Research* **31** (4), 387–394.

Kilmann, R. and Thomas, K. (1975) 'Interpersonal conflict-handling behaviour as a reflection of Jungian personality dimensions'. *Psychological Reports* **37**, 971–980.

Kirkpatrick, D. L. (1960) 'Techniques for evaluating training programmes'. *American Society for Training and Development* **14** (1), 13–32.

Korman, A. (1974) *The Psychology of Motivation*. Englewood Cliffs, New Jersey: Prentice-Hall.

Langford, D., Hancock, M., Fellows, R. and Gale, A. (1995) *Human Resources Management in Construction*. Harlow: Longman.

Langford, D. and Male, S. (1991) *Strategic Management in Construction*. Aldershot: Gower.

Lansley, P. (1981) 'Maintaining the company's workload in a changing market'. *Proceedings of the Chartered Institute of Building: Annual Estimating Seminar*. Ascot: Chartered Institute of Building.

Lansley, P., Sadler, P. and Webb, T. (1975) 'Managing for success in the building industry'. *Building Technology and Management* **13** (7), 21–3.

Latham, M. (1994) *Constructing the Team*. London: HMSO.

LeDoux, J. (1998) *The Emotional Brain*. New York: Phoenix.

Lee, F. (1997) 'When the going gets tough, do the tough ask for help? Help seeking and power motivation in organizations'. *Organizational Behaviour and Human Decision Processes* **72** (3), 336–363.

Leonard-Barton, D. (1995) *Wellspring of Knowledge: Building and Sustaining the Sources of Innovation*. Boston, MA. Harvard Business School Press.

Leonard, D. and Strauss, S. (1997) 'Putting your company's whole brain to work'. *Harvard Business Review*, July–August, 111–121.

Likert, R. (1961) *New Patterns of Management*. New York: McGraw-Hill.

Likert, R. (1987) *New Patterns of Management*. London: Garland.

Lin, F. R. and Shaw, M. J. (1998) 'Re-engineering the order fulfilment process in supply chain networks'. *International Journal of Flexible Manufacturing Systems* **10**, 197–299.

Littlepage, G. E. and Silbiger, H. (1992) 'Recognition of expertise in decision-making groups: Effects of group size and participation patterns'. *Small Group Research* **22**, 344–355.

Littler, C. R. and Innes, P. (2003) 'Downsizing and deknowledging the firm'. *Work, Employment and Society* **17** (1), 73–100.

Lloyd, S. R. (1988) *How to Develop Assertiveness*. London: Kogan Page.

LMU (2003) *Mind maps, Skills for Learning*. Learning and Information Services, Leeds Metropolitan University.

Locke, M. (2002) 'Value management in collaborative agreements'. *HKIVM Fifth International Conference*, May.

Loosemore, M. (1996) 'Crisis management in building projects: A longitudinal investigation of communication behaviour and patterns within a grounded framework'. Unpublished Ph.D thesis, Department of Construction Management and Engineering, University of Reading.

Loosemore, M., Nguyen, B. T. and Denis, N. (2000) 'An investigation into the merits of encouraging conflict in the construction industry'. *Construction Management and Economics* **18**, 447–456.

Love, P. E. D. and Haynes, N. S. (2001) 'Construction managers' expectations and observations of graduates'. *Journal of Management Psychology* **16** (8), 579–593.

Love, P. E. D. and Li, H. (1998) 'From BPR to CPR – conceptualising re-engineering in construction'. *Business Process Management Journal* **4** (4), 291–305.

Lowe, R. J. and Bell, M. (2000) 'Building regulation and sustainable housing Part 2: technical issues'. *Structural Survey* **18** (2), 77–88.

Lowe, R. J., Bell, M. and Roberts, D. (2003a) 'Developing future energy performance standards for UK housing: the St. Nicholas Court project – Part 1'. *Structural Survey* **21** (4).

Lowe, R. J., Bell, M. and Roberts, D. (2003b) 'Developing future energy performance standards for UK housing: the St. Nicholas Court project – Part 2'. *Structural Survey*, **21** (5).

Macan, T. A. (1994) 'Time management: test a process model'. *Journal of Applied Psychology* **79** (3), 381–391.

Maddux, R. (1988) *Successful Negotiation*. London: Kogan Page.

Madique, M. A. (1980) 'Entrepreneur champions and technological innovation'. *Sloan Management Review* **21** (2), 59–76.

Malhotra, Y. (2000) 'Knowledge management for e-business management'. *Information Strategy: The Executive's Journal* **16** (4), 5–16.

Maslow, A. H. (1954) *Motivation and Personality*. New York: Harper & Row.

Maude, B. (1977) *Communication at Work*. London: Business Books.

May, G. H. (1996) *The Future is Ours: Foreseeing, Managing and Creating the Future*. Adamantine Press.

Maylor, H. (2002) *Project Management*. Harlow: Financial Times/Prentice-Hall.

McCall, G. J. and Simmons, J. L. (1966) *Identities and Interactions*. New York: Free Press.

McCampbell, A. S., Clare, L. M. and Gitters, S. H. (1999) 'Knowledge management: the new challenge for the 21st century'. *Journal of Knowledge Management* **3** (3), 172–179.

McClelland, D. C. (1961) *The Achieving Society*. Princeton, New Jersey: Van Nostrand Reinhold.

McCroskey, J. C. (1977) 'Oral communication apprehension: a summary of recent theory and research'. *Human Communication Research* **4**, 78–96.

McCroskey, J. C. (1997) 'Willingness to communicate, communication apprehension, and self-perceived communication competence: conceptualizations and perspectives'. In Daly, J. A., McCroskey, J. C., Ayres, J., Hopf, T. and Ayres, D. M. *Avoiding Communication: Shyness, Reticence and Communication Apprehension*. New Jersey: Hampton Press, 75–108.

McCroskey, J. C. and Richmond, V. P. (1990) Willingness to communicate: A cognitive view. *Journal of Social Behavior and Personality* **5**, 19–37.

Miller, E. and Rice, A. K. (1967) *Systems of Organisation*. London: Tavistock Publications.

Miller, G. (1966) *Psychology: The Science of Mental Life*. Harmondsworth: Penguin Books.

Mintzberg, H. (1973) *The Nature of Managerial Work*. New York: Harper & Row.

Mintzberg, H. (1976) 'The manager's job: folklore and fact'. *Building Technology and Management* **14** (1), 6–13.

Mitchell, E. (1989) 'The training of the police: new thinking'. *Journal of the Royal Society of Arts* **137**, (5396), 501–512.

Moran, P. and Ghoshal, S. (1999) 'Markets, firms, and the process of economic development', *Academy of Management Review* **24** (3), 390–412.

Morrison, E. W. (1993) 'Newcomer information seeking: exploring types, models, sources and outcomes'. *Academy of Management Journal* **36** (3), 556–589.

Morse, J. J. and Lorsch, J. W. (1970) 'Beyond theory Y'. *Harvard Business Review* **48** (3), 61–8.

Moscovici, S. and Zavalloni, M. (1969) 'The group as a polarizer of attitudes'. *Journal of Personality and Social Psychology* **12**, 125–35.

Mowrer, O. H. (1950) *Learning Theory and Personality Dynamics: Selected Papers.* New York: Ronald Press.

Mullins, L. J. (2002) *Management and organisational behaviour, 6th* edn. Harlow: Financial Times–Prentice Hall.

Murdoch, J. and Hughes, W. (2000) *Construction Contracts: Law and Management.* London: E & F N Spon.

Murphy, L. R. (1984) 'Occupational stress management: a review and appraisal'. *Journal of Occupational Psychology* **57** (1), 1–15.

Murphy, L. R. (1995) 'Managing job stress: an employee assistance/human resource management partnership'. *Personnel Review* **24** (1), 41–50.

Murphy, N. (1980) 'Image of the industry'. *Building* **238** (31).

Murray, H. A. (1938) *Explorations in Personality.* New York: Oxford University Press.

Nattrass, S. (1995) *Construction (Design and Management) Regulations: HSE's Implementation Strategy.* London: HSE.

Nolan, V. (1987) *Teamwork.* London: Sphere Books.

Nonaka, I. and Takeuchi, H. (1995) *The Knowledge-Creating Company*, Oxford University Press.

Norman, D. (1978) 'Overview'. *Cognitive Psychology.* Milton Keynes: The Open University Press.

Northledge, A. (1993) *The Good Study Guide.* Open University Press.

Novelli, L. Jr and Taylor, S. (1993) 'The context for leadership in 21st-century organisations'. *American Behavioral Scientist* **37** (1), 139–47.

OGC (2002) Procurement Guidance No.3 [http://www.ogc.gov.uk/sdtoolkit/reference/achievingiguide3.html].

O'Neil, P. (2002) 'Conflict Management'. In Stevens, M. *Project Management Pathways.* High Wycombe: The Association for Project Management, Section 73, 1–18.

Ostmann, A. (1992) 'On the relationship between formal conflict structure and the social field'. *Small Group Research* **23** (1), 26–39.

Palmer, A. (1990) *A Critique of Value Management.* Technical Information Sheet No. 124, CIOB, Ascot.

Parnes, S. J. (ed.) (1992) *Source Book for Creative Problem Solving.* Buffalo, New York: The Creative Education Foundation.

Pask, G. and Scott, B. C. E. (1972) 'Learning strategies and individual competence'. *International Journal of Man–Machine Studies* **4**, 217–53.

Patchen, M. (1993) 'Reciprocity of coercion and co-operation'. In Felson, R. B. and Tedeschi, J. T. *Aggression and Violence: Social Interactionist Perspectives.* Washington: American Psychological Association, 119–144.

Patel, M. B., McCarthy, T. J., Morris, P. W. G. and Elhag, T. M. S. (2000) 'The role of IT in capturing and managing knowledge for organisational learning on construction projects'. In *Proceedings of CIT 2000*, 28–30 June, Reykjavik, Iceland, 674–685.

Payne, R. (1984) 'Organisational behaviour'. In Cooper, C. L. and Makin, P. (eds) *Psychology for Managers.* Leicester and London: British Psychological Society and Macmillan Press, 9–51.

Payne, R., Fineman, S. and Jackson, P. (1982) 'An interactionist approach to measuring anxiety at work'. *Journal of Occupational Psychology* **55** (1), 13–25.

Pearce, P. (1992) *Construction Marketing: a professional approach.* London: Thomas Telford.

Pedler, M. and Boydell, T. (1985) *Managing Yourself.* London: Collins.

Pedler, M., Burgoyne, J. and Boydell, T. (1986) *A Manager's Guide to Self-development.* London: McGraw-Hill.

Pemberton, C. and Herriot, P. (1994) 'Inhumane resources'. *The Observer*, 4 December.

Penrose, E. T. (1959) *The Theory of Growth of the Firm.* Oxford: Basil Blackwell.

Perry, J. G. and Hayes, R. W. (1985) 'Risk and its Management in Construction Projects'. In *Proceedings of the ICE*, Part 1, 78, June, 499–521.

Peters, T. (1989) *Thriving on Chaos.* London: Pan Books.

Peters, T. (1992) *Liberation Management: Necessary Disorganization for the Nanosecond Nineties.* New York: Alfred A. Knopf.

Peters, T. and Austin, N. (1985) *A Passion for Excellence.* London: Collins.

Petit, T. (1967) 'A behavioural theory of management'. *Journal of the Academy of Management* **1**, 341–50.

Pettinger, R. (1998) *Construction Marketing: Strategies for Success.* Macmillan. UK/USA.

Pheng, L. S. (1999) 'The extension of construction partnering for relationship marketing'. *Marketing Intelligence & Planning* **17** (3).

Pickrell, S., Garnett, N. and Baldwin, J. (1997) *Measuring up: a practical guide to benchmarking in construction.* Construction Research Communications Ltd. Garston, UK.

Porter, L. W. and Lawler, E. E. (1968) *Managerial Attitudes and Performance.* Homewood: Irwin-Dorsey.

Prahalad, C. K. and Hamel, G. (1990) 'The core competence of the corporation'. *Harvard Business Review* **68** (3), 79–91.

Preece, C. and Male, S. (1997) Promotional literature for competitive advantage in UK construction firms, *Construction Management and Economics* **15**, 59–69.

Prince, G. M. (1995) 'Synectics'. In Prince, G. M. and Logan-Prince, K. (eds) *Mind-Free.* 13th ed. Mass. USA: Mind-Free Group, Inc.

Pruitt, D. G. and Rubin, J. Z. (1986) *Social Conflict: Escalation, Stalemate and Settlement.* New York: Random House.

Quinn, J. B. (1985) 'Managing innovation: controlled chaos'. *Harvard Business Review* **63**, May–June.

Rackham, N. (1977) *Behaviour Analysis in Training.* Maidenhead: McGraw-Hill.

Radford, J. and Govier, E. (1980) *A Textbook of Psychology.* London: Sheldon Press.

Rahim, M. A. (1983) 'A measure of styles of handling interpersonal conflict. *Academy of Management Journal* **26** (2), 368–376.

Respect for People (2002) *A Framework for Action, The Report of Rethinking Construction's Respect for People Working Group.* Rethinking Construction Ltd, October.

Rethinking Construction (2002) *Accelerating Change.* A report by the Strategic Forum for Construction Chaired by Sir John Egan, September, UK.

Ribeiro, F. L. and Henriques, P. G. (2001) 'How Knowledge Can Improve E-Business In

Construction'. In *Proceedings of the Second International Postgraduate Research Conference in the Built and Human Environment*, University of Salford. UK: Blackwell Publishing.

RICS (1998) 'Improving local services through best value'. *Public Policy* 5 May [http://www.rics.org/].

RICS (2000) *The Management of Risk: An Information Paper*. London: RICS Business Services Ltd.

RICS (2003) *Contracts-in-Use Survey: A Survey of Building Contracts in Use During 2001*. DLE.

Roberts, P. (1994) 'Environmental sustainability and business'. In Williams, C. and Haughton, G. (eds) *Perspectives Towards Sustainable Environmental Development*. Aldershot: Avebury Studies in Green Research.

Rogers, Carl R. (1951) *Client Centred Therapy*. Boston: Houghton Mifflin.

Rogers, E. M. (1983) *Diffusion of Innovations*, 3rd edn. New York: Free Press.

Rothwell, R. (1992) 'Successful industrial innovation: critical success factors for the 1990s'. *R&D Management* **22** (3), 221–239.

Rothwell, R. and Dodgson, M. (1994) 'Innovation and Size of Firm'. In Dodgson, M. and Rothwell, R. (eds) *The Handbook of Industrial Innovation*. Edward Elgar.

Ruikar, K., Anumba, C. J. and Carrillo, P. M. (2003) 'Reengineering construction business processes through electronic commerce'. *The TQM Magazine* **15** (3), 197–212.

Saad, M., Jones, M. and James, P. (2002) 'A review of the progress towards the adoption of supply chain management (SCM) relationships in construction'. *European Journal of Purchasing Management* **8**, 173–183.

Sadgrove, K. (1994) *ISO 9000/BS 5750 Made Easy: a practical guide to quality*. London: Kogan Page.

Salancik, G. R. and Pfeffer, J. (1977) 'An examination of need satisfaction models of job attitudes' *Administrative Science Quarterly* **22**, 427–56.

Scally, M. and Hopson, B. (1979) *A Model of Helping and Counselling: Indications for Training*. Leeds: Counselling and Careers Development Unit, Leeds University.

Scarborough, H., Swan, J. and Preston, J. (1999) *Knowledge Management: A Literature Review – Issues in People Management*. London: Institute of Personnel and Development.

Schachter, S. (1959) *The Psychology of Affiliation*. Stanford C. A.: Stanford University Press.

Schmidt, W. H. and Finnigan, J. P. (1992) *The Race without a Finish Line: America's quest for total quality*. San Francisco: Jossey-Bass Publishers.

Selwyn, N. (1980) *Law of Employment*. Butterworth.

Shepherd, C. R. (1964) *Small Groups: Some Sociological Perspectives*. New York: Chandler Publishing.

Shimmin, S., Corbett, J. and McHugh, D. (1980) 'Human behaviour: some aspects of risk-taking in the construction industry'. In *Safe Construction for the Future* (proceedings of a conference). London: Institution of Civil Engineers.

Skinner, B. F. (1953) *Science and Human Behavior*. New York: Macmillan.

Slocombe, T. E. and Bluedorn, A. C. (1999) 'Organizational behavior implications of the congruence between polychronicity and experienced work-unit polychronicity'. *Journal of Organizational Behavior* **20** (1), 75–99.

Smith, A. J. and Piper, J. A. (1990) Evaluation techniques and instruments. *Journal of European Industrial Training* **14** (8), 5–24.

Smith, P. B. (1984) 'The effectiveness of Japanese styles of management: A review and critique'. *Journal of Occupational Psychology* **57** (2), 121–36.

Smith, R. (1993) *Psychology*. Minneapolis: West Publishing.

Smode, A. (1958) 'Learning and performance in a tracking task under two levels of achievement information feedback'. *Journal of Experimental Psychology* **56**, 297–304.

Snyder, R. A. and Williams, R. R. (1982) 'Self theory: an integrative theory of work motivation'. *Journal of Occupational Psychology* **55** (4), 257–67.

Srivastava, A. (1996) *Widening Access: Women in Construction Higher Education*. PhD thesis: Leeds Metropolitan University.

Srivastava, A. and Fryer, B. (1991) 'Widening access: women in construction'. In *Proceedings of the seventh annual conference*. Association of Researchers in Construction Management.

Standing, N. (2001) *Value Management Incentive Programme*. Thomas Telford.

Stein, M. (2002) Personal communication.

Stephenson, J. and Weil, S. (eds) (1992) *Quality in Learning*. London: Kogan Page.

Stewart, A. M. (1994) *Empowering People*. Corby and London: IM and Pitman Publishing.

Stewart, R. (1988) *Managers and Their Jobs*. Basingstoke: Macmillan.

Stewart, R. (1997) *The Reality of Management*. Oxford: Butterworth-Heinemann.

Stewart, T. A. (1997) *Intellectual Capital: The New Wealth of Organisations*. London: Nicholas Brealey Publishing.

Stoner, J., Freeman, A. and Gilbert, D. (1995) *Management*. Englewood Cliffs, New Jersey: Prentice-Hall.

Sturges, J. L., Bates, M. and Abdullah, N. (1997) 'CAPM, an under-utilised resource within the UK construction industry'. In *Proceedings of the First International Conference on Construction Industry Development: Building the Future Together*, Singapore.

Sullivan, P. H. (1999) 'Profiting from intellectual capital'. *Journal of Knowledge Management* **3** (2), 132–142.

Tannenbaum, R. and Schmidt, W. H. (1973) 'How to choose a leadership pattern'. *Harvard Business Review* **51** (3), 162–80.

Tatum, C. B. (1987) 'Process of innovation in construction firms'. *Journal of Construction Engineering Management* **113** (4), 648–663.

Taylor, A., Sluckin, W. *et al.* (1982) *Introducing Psychology*. Harmondsworth: Penguin Books.

Teece, D. and Pisano, G. (1994) 'The dynamic capabilities of firms: an introduction'. *Industrial and Corporate Change* **3**, 537–556.

Teece, D., Pisano, G. and Sheun, A. (1997) 'Dynamic capabilities and strategic management'. *Strategic Management Journal* **18** (7), 509–533.

Thomas, K. W. (1976) 'Conflict and conflict management'. In Dunnette, M. D. (ed.) *Handbook of Industrial and Organizational Psychology*. Chicago: Rand McNally, 889–935.

Thomason, G. (1994) 'Management styles for the twenty-first century', *Professional Manager* **3** (5), 4.

Thoms, P. and Pinto, J. K. (1999) 'Project leadership: a question of timing'. *Project Management Journal* **30** (1), 19–26.

Toakley, A. R. and Marosszeky, M. (2003) 'Towards total project quality – a review of research needs'. *Engineering, Construction and Architectural Management* **10**, (3), 219–228.

Tolman, E. C. (1932) *Purposive Behavior in Animals and Men.* Appleton-Century-Crofts (reprinted 1967).

Torrance, E. P. (1995) *Why Fly? A Philosophy of Creativity.* Norwood, New Jersey: Ablex.

Turban, E., Lee, J., King, D. and Chung, M. H. (2000) *Electronic Commerce: A Managerial Perspective.* New Jersey: Prentice Hall.

UKOnline (2002) UK Business 'more strategic approach to e-business. [http://www.ukonlineforbusiness/gov.uk/cms/template/news-details.jsp?id=227567] [Accessed 4th February 2003].

VanDemark, N. L. (1991) *Breaking the Barriers to Everyday Creativity.* Buffalo, New York: The Creative Education Foundation.

Van de Ven, A. H., Angle, H. L. and Poole, M. S. (eds) (1989) *Research on the Management of Innovation: The Minnesota Studies.* New York: Harper & Row Publishers.

VanGundy, A. (1988) *Techniques of Structured Problem Solving.* New York: Van Nostrand Reinhold.

VanGundy, A. (1992) *Idea Power.* New York: American Management Associations, Publications Group.

Von Krogh, G., Ichijo, K. and Takeuchi, H. (2000) *Enabling Knowledge Creation.* Oxford University Press.

Vroom, V. H. (1964) *Work and Motivation.* New York: Wiley.

Vroom, V. H. and Yetton, P. W. (1973) *Leadership and Decision-making.* Pittsburgh: University of Pittsburgh Press.

Walker, D. H. T. (2000) 'Client/customer or stakeholder focus? ISO 14000 EMS as a construction industry case study'. *The TQM Magazine* **12** (1), 18–26.

Wallace, W. A. (1987) *The influence of design team communication content upon the architectural decision making process in the pre-contract design stages.* Unpublished PhD Thesis, Department of Building, Heriot-Watt University.

Warr, P. (ed.) (1978) *Psychology at Work.* Harmondsworth: Penguin Books.

Wason, P. C. (1978) 'Hypothesis testing and reasoning' Unit 25, *Cognitive Psychology.* Milton Keynes: The Open University Press.

Weber, R. J. (1971) Effects of videotape feedback on task group behavior. *Proceedings of the Annual Convention of the American Psychological Association*, 79th Annual Convention **6** (2), 499–500.

Weisberg, R. W. (1993) *Creativity: Beyond the myth of genius.* New York: Freeman.

Weitzman, M. L. (1996) *Recombinant Growth.* Cambridge, MA: Harvard University Press.

West, M. (1997) *Developing Creativity in Organizations.* Leicester: BPS Books.

Wheelen, T. L. and Hunger, J. D. (2002) *Strategic Management and Business Policy.* Prentice Hall.

White, M. and Trevor, M. (1983) *Under Japanese Management.* London: Heinemann.

White, R. (1959) 'Motivation reconsidered: the concept of competence'. *Psychological Review* **66**, 297–333.

Winch, G. (1998) 'Zephyrs of creative destruction: understanding the management of innovation in construction'. *Building Research and Information* **26** (4), 268–279.

Winch, G. M. (2000) 'Institutional reform in British construction: partnering and private finance'. *Building Research and Information* **18** (2), 141–155.

Winch, G. M. (2001) 'Processes, maps and protocols: understanding the shape of the construction process'. *Construction Management and Economics* **19**, 519–531.

Wolfe, R. A. (1994) 'Organizational innovation: review, critique and suggested research directions'. *Journal of Management Studies* **31** (3), 405–431.

World Commission on Environment and Development (1987) *Our Common Future*. Oxford, UK: Oxford University Press.

Wood, G. D. (2001) 'Management of the design process'. In Ellis, R. C. T. and Wood, G. D. *Project Management (Level M)*. Leeds Metropolitan University.

Wood, G. D. and Ellis, R. C. T. (2003) 'Risk management practices of leading UK cost consultants'. *Engineering, Construction and Architectural Management* **10** (4).

Woodman, R. W., Sawyer, J. E. and Griffin, R. W. (1993) 'Toward a theory of organisational creativity'. *Academy of Management Review* **18** (2), 293–321.

Woodward, J. (1958) 'Management and technology'. *Problems of Progress in Industry 3*. London: HMSO.

Woodward, J. (1965) *Industrial Organization: Theory and Practice*. Oxford: Oxford University Press.

Wright, A. (1998) 'Counselling skills Part 1 – can you do without them?'. *Industrial and Commercial Training* **30** (3), 107–109.

Yisa, S. B., Ndekugri, I. and Ambrose, B. (1996) 'A review of changes in the UK construction industry. Their implications for the marketing of construction services'. *European Journal of Marketing* **30** (3), 47–64.

Zaltman, G., Duncan, R. and Holbek, J. (1973) *Innovation and Organizations*. New York: Wiley.

Zhang, Z. H. and Shen, L. Y. (2000) 'A framework for implementing ISO 14000 in Construction'. *Environmental Management and Health.* **11** (2), 139–149.

Index

ability, 138
ACAS, 302, 303
accommodating, 102
Adair, J., 126, 127, 131
Adams, E.C. and Freeman, C., 216
adjudication, 105
administration and records, 239
advertisements, 247, 248
alertness, 142
Alexander, G., 272
Allied Trades Joint Industrial Council, 296
alternative dispute resolution, 104
Anderson, C. M., Riddle, B. L. and
        Martin, M. M., 70
Ansoff, H. I., 155
anxiety, 142
APM, 30, 202
arbitration, 106
Arbitration Act, 106
Argyle, M., 87, 113, 117
Arsenault, A. and Dolan, S., 143
artificial intelligence, 159–160
asbestos, 276
Asch, S., 115
assertiveness, 86
attitude change function, 67
auditing
    external, 312
    internal, 312
auditing of suppliers and subcontractors, 312
autonomous work group, 150
Averill, J. R., 72
avoiding, 101

Baden Hellard, R., 124, 308, 311

Bales, R. F., 65, 70, 100, 118, 119, 121
Ball, P., 311
Barnes, S. and Hunt, B., 218
Barrett, P., 306
Battikha, M.G., 306
Belbin, R. M., 93, 113, 127
Bell, M. and Lowe, R. J., 35
benchmarking, 307–8
Bennett, J. and Jayes, S., 225
Bennett, R. and Gabriel, H., 215
Bennett, S., 195
Bennis, W., 48
Bennis, W., Parikh, J. and Lessem, R., 315
best value, 195
Bhatt, G.D., 216
BITF, 35
Blake, R. R. and Mouton, J. S., 56, 50, 99
Bogdanov, J., 218
brainstorming
    reverse, 170
    techniques, 168–9
BRE, 308, 316
Bresnen, M. and Marshall, N., 225
British Quality Award, 308
brochures, 183
Brown, R., 163
Brown, W. and Jaques, E., 4
Brundtland commission, 35
Bruner, J. S., 146, 147, 148
BS5750, 306, 308, 309
BSRIA, 316
Burns, T. and Stalker, G. M., 11
business and the environment, 315
Business Impact Task Force, 34
business process re-engineering, 186

Byrne, R., 215

Cannon, T., 218
Capers, B. and Lipton, C., 72, 165
CAPM, 18
Carnegie, D., 84
CBPP, 181, 195, 198
CDM regulations, 201, 271, 272, 273, 282, 317
change, 32
    implementing, 176
    people's attitudes, 187
    planning organisations, 175
characteristics
    effective teams, 126
    leader, 49–50
    organisations, 36–42
Christopher, M., 222
CIC, 207
Cicmil, S. J. K., 32
CIOB, 21, 197
CIRIA, 15, 195, 202, 306, 309, 316–18
CITB, 258, 267
Clarke, N., 120
client expectations, 96–7
Cline, R. J. W., 65, 72, 162, 163, 165
closed questions, 71
cognitive consistency theories, 145
Cole-Gomolski, B., 215
collaborative
    agreements, 196
    workshops, 204
collecting information, 157
collective bargaining, 295–97
commercial employment agencies, 247
communication
    direction of, 68–9
    dominance, 70–71
    downward, 69
    failure, 70
    functions of, 66
    group, 65
    influence, 84
    lateral, 69
    methods, 74–8
    process, 63–8
    skills, 63
    spoken, 74

subcontractors, 69
structure, 68
training, 64
upward, 69
written, 77
competence models, 147
compromise, 101
computer-aided learning, 261
conceptual skill, 23
conflict
    change, 97
    disputes, 104–6
    dysfunctional, 93–5
    emergence and development, 95
    functional, 93–5
    management, 93–110, 102, 167
    objectives, 95
    processing, 64
    resolution, 97
Constable, J. and McCormick, R., 265
Construction Best Practice Programme, 195
Construction Confederation, 269, 292–3, 296
Construction Industry Advisory Committee,
    285–6
construction personnel, 232–3
Construction Quality Forum, 308
Construction Regulations, 286–7
Construction Round Table, 310
Construction Skills Certification Scheme, 269,
    293
contingency
    management, 9
    theory, 6, 16
contract sum development, 201
contracts of employment, 299
contractual relationships, 124
contributory negligence, 281
controlling, 20
Cooper, C. L., 143, 148
corporate identity, 183
corporate social responsibility, 35
COSHH regulations, 276, 277, 289
counselling, 143, 238
Cox, A. and Ireland, V., 223–4
Cox, A. and Thompson, I., 225
Cox, C. J. and Cooper, C. L., 87
CPD, 265–7

creative problem-solving, 168, 191
crises and breakdowns, 33
CRISP, 207
cultures and multiculturalism, 12
curiosity, 147

Dainty, A. R. J., Bryman, A. and Price, A. D. F.,
    28
Dainty, A.R.J., 301, 302
Davis Langdon, 200
Davis, H. and Scase, R., 190
Day, D., 129
Dearlove, D., 217
De Bono, E., 167, 191, 199
debate motions, 31
decision making, 24
decisions
    administrative, 156
    non-programmed, 156
    operating, 155
    programmed, 156
    reaching and acting on, 158
    strategic, 155–6
Defence Estates Organisation, 198
Defence Housing Executive, 227
Delegation, 124, 153
Delphi method, 170
Demming, E., 307, 311
developing
    creative thinkers, 191
    group performance, 112
    organisations, 33
    people, 17
    staff, 21
development activities, 259
development programmes
    evaluating, 259
    planning, 258
direct labour, 149
directing, 19–20
Disability Discrimination Act, 302
disagreement, 72
disciplinary procedures, 302–3
discrimination and equal opportunities, 300
disjunction, 73
dismissal, 303–4
disputes procedure, 303

dominating, 100
Donald, M., 210
Drake, C.D. and Wright, F.B., 283
Drennan, D., 87
drive reduction theory, 144
Drucker, P. E., 17, 37, 158, 215
Drummond, H., 313
DTI, 35, 207
dual process hypothesis, 159
dynamic engagement, 12–13

e-business initiatives, 218–19
EC Directive, 272, 287, 289, 319
education and training, 257
Edvinsson, L., 209
Egan, J., 33, 173, 222
Egbu, C. and Botterill, K., 218
Egbu, C. O., 24, 207–9, 212, 218
Elliman, T. and Orange, G., 187
Ellis, D. G. and Fisher, B. A., 71
Ellis, R. C. T and Wood, G. D., 196, 199, 200,
    203, 259, 261
Emmitt, S. and Gorse, C. A., 66, 94, 100, 104,
    105
empathy, 85
employee
    participation, 298
    performance, 135–44
    remuneration, 237–8
employer
    associations, 292
    duties, 283
    federations, 292
employment and workplace relations, 299–304
employment planning, 241
Employment Protection Act, 299
empowerment, 27–9
empowerment culture, 129
encouraging
    dialogue, 189
    interaction, 116
enquiries and contracts, 183
Environment Agency, 317
environmental
    action planning, 319
    impact assessment, 315–20
    management, 316, 317

objectives, 35–6
scanning, 178
Environmental Protection Act, 317
ergonomics, 151
ethics and social responsibility, 12
European Commission, 35
Evans, H., 79
Evans, M. G., 57
Evenden, R. and Anderson, G., 240
expectancy theories, 145
experiential learning, 260
expert determination, 105
express and implied terms, 300
external labour supply, 243

Fahey, L., Srivastava, R., Sharon, J. S. and
    Smith, D. E., 218
Fairclough Report, 2
Farmer, S. M. and Roth, J., 103
Farnham, D., 298
Fayol, H., 2
Federation of Environmental Trade
    Associations, 293
Federation of Master Builders, 296
feedback, 7, 20
    extrusive, 137
    intensive, 137
Festinger, L., 145
Fiedler, F. E., 55
Field, D., 281
financial incentives, 149
Finnigan, J. P., 314
Fitts, P. M. and Posner, M. I., 136, 139, 142
Flanagan, R. and Norman, G., 203
Flowers, R., 83
forecasting and budgeting, 241–2
Foundation for Management Education, 264
Francis, S., Shemmings, S. and Taylor, P., 287,
    318
Frey, L. R., 65
Fryer, B., 15, 20, 87, 117, 125, 147, 151, 264, 316,
    318
Fryer, M., 5, 190, 191, 192
future studies, 174

Galuinic, D. C. and Rodan, S. A., 210
Gameson, R. N., 71, 96

Gann, D. M., 209
Gardiner, P. D. and Simmons, J. E. L., 93, 97,
    98
Garnett, N. and Pickrell, S., 180
Gayle, A., 301
Gibb, J. R., 71, 127
globalisation, 12
GMB, 295
goal-setting theory, 148
Godfrey, K., 202
Gordon, W. J. J., 169
Gorse, C. A. 65, 71, 94, 95, 100, 102, 118
Gorse, C. A. and Whitehead, P., 166
Gould, J. D., 137
Grant, R. M., 210
graphic and numerical, 81–2
Greed, C., 301
grievance procedures, 303
Griffith, A., 311, 317, 318, 319
group
    behaviour, 117–22, 164
    core sureness, 113
    decision making, 161–7
    polarisation, 163
    think, 163, 165
Grundy, T., 177

Hackman, J. R., 164
Hall, R., 209
Hammer, M. and Champy, J., 13, 186
Handy, C., 13, 38, 50, 53, 56, 95, 264, 265
Hannagan, T., 13, 185
Hare, A. P., 70
Harris, F. and McCaffer, R., 149
Harrison, F. L., 45
Harry, M., 83
Hartley, P., 163, 165
Harvey-Jones, J., 14
Haslett, B. B. and Ruebush, J., 70
Hastings, C., Bixby, P. and Chaudhry-Lawton,
    R., 125, 128, 129
Hawkins, K. and Power, C. B., 71
Hawley, R., 191
Hax, A. and Majluf, N., 178
health and safety administration, 284
Health and Safety at Work Act, 269, 277, 281,
    282, 317

health and safety, 275–9
  commission, 284
  common law, 280
  effective communications, 279–80
  files, 273
  plans, 273
  safety inspectors, 284–5
  safety management, 279
  team approach, 272–3
help seeking, 72
Hersey, P., Blanchard, K. and Johnson, D. E.,
      56
Hertin, J., Berkhout, F., Gann, D. M. and
      Barlow, J., 35
Hewison, R., 14
high-risk activities, 273–5
  demolition, 274
  excavations, 274
  falsework, 275
  maintenance, 275
  painting, 275
  roof work, 275
  scaffolding, 274
  site transport, 275
  steel erection, 274
Highways Agency, 270
Hirokawa, R. Y. and Poole, M. S., 164, 165
Hirowaka, R. Y., Erbert, L. and Hurst, A., 165
HM Inspectorate of Pollution, 318
HM Treasury, 202, 270
Hollander, E. P., 57
Hollander, E. P. and Julian, J. W., 60
Hopson, B., 239
Horne, D., 14
Housing Corporation, 195
Housing Grants and Construction Regeneration
      Act, 104
HSE, 269, 272, 273, 276, 280
Hudson, L., 167
Hull, C. L., 144
human performance, 136
human resources management, 232
human skill, 23
Hunt, J., 161, 162

ICE, 198, 202
identifying development needs, 258

improvement notice, 285
individuals, 122–3
induction, 159
industrial
  democracy, 298
  relations, 291–305
Industrial Training Act, 21
Industrial Training Boards, 21
industry action, 309
information communication technologies, 217
information
  function, 67
  handling, 25
  management, 82–3
  technology, 82, 143
innovation strategies of organisations, 208
Innovative Teamwork Programme, 131
Institute of Management, 264, 319
Institute of Public Policy Research, 230
instrumental
  conditioning, 146
  function, 67
interaction analysis, 117
intermediate stress, 140
ISO 1400, 318
ISO 9000, 306, 309, 312, 317
ISO 9004, 317, 318

Jacobs, A., Jacobs, M., Cavior, N. and Burke, J.,
      65
Jahoda, M., 189
Jarboe, S., 165
JCT 98 with Contractor Design, 200
job
  centres, 247
  description, 38, 41, 245
  design, 143, 150
  enlargement, 150
  enrichment, 150
  evaluation, 238
  plans, 199
  rotation, 151
  specifications, 245
  stress, 140
Johari window, 240
Joyce, R., 286
just in time, 222

Kahn, R. L., 134
Kalakota, R.M.R., 218, 220
Kanter, R. M., 14
Katz, R., 23
Keel, D.K., 264
Kelly, J. and Male, S., 197, 198, 197
Kelly, J. and Poynter-Brown, R., 197
Kennedy, C., 13
Key Performance Indicators, 139
Keyton, J., 65
Kilmann, R. and Thomas, K., 100, 102
Kirkpatrick, D. L., 259
knowledge
    innovations, 209, 214–16
    integration, 97
    sharing, 214
Korman, A., 145
KPI Report, 139

labour turnover analysis, 242
Langford, D., Hancock, M., Fellows, R. and
        Gale, A., 292
Lansley, P., Sadler, P. and Webb T., 11
lateral thinking, 167
Latham, M., 2, 33, 222, 310, 311
leader
    authoritarian, 50
    competence, 59
    decision-making, 52
    setting, 55
    situation, 53–7
    style continuum, 51
    task, 54
leadership
    democratic, 50
    employee-centred, 51
    formal, 59, 171
    functions, 127
    goal setting, 58–9
    informal, 59, 171
    path goal, 57
    situational, 55–6
    structuring, 50
    style, 50–53
    supportive, 50, 56
Learndirect, 21
lectures, 262

LeDoux, J., 63
Lee, F., 71, 72
Leonard, D. and Strauss, S., 216
Leonard-Barton, D., 210
liability
    occupiers, 282
    personal, 281
    vicarious, 281–2
Likert, R., 51
Lin, F.R. and Shaw, M.J., 223
line and staff organisations, 42–3
listening, 85–6
litigation, 106
Littlepage, G. E. and Silbiger, H., 170
Littler, C. R. and Innes, P., 39
Lloyd, S. R., 86
Locke, E., 148
Locke, M., 197, 204
Loosemore, M., 97
Loosemore, M., Nguyen, B. T. and Denis, N., 103
Love, P. E. D., 23, 186
Lowe, R. J., 35

Macan, T. A., 151
Mace, C. A., 136
Madden, J. F., 301
Maddocks, R., 297
Madique, M. A., 210
Major Contractors Group, 269
Malhotra, Y., 219
management
    classical, 2
    information systems, 83
    management development, 264–5
        innovation, 266
        strategic, 265
    power, 25
    selection consultants, 247
    skills, 23–5
    social sciences, 4–6
    tasks, 17–22
    time, 22
    training, 143
managing
    change, 173–94
    conflict, 99–104
    creativity, 190–2

innovation, 207–21
knowledge, 216–18
quality and environmental impact, 306–21
supply chains, 222–31
Management Charter Initiative, 265
market research, 181
marketing, 180–4
audit, 184
strategy, 182
Maslow, A. H., 145
matrix organisations, 43–6
Maude, B., 78
May, G. H., 174
Maylor, H., 33
Mayo, E., 4
McCampbell, A. S., Clare, L .M. and
Gitters, S. H., 217
McClelland, D. C., 145, 265
McCroskey, J. C., 70
measuring innovation success, 213–14
mediation, 105
meetings
action points, 75–6
clients, 183
project, 76
site, 76–7
mentoring, 263–4
merit rating techniques, 238
method choice, 73
Miller, E. and Rice, A. K., 6
Miller, G., 144
Mintzberg, H., 4, 22
MoD, 227
Monte Carlo simulation, 203
Moran, P. and Ghoshal, S., 210
Morrison, E. W., 72
Morse, J. J. and Lorsch, J. W., 11
Moscovici, S. and Zavalloni, M., 163
motivation, 144–9
human performance, 134–53
intrinsic, 146
job satisfaction, 148
Movement for Innovation, 196
Mowrer, O. H., 144
Mullins, L. J., 38, 113, 243
multiple observation methods, 121
Murdoch, J. and Hughes, W., 29, 30

Murphy, L. R., 141, 143
Murray, H. A., 145
mutual recognition, 29

National Joint Council for Building Industry,
293, 295
National Rivers Authority, 318
National Training Organisation, 267
National Working Rule Agreements, 296–7
negative socio-emotional interaction, 65
negotiation, 105, 297
NHBC, 310
NHS plan, 230
Noise at Work Regulations, 289
Nolan, V., 126, 131, 132
Nonaka, I. and Takeuchi, H., 209
Norman, D., 159
norms, 114–15
Northledge, A., 31
Novelli, L. Jr and Taylor, S., 48

O'Neil, P., 93, 100
objectives
economic, 34
organisational, 34
policies, 33
occupational health hazards, 276
OGC, 195
optimal stress, 142
organisation
centralisation, 39–40
charts, 41
committees, 42
decentralisation, 39–40
downsizing, 39
flexibility, 40–41
hierarchy, 38–9
issues, 41
manuals, 41
paperwork, 42
procedures, 41
product-based, 42
rigidity, 40–41
size, 40
specialisation, 37–8
structure, 36–7
survival, 36

organisational
  activities, 33–4
  area based, 42
  change process, 174
  climate, 175
  development, 184–7
  environments, 12
  innovations, 210–12
  review, 257
organising, 19
Ostmann, A., 93
overloading, 73

Palmer, A., 197
Parnes, S. J., 157
partnering, 222–7, 316
  commitment, 226
  early involvement, 226
  equity, 226
  evaluation, 227
  implementation, 227
  mutual goals, 226
  timeliness, 227
Patchen, M., 72
Patel, M. B., McCarthy, T. J., Morris, P. W. G.
    and Elhag, T. M. S., 216
Payne, R., Fineman, S. and Jackson, P., 143
Pearce, P., 182
Pemberton, C. and Herriot, P., 232
Penrose, E. T., 209
people management, 13
performance appraisal, 37, 253–7
Perry, J. G. and Hayes, R. W., 202
personal and corporate relationships, 183
personal digital assistants, 152
personal
  power, 26
  skills, 86–8
Personal Protective Equipment at Work
    Regulations, 289
personality of individual behaviour, 111
personnel
  action programme, 243
  function, 233
  management, 232–40
  management tasks, 233–5
  policy employment, 236

policy, 235–7
  general, 235–6
  health and safety, 237
  industrial relations, 236
  remuneration and employee services, 237
  staff development, 236
selection
  curriculum vitae, 250
  group methods, 252
  interviewing, 250–51
  measures of performance, 251
  tests and questionnaires, 251
  work try outs, 253
specification, 246
statement, 235
persuasion, 188
Peters, T., 13, 310, 311, 313,
Petit, T., 7
Pettinger, R., 181
Pheng, L. S., 181
physical health hazards, 277–9
  dusts, 277–8
  radiation, 279
  skin troubles, 278–9
  toxic fumes, 278
  vibration and noise, 278
plain talking, 78–81
plan implementation, 179
planning
  business, 178
  corporate, 178
  functional, 178
  projects, 243–4
  strategic, 178–9
Porter, L. W. and Lawler, E. E., 145
portfolio management, 13–14
position power, 26
positioning, 182
positive socio-emotional interaction, 65
post-industrial era, 13
post-industrial society, 13
post-modern management, 14
Prahalad, C.K. and Hamel, G., 215
Preece, C. and Male, S., 184
present labour force, 242–3
prime contracting, 227–8
Prince, G. M., 131

priorities, 157
Private Finance Initiative, 197, 228, 229
problem definition, 157
problem solving
  decision making stages, 157, 172
  human reasoning, 158
  suspending judgement, 167
  techniques, 169
problem types, 155–6
process change, 187
productivity bargaining, 238
professionalism, 29–30
profit, 34
prohibition notice, 285
project
  information, 152
  organisation, 44
propaganda, 188
protective equipment, 289
Pruitt, D. G. and Rubin, J. Z., 99
Public Private Partnerships, 228–31

QUAD, 89, 167
quality and environmental management, 15
quality
  attitudes, 311
  benefits, 310
  control, 307
  culture, 313
  definition, 307
  in service organisations, 313
  management, 308–11
  manual, 313

Rackham, N., 118
Rackham, N., Honey, P. and Colbert, M., 118
Rahim, M. A., 100
recognition, 84–5
recruitment
  procedures, 246
  web sites, 248, 249
reports, 77–8
resource power, 25—6
Revans, R., 131, 264
review and control performance, 180
reward power, 25–6
Ribeiro, F. L. and Henriques, P. G., 218

RICS, 21, 196, 202
risk
  analysis, 203
  identification, 203
  management, 201
  registers, 203
  response continuum, 204
risky shift, 163
Robbens Committee, 282
Roberts, P., 15, 319, 320
Rogers, C. R., 238
Rogers, E. M., 210
role
  ambiguity, 117
  behaviour, 116
  clarity, 135
  conflict, 117
  expectations, 116
  play, 262–3
  power, 26
  related function, 68
  set, 117
  storming, 170
roles, 116–17
Rothwell, R., 207, 210
Ruikar, K., Anumba, C. F. and Carrillo, P. M.,
  187

Saad, M., Jones, M. and James, P., 223
Sadgrove, K., 307, 313
safety, 269–75
  attitudes, 271–2
  policy, 283–4
  regulations and committees, 287–9
Scarborough, H., Swan, J. and Preston, J., 217
scenario building, 192
Schachter, S., 145
Schein, E., 177
Schmidt, N. H. and Finnigan, J. P., 48, 307
scientific management, 2
Sector Skills Councils, 267
self-managed teams, 20, 28–9
Selwyn, N., 304
Shepherd, C. R., 65
Shimmin, S., Corbett, J. and McHugh, D., 272
Simon, H. A., 159
situational management, 9

skills
  analysis, 242
  development, 138
Skinner, F. F., 146
Slocombe, T. E. and Blectorn, A. C., 151
SMART, 189
Smith, A. J. and Piper, J. A., 259
Smith, P. B., 51
Smode, A., 138
SMOT, 121
Snyder, R. A. and Williams, R. R., 145
social relationships function, 67
specialisations, 3
Srivastava, A., 300–302
staff capabilities, 40
staff development
  approaches, 259–60
  methods, 261–3
Standing, N., 196
Stein, M., 190
Stewart, A. M., 130
Stewart, R., 1, 17, 28, 209
Stoner, J., Freeman, A. and Gilbert, D., 12, 28, 39, 113
Strategic Defence Review, 227
strategic
  management, 177
  planning, 237
stress, 142–3
subcontract labour, 149
subcontractors, 143
subordinates, 54
succession plan, 242
Sullivan, P. H., 215, 217
supply chain management, 222–4
sustainable development, 35, 123, 317
SWOT analysis, 182
SYMLOG, 120, 121
systems
  management, 6–9
  theory, 6

Tannenbaum, R. and Schmidt, W. H., 52
targets and performance standards, 143
tasks
  closed, 159
  generative, 159

socio-emotional roles, 59
task-based interaction, 66
Tatum, C. B., 210, 216
Tavistock Institute, 5, 6
team leadership, 127–30
teamwork, 123–5, 131
  roles, 127, 128
  training, 130–32
team building, 124
Teece, D. and Pisano, G., 210, 216
termination of contract, 304
terms and conditions, 299
TGWU, 293
Thomas, K. W., 100
Thomason, G., 48
Thoms, P. and Pinto, J. K., 33
time management, 151–3
Toakley, A. R. and Marosszeky, M., 306
Tolman, E. C., 57
Torrance, E. P., 190
total quality management systems, 28
TQM, 298, 307, 314
Trade Union and Labour Relations Act, 295
trade unions, 293
training
  plans, 242
  team leadership, 130–32
Turban, E., Lee, J., King, D. and Chung, M. H., 218

UCATT, 293
UK-Online, 218
unanimous agreement, 162
unique selling proposition, 182
University for Industry, 21
Urwick, L., 2

value management, 195–206
value engineering, 197
value stream mapping, 197
VanDemark, N. L., 191, 192
Van de Ven, A. H., Angle, H. L. and Poole, M. S., 210, 216
VanGundy, A., 168
Vroom, V. H., 52, 145

Walker, D. H. T., 181

Wallace, W. A., 70, 95, 97
Wason, P. C., 146, 159
Waste Regulation Authorities, 318
Weber, R. J., 166
Weisberg, R. W., 191
Weitzman, M. L., 210
West, M., 192
Wheelan, T. L and Hunger, J. D., 178, 180
White, M. and Trevor, M., 51
White, R., 145
Winch, G. M., 175, 209
Winslow Taylor, F., 2
Wolfe, R. A., 207
women in construction, 301

Wood, G. D., 44
Woodman, R. W., Soyer, J. E. and
      Griffin, R. W., 215
Woodward, J., 10
work-based learning, 260, 263
work breakdown, 152
Workplace Regulations, 287
World Commission on Environment and
      Development, 35
Wright, A., 238

Yisa, S. B., Ndekugri, I. and Ambrose, B., 181

Zaltman, G., Duncan, R. and Holbek, J., 216